UNDOING PLACE?

UNDOING PLACE?

A GEOGRAPHICAL READER

Edited by
Linda McDowell
Department of Geography,
University of Cambridge, UK

A member of the Hodder Headline Group
LONDON • NEW YORK • SYDNEY • AUCKLAND

First published in Great Britain in 1997 by
Arnold, a member of the Hodder Headline Group
338 Euston Road, London NW1 3BH
175 Fifth Avenue, New York, NY 10010

Copublished in the US, Central and South America by
John Wiley & Sons, Inc.,
605 Third Avenue,
New York, NY 10158–0012

British Library Cataloguing in Publication Data
A catalogue entry for this book is available from the British Library

Library of Congress Cataloging-in-Publication Data
[copy to come]

ISBN 0 340 67746 5 (pb)
ISBN 0 470 23640 X (Wiley)

ISBN 0 340 67747 3 (hb)
ISBN 0 470 23639 6 (Wiley)

Typeset in 10/12 Times by Photoprint, Torquay, Devon
Printed and bound in Great Britain by J W Arrowsmith Ltd

This one is for the women in my life:
my mother, Olive, my sisters, Judith and Kate,
and my daughter, Sarah.

CONTENTS

ACKNOWLEDGEMENTS

As this reader is based around a third year course that I teach in the geography department at Cambridge, I should like to thank all those students who have worked with me on it, seemed to enjoy it, indulged my reminiscences of the 1960s and more importantly taught me a lot about the 1990s. Hugh McDowell did the photocopying again, so thanks again too. And Laura McKelvie continued to be fine supportive editor. Finally I should like to thank all the authors for agreeing to being reprinted here, especially those who never knew they were reaching geographers.

The editor and publishers would like to thank the following for permission to use copyright material in this book.

The authors and Routledge Ltd. for ' "It's all falling apart here": coming to terms with the future in Teeside' by Huw Benyon, Ray Hudson, Jim Lewis, David Sadler and Alan Townsend in P. Cooke (ed.) *Localities: The Changing Face of Urban Britain* (Unwin Hyman, 1988); the author and Reed Consumer Books Ltd. for 'Space and power' in *Goliath: Britain's Dangerous Places* by B. Campbell (Methuen, 1993); the author and HarperCollins Publishers for ' "Leave it to Beaver" and "Ozzie and Harriet": American families in the 1950s' in *The Way We Never Were: American Families and the Nostalgia Trap* by Stephanie Coontz, copyright © 1992 by Basic Books, a division of HarperCollins Publishers, Inc.; the author and Pion Ltd. for 'West Hollywood as symbol: the significance of place in the construction of a gay identity' by B. Forest in *Environment and Planning D: Society and Space*, **13** (1995); the author and Routledge Ltd. for ' "The whisper wakes, the shudder plays": "race", nation and ethnic absolutism' in *There Ain't No Black in the Union Jack* by Paul Gilroy (Unwin Hyman, 1987); the author and Blackwell Publishers Inc. for 'The magic of the mall' by Jon Goss in *Annals of the Association of American Geographers* **83**, 1 (1993); the author and Lawrence and Wishart Ltd for 'Cultural identity and diaspora' by Stuart Hall in J. Rutherford (ed.) *Identity: Community, Culture, Difference* (Lawrence and Wishart Ltd., 1990); the author and South End Press for 'Homeplace: a site of resistance' in *Yearning: Race, Gender and Cultural Politics* by bell hooks (South End Press, 1991); the authors and Sage Publications Inc. for 'Women in outlaw motorcycle gangs' by Columbus B. Hopper and Johnny Moore in *Journal of Contemporary Ethnography*, **18**, 4, pp. 363–87 © 1990 Sage Publications Inc.; the author for 'Roast beef and reggae music: the passing of whiteness' by Diana

Jeater in *New Formations* **18** (1992); the author and Verso for 'O life unlike to ours! Go for it! New Age travellers' in *Senseless Acts of Beauty* by George McKay (Verso, 1996); the author and Pion Ltd. for 'The political place of locality studies' by Doreen Massey in *Environment and Planning A*, **23** (1991); the author and the American Anthropological Association for 'Future travel: anthropology and cultural distance in an age of virtual reality or, a past seen from a possible future' by Christoper Pinney in *Visual Anthropology Review* **8**, 1 (1992); the author and Routledge Ltd. for 'Beyond the modern home: shifting the parameters of residence' by Tim Putnam in J. Bird, B. Curtis, T. Putnam, G. Robertson and L. Tickner (eds) *Mapping the Futures: Local Cultures and Global Change* (1993); the author and Routledge Ltd. for 'Tradition and translation: national culture in its global context' by Kevin Robins in John Corner and Sylvia Harvey (eds) *Enterprise and Heritage: Crosscurrents of National Culture* (Routledge, 1991); the author and Routledge Ltd. for 'Mods, rockers and turf gangs' in *Places on the Margin: Alternative Geographies of Modernity* by R. Shields (Routledge, 1991); the author and Princeton Architectural Press for 'The suburban home companion: television and the neighborhood ideal in postwar America' by Lynn Spigel in B. Colomina (ed.) *Sexuality and Space* (Princeton Architectural Press, 1992); the author and Pion Ltd. for 'Ambivalent attachments to place in London: twelve Barbadian families' by J. Western in *Environment and Planning D: Society and Space*, **11** (1993); the author and Routledge Ltd. for 'Racism, black masculinity and the politics of space' by Sallie Westwood in J. Hearn and D. Morgan (eds) *Men, Masculinities and Social Theory* (Unwin Hyman, 1990); the author and Routledge Ltd. for 'On the road again: metaphors of travel in cultural criticism' by Janet Wolff in *Cultural Studies* **7** (1993); the author for 'The ghosting of the inner city' in *On Living in an Old Country: the National Past in Contemporary Britain* by Patrick Wright (Verso, 1985); the author and Lawrence and Wishart Ltd. for 'Together in difference: transforming the logic of group political conflict' by I. M. Young in J. Squires (ed.) *Principled Positions* (Lawrence and Wishart Ltd., 1993); and the author and the University of California Press for 'Disney World: the power of facade/the facade of power' in *Landscapes of Power: From Detroit to Disney World* by Sharon Zukin, copyright © 1991 The Regents of the University of California.

Every effort has been made to trace copyright holders of material reproduced in this book. Any rights not acknowledged here will be acknowledged in subsequent printings if notice is given to the publishers.

INTRODUCTION: RETHINKING PLACE

In the last few years human geography has been transformed. Not only has the subject itself become immeasurably more interesting (in my view at least) as new ways of looking at and explaining the enormous socio-spatial changes of recent years have produced a vivid, vibrant and diverse discipline, but geography also finds itself, somewhat unaccustomedly, at the centre of debates in the social sciences and humanities about space–time compression, the expansion of mass and niche consumer markets, the impact of migration and the development of diasporic identities, about the effects of global communications on local cultures, about the production and symbolism of imagined or spectacular places. As, in the final years of the twentieth century, places in the world become increasingly interconnected and as global capitalism seems like an unstoppable juggernaut, erasing all that is local and particular in its path, there is increasing anxiety among many commentators about both the 'end of history' and the 'end of geography'. Somewhat paradoxically, there has, at the same time, been a reassertion of the significance of local or regional identities: in the turn back to local languages in the peripheral parts of Great Britain, in the rise of new forms of regional and ethnic nationalism and in the shattering of former communist states into multiple parts and new nations.

The principal aim of this reader is to uncover and assess some of these changes, to examine the ways in which the relationships between places and individual and group identities have changed over time. This is clearly a large aim so I have restricted my focus to issues that generally fall under the heading of social geography – that is to areas of daily life that are outside the social relations of waged and unwaged labour, although this division can never be watertight and the spatial distribution of the social classes has long been a key issue in social geography. I have concentrated on the 'advanced industrial west', in the main Great Britain and the USA, and on the period between the end of the Second World War and the end of the century.

The main argument of what follows is that there is a reciprocal relationship between the constitution of places and people. Thus there is a dual focus on how places are given meaning and how people are constituted through place (as well as how they perceive and consume place in everyday social interactions). Further, the approach taken in the reader is to unite the social and the cultural in the sense that the significance of both material social relations and symbolic meanings in the construction of place are examined. The organisation of space, in the sense of devising, channelling and controlling social interactions, and the construction of places, in the sense of known and definable areas, is a key way in which groups and collectivities create a shared, particular and distinctive identity. (Think of what it means to claim a place-based identity: I am from Stockport (a former textile town on the Mersey), I am a Northerner (a warm-hearted outgoing, plain speaking sort of a person, or so you might assume!). I am English (an identity tied to a real place but also to a myth of nationality – the bulldog breed, or less positively a Little Englander).) So places are both concrete and symbolic. They are literally and metaphorically made up: of buildings, field systems, roads and railways as well as of myths and legends, statues and ceremonies that link people to a place. And places are by very definition, exclusive. They define themselves and their inhabitants as 'different from', although this is not to deny the multiple senses and meanings of place constructed by the co-inhabitants of any place. The meaning of place varies depending on the age, class, gender, status and point of view of its occupants, but nevertheless boundaries and exclusion are essential characteristics of place formation.

The aim of the readings is to examine critically the argument that in the period of postmodernity (post-Fordism, high modernity or what-ever we choose to term it) the close associations between place (at the scale of the home, the locality or the community, the region and the nation state) and a set of social divisions including gender, class and nationality or ethnicity are either breaking down or are being reconsti-tuted in radically different ways. In the recent past, it has been the association between social class and place that seems to have been seen as most under threat or subject to change. Writing in 1979, the sociologist Anthony Giddens succinctly noted the way in which social divisions and spatial structures traditionally were connected in a capitalist society like Britain:

> In a class society, spatial division is a major feature of class differentiation. In a fairly crude, but nevertheless sociologically significant sense, classes tend to be regionally concentrated. One can easily instance the contrasts between the north and south of England, or west and east Scotland, to make the point. Such spatial differentiations have always to be regarded as time-space formations in the terms of social theory. Thus one of the important features is the sedimentation of divergent regional 'class cultures' over time:

but, as he added:

> class cultures which today, of course, are partly dissolved by new modes of transcending time-space distances (Giddens, 1979).

It is with the causes and consequences, and the timing, of this supposed transcendence that this reader is concerned, with the dis-solution of older ways of living and with the development of new time–space formations. Since Giddens made his astute observation, the pace of change and academic interest in it has magnified enormously. A huge range of changes, not only the 'new modes of transcendence', those technologies of 'space–time compression' (Harvey, 1989) but also global and local social, political and economic changes at a range of spatial scales have fundamentally restructured the links between people and place. The collapse of communism and the rise of new nation states has transformed the map of central Europe. Within old industrial nations, single-class communities based on the dominance of a particular manufacturing industry, for example, have been rup- tured by the rise of the service economy and new forms of working. At a finer spatial scale the association of women with the private domain of the home and men with the workplace has also been torn asunder by women's entry into the labour force and at the national scale, global migration and the rise of regional movements has disrupted the links between territory and a people. For some, this period is one of fear, in which all that is solid has seemed to melt into air and there is a nostalgic reference back to the 1950s – the period in which it now seems that there was a set of clear relationships between people and place, a period when people knew their place, not only in the sense of deference but in a sense of belonging to a place, be it the North or England or wherever.

I want to show how our current concern with disruption and displace- ment is based on a particular construction of the past as an imaginary construct or 'structure of feeling', to use Raymond Williams' term. We now have a view of the 1950s as a period of relative geographic immobility and of certainty, an idea that this was the end of the stability of the modern period compared with the restless instability of the present. I shall argue that this view of the past, or at least of the 1950s, is as one-dimensional as an idea of the present as one in which links between place and identity have been completely recast. Increasingly the 1950s are being seen by social and cultural theorists as a key period in understanding more recent shifts in the relationships between iden- tity, meaning and place. Thus the reader is organised around the strong belief that not only is comparative analysis an essential prerequisite for understanding the nature of the relationships between identity and place but also that the preconditions and origins of the contemporary set of socio-spatial changes lie in the 1950s where, as Ross (1995) sug- gests we shall find the prehistory (and I would add, the pregeography) of postmodernity.

In my own department, where I teach a third year undergraduate course in the social geography of the post-war UK and USA, I address these issues about the changing relationships between identity and place through the lens of a set of specific questions. I have used these questions to structure this reader too.

- How are social characteristics (class, race, gender, etc.) mapped out on the ground? How has the geography of these distributions changed in the post-war decades?'
- Can we still – could we in the 1950s – distinguish 'communities' (spatially defined and discrete, locally based or, perhaps, imaginary)? What are the sets of meanings attached to these communities or localities and how, in turn, do local factors – a sense of place – affect what it means to be black or gay, a woman or working class? Does

geography affect our sense of ourselves? Does it matter where we were born or where we live now?

- How are the relationships between identity and place being altered in an increasingly interconnected world? Is there any 'authentic' sense of place now or are we increasingly 'placeless'? What has changed between the 1950s and the 1990s?

The retrospective view of the 1950s as a purportedly settled period with strong links between place and identity has not only affected our view of that period and of the present but has also structured geographic concepts such as 'locality' which have remained a key way of thinking about place right through to the end of the 1980s. Thus the decision to look backwards has a theoretical justification too. In the individual sections, the aim is not merely to counterpose the two periods as 'structures of feeling' but also to draw out the counter or oppositional movements in each decade and show how the supposed stability of one and the mobility of the other is exaggerated.

Changes in the conceptual definition and geographical analysis of the significance of place are an important part of the post-war shifts. One of the most important changes has been the growing awareness of the significance of what we might broadly term culture (i.e. issues of meaning, symbolism and representation) in understanding socio-spatial structures. Thus for example, recent geographical approaches to such key issues as community or landscape have emphasised the construction of place and place-based identities as both a material and symbolic product, while also recognising the significance of past changes. Thus we might define places as living histories of past and current social and cultural relations. I shall try to capture some of the flavour of this new approach. This definition – the recognition that places are constructed through social relationships – has also led to a far more dynamic conceptualisation of place. Instead of the former tendency to define a place through establishing a spatial boundary, it is now recognised that the significance of place depends on the issue under consideration and the sets of social relationships that are relevant to that issue. There has also been a further shift in the definition of place and the subject matter of social geography and that is the growing emphasis on the significance of the social relations of consumption in geographical analysis. As John Urry (1995) has suggested 'places are increasingly being restructured as centres for consumption, as providing the context within which goods and services are compared, evaluated, purchased and used' (p. 1). This reflects the changes in living standards and everyday life in the post-war period in which the development of mass consumption and later niche marketing altered people's relationships to their homes and neighbourhood, as well as facilitating escape to spaces and places of pleasure as mass transport improved and increased both its capacity and range. Growing numbers of people became able to visit and consume greater numbers of places. Urry thus suggests two further ways in which place and consumption are linked: first that places are consumed visually by visitors and second that the amenities and facilities in these places may themselves be consumed, in the sense of used up. Finally, Urry suggests a fourth

link between place, consumption and identity. Through the development of local interests, the romantic rediscovery of local or 'folk'
culture, preservation societies or action groups, repeated visits perhaps, 'it is possible for localities to consume one's identity, so that such
places become almost literally *all-consuming* places' (p. 2, original
emphasis).

The temporal comparison

As I suggest above, one of the aims of the reader is to reflect on the
past through the eyes of the present. I have indicated that it is to the
1950s that contemporary commentators now refer when they variously
lament or celebrate the breakup of a perceived stability, a social order
in which the social classes, women and various other unruly Others
knew their place. This temporal reference is not, however, tightly
restricted to the years between 1950 and 1959 but extends both back to
the post-war social compact and forward into some date in the 1960s,
when the challenges to the Establishment became uncontainable and
the hedonism of the decade – the swinging sixties – and the politics of
protest – the anti-Vietnam war protests, the Women's Movement, the
'events' of Paris 1968 – seemed to herald the dawning of a new age. In
economic terms the death knell of the old era came a few years later,
with the oil crisis and the recession which also punctured the optimism
of the new age as well as marking the automatic assumption of growth
in post-war western democracies. Thus David Harvey refers to 'around
1973' as the pivot of change and Raymond Williams similarly suggests
the early 1970s as the end of the older order. Interestingly, Marshall
Berman suggests two moments and two rather different social movements that led to the 'postmodern': the first political as well as cultural,
located in the USA, and a decade before the period chosen by Harvey
and Williams; the second an intellectual movement among Continental
theorists.

> The first wave of postmodernism emerged around 1960 in America's uni-
> versities and Bohemian enclaves. It sprang from the people who invented
> happenings, assemblages, environments and the art that would come to be
> called Pop – people who without knowing it were inventing the 1960s. For
> the most part, they were too busy to worry about labels. But they were at
> least occasionally willing to answer to the label 'postmodern' because they
> all deplored the cultural orthodoxy that, in the 1950s, seemed to pre-empt
> the label of modernism . . . Nothing would have appalled the 1950s trustees
> of culture more than the idea that serious art could make you laugh. . . .
>   The new faces of the early sixties were more active politically, and more
> militant in the demands they made on life, than were the modernists of the
> cold war years. At the same time they were in love with the world they
> wanted to change. . . .
>   If the first wave of postmoderns was composed of the people who
> invented the 1960s, the second wave, still flowing today, is a strange
> combination of people who were born too early to participate actively in the
> 1960s, and people who were born too late and missed the sixties. This
> postmodernism was created by Parisian academics who spent their whole

lives as members of the enviably privileged French mandarin caste. For two minutes, in May 1968 their lives were transfigured, a terrible beauty was born; in two minutes more, all their hopes were dead. The postmodernisms of the past twenty years grew out of this trauma, and also out of a collective refusal to confront it.

Instead the Left Bank exploded with all the feverish rhetoric and sectarian fanaticism that typify radical politics at its worst, combined with a total abdication of concern for political issues and relationships in the grubby real world. (Indeed it was typical of Parisian postmodernism to say it made no sense even to talk about a real world: there was 'nothing outside the text' as Jacques Derrida liked to say.) Derrida, Roland Barthes, Jacques Lacan, Michel Foucault, Jean Baudrillard, and all their legions of followers, appropriated the whole modernist language of radical breakthrough, wrenched it out of its moral and political context, and transformed it into a purely aesthetic language game. Eros, revolution, terrorism, diabolical possession, apocalypse, were now simply ways of playing with words and signifiers and texts. As such they could be experienced and enjoyed – jouir, jouissance, Roland Barthes' favourite words – without engaging in any action, taking risks or paying any human costs. If modernism found both its fulfilment and its ruin in the streets, postmodernism saved its devotees the trouble of ever having to go out at all. (Berman, 1991, pp. 43–4)

These are jaundiced words and it is important to remember that some of these thinkers that Berman lumps together as postmodern may not only disagree with the label but were also active in French politics, most notably Michel Foucault. But it is a common claim that the postmodern emphasis on texts and language, on multiplicity and difference was politically disabling. I shall return to these arguments in Section Six. The huge impact of postmodern thought on geographies, theories and methods is discussed in the companion reader by Trevor Barnes and Derek Gregory (1997) *Reading Human Geography: The Poetics and Politics of Inquiry.*

Here I concentrate on how the socio-spatial changes of the period of changes have affected people's attachment to place. I want to look back at the years between the end of the Second World War and the early to mid 1970s through the lens of the 1990s, when a dour, moralistic *fin de siècle* sullenness seems to have descended on the west and a series of backward-looking appeals to an imaginary set of moral values and family life are common. The current political rhetoric, from both the left and the right, of belonging to a family, community and nation is profoundly conservative. These years between about the mid 1950s and the end of the 1960s were, in Britain at least, years of extraordinary transformation. A poorly housed and under-educated population experienced the effects of modernisation, economic growth and rising living standards under a Keynesian welfare state. The coincidence of a vast range of socio-economic changes challenged the stability of the 'old order': new ways of living, growing access to consumer durables, migration into the more prosperous areas of the country, a new generation of students who gained access to the higher education institutions that previously had been restricted to a tiny élite, social mobility for newly educated through movement into new white collar jobs in the public and private sector, as well as a growing workforce of former colonial peoples meant the 'racialisation' of the working class as well as its feminisation, recutting 'race', class and gender divisions and challenging the separation of public and private spheres.

Similar processes were apparent in other parts of western Europe, most notably France and the former West Germany. In the United States, the end of the Second World War marked rather a resumption of modernisation than its beginning. The possession of new consumer durables had begun in that nation in the 1930s rather than the 1950s, and of course the experience of migration was different. As Kirsten Ross has noted in her book about the modernisation of France in the 1950s, Henri Lefebvre was an astute commentator on the contrasts between the experiences of his country and that of the USA. 'Contrasting the French experience to the slow, steady, "rational" modernization of American society that transpired throughout the twentieth century, Lefebvre evoked the almost cargo-cult-like, sudden descent of large appliances into war-torn French households and streets in the wake of the Marshall Plan. Before the war, it seemed, no one had a refrigerator; after the war, it seemed, everyone did' (Ross, 1995 p. 4). In Britain, the proportion of working-class households owning a refrigerator rose from 4 per cent in 1951 to twenty times that by 1971. And, as Aglietta has pointed out in his analysis of Fordism, and others have in their work on the post-war spread of the suburbs, this era was dominated by two mass consumption products: standardised housing and the motor car (the racier US term 'the automobile' is better) which allowed the increasing separation of the home and the workplace and necessitated the growing individualisation of consumption. Here the experiences of the US and western Europe are parallel. So for all households, but especially for the working class, the 1950s and 1960s were decades dominated by things. Born in 1949, I grew up in these decades and the passing of the years is marked for me by the gradual acquisition of goods. First a move to a house with an indoor lavatory, to a street with electric lighting in 1954, then a fridge, a washing machine, later a phone, then central heating, first solid fuel, which still entailed carrying coal into the house, but later gas-fired; but not a car until after I became an adult and left home. My own children seeing all these things as basic necessities cannot understand the difference they made to the standard of living for most people, and for women in particular, in the immediate post-war decades.

In the more recent past, however, it has been argued that these consumption goods have become more than a way of raising living standards and improving the circumstances of everyday life, but have begun to determine everyday life. The social relations of consumption, rather than of the workplace, have become the key determinant of identity. Through the media and advertising and market dynamics, a constant search for novelty, new styles and new sensations has become predominant and, it is argued by postmodern theorists and cultural analysts, life itself became redefined as an act of consumption or a work of art, what Featherstone (1991) has termed 'the aestheticization of everyday life'. Images, aesthetics and cultural practices are central in the identification of a shift from modernity to postmodernity in western societies. This new emphasis has been variously explained. While critics such as Harvey and Jameson see it as a necessary consequence of changes in the mode of capitalist organisation, 'the cultural logic of late capitalism' (Jameson, 1984), other theorists, Urry among them, dispute the neccessity of the connections. There is general agreement, however, that these cultural and economic changes are 'bound up with the emergence of new dominant ways in which we experience space and time' (Harvey, 1989, p. vii). Technological innovations and new means of transport from the telephone and the train to the internet and the supersonic aircraft have, according to Harvey and many others, speeded up the 'annihilation of space by time', reducing the friction of distance and facilitating almost immediate communication over increasingly large distances. This, in the words of the urban theorist Manuel Castells (1989) has resulted in the replacement of the modern spaces of places with postmodern spaces of flows, in deterritorialisation, and the detachment of individual and group identity from local places in network societies (Castells, 1997a and b). I have captured my unease with these arguments in the question mark in the reader's title. Place and the local still matter and millions of people spend the vast majority of their daily lives in a restricted spatial sphere. Further it is clear that place in a symbolic sense is a crucial part of contemporary identities. We may instead be witnessing a reterritorialisation, in which new geographies and communities based on different forms of collective identification, which may or may not be place-based, are developing. But there is little doubt that in the contemporary era the restructuring of space and time places geographic questions right at the forefront of intellectual work.

In trying to think through the issues raised by postmodern shifts in the nature of space and time, I have found the parallels in the work of David Harvey (1989) and that of Marc Auge, a French anthropologist who is perhaps less well-known to geographers than is David Harvey, extremely helpful. Auge (1996) suggests that the key experience in the transformation of space and time is that of excess or superabundance. He prefers the term supermodernity to postmodernity as it more accurately represents what is happening: a speeding up or multiplication rather than a complete or radical transformation. Indeed, Harvey's argument in *The Condition of Postmodernity* is similar despite using a different term.

Let us look at the arguments of Auge. First he suggests that our perception and use of time has changed. 'The idea of progress, which implied an afterward explainable in terms of what had gone before, has run aground, so to speak on the shoals of the twentieth century, following the departure of the hopes or illusions that had accompanied the ocean crossing of the nineteenth' (p. 24). There are a number of reasons for disillusionment with a linear idea of history:

> the atrocities of the world wars, totalitarianisms and genocidal policies, which (to say the least) do not indicate much moral progress on the part of humanity; the end of the grand narratives, the great systems of interpretations that aspired to map the evolution of the whole of humanity, but did not succeed, along with the deviation or obliteration of the political

systems officially based on some of them; in sum a doubt as to whether history carries any meaning (p. 24).

To these doubts about meaning, Auge adds a speed up in time, the acceleration of history in which the events of the recent past become history almost immediately: 'Nowadays the recent past – "the sixties", "the seventies", now "the eighties" – becomes history as soon as it is lived' (p. 26). Thus there is, Auge suggests, an overabundance of events and of information about them.

This magnification or excess of history is paralleled by the second change: an accelerated transformation of space – 'the excess of space is correlative with the shrinking of the planet' (p. 31). This shrinking is the result not only of travel by rapid means of transport that brings any capital city within a few hours of any other (as long as you can afford the fare, of course) but also because of the ways in which new information technologies have altered the relations between space and time, proximity and distance, public and private events. 'In the privacy of our homes, images of all sorts, relayed by satellites and caught by the aerials that bristle on the roofs of our remotest hamlets, can give us an instant, sometimes simultaneous vision of an event taking place on the other side of the planet' (p. 31). The screen also makes familiar to us world actors in a range of domains. News is mixed with entertainment, sport with politics. We may not know the participants in these events but we recognise them. Images replace knowledge. The result, Auge argues, is that '[w]e can say of these universes, which are themselves broadly fictional, that they are essentially universes of recognition . . . closed universes where everything is a sign; . . . totalities which are partially fictional but which are effective' (p. 33). Here is the reason, I think, why geographers are increasingly dealing with questions of meaning and representation, fictional universes in Auge's words, and why Derrida's argument that 'there is nothing outside the text' appeals to some, although notice that Auge is more cautious, referring instead to partial fictions. The effects of television on the meaning of space and place, for example, are discussed in the extract by Spigel in Section One of this reader and in Section Five I examine imagined and fictional places and the impact of cyberspace on geographic 'reality'.

In his consideration of space, Auge argues that, like time, the effect of the spatial overabundance of the present is less to subvert than to complicate an understanding of space, 'for soils and territories still exist, not just in the reality of facts on the ground but even more in that of individual and collective awareness' (p. 35). Here too we see a reason to turn to cultural questions and imagined spaces; to examine people's desire for a territorial identity and their willingness to construct and act on myths of belonging. In considering 'real' places and spaces – new urban concentrations, new means of transport – Auge interestingly suggests that there is a growing proliferation of what he terms 'non places' associated with accelerated circulation of people, goods and ideas, rather than traditional notions of place as 'the idea of a culture localised in time and space'. In Sections Two and Three I have included readings that examine the decline of localised place, as well as shifts in the relationship of belonging and territory, and the rising significance of 'imaged and imaginary references' (Auge, p. 34).

Auge's discussion of space has clear parallels with Harvey's notion of space–time compression. Auge also points, as geographers have, to the paradox of globalisation ('when it becomes possible to think in terms of the unity of terrestrial space' (p. 34)) and the assertion of local particularisms or what is sometimes, somewhat inelegantly, termed glocalisation. Increasingly, local voices are raised: 'clamour from those who want to stay at home in peace, clamour from those who want to find a mother country. As if the conservatism of the former and the messianism of the latter were condemned to speak the same language: that of land and roots' (Auge, 1996 p. 35). Auge's recognition of the conservatism of this language of home is important. Harvey is more ambivalent about the claims, although he too fears the conservatism of localism. In the final section, I have included an article by Doreen Massey which asserts the progressive potential in local action and the 'locality approach' by geographers.

Ambivalence, however, is a key theme in assessments of the implications and consequences of all the social trends that go to constitute postmodernity. A sense of loss for the past is as common as a celebration of current and future change. Some commentators have emphasised the gain/pleasure/freedom that has resulted from the challenges to old forms of social organisation and community, the ability for ever growing numbers of people to break out of the old bounds and bonds that tied people to place, be they the ties of the family and kinship, of class-based communities, or of a white nation. These place ties have been broken by changes in family and household forms, by a challenge to patriarchal gender relations, from women's entry into waged work, and growing male employment consequent on regional restructuring, by migration, by new ethnicities, and new forms of politics. This celebratory stance is nicely summed up in the phrase the 'new times' (Hall and Jacques, 1990) whereas the despair felt by others is expressed in the angry response by Sivanandan (1990) who referred to the 'hokum of the new times' when all that is solid (class politics, masculine certainty) has melted into air. But for those who were excluded from mainstream society – that diverse cast of 'Others' that includes women, people of colour, the perverse and the mobile – the move towards a radical decentring of old certainties has constructed a space from which to speak. It is hardly surprising that those whose authority is so challenged find the present less comfortable. Interestingly Robert Young (1995) has recently suggested that the 1990s reactions: the desire to recreate moral and family values, fears of diversity, and anxiety about relativism in Britain and the US mark societies that are uneasy and uncertain. Difference and diversity, he suggests, are only tolerated by a society that is confident and outward-looking.

Selecting the articles and extracts to include

Selecting the readings for any reader is always difficult. I am lucky in that I have been able to develop a reader with a specific focus, united by a common theme and periodisation and did not have to discuss either what is social geography or outline the history of its shifting interests. That has been done for me, and you, by Chris Hamnett (1996) in his companion reader to this volume, *Social Geography: A Reader*, where concepts of race, class and gender are defined in some of the classics and other pieces by geographers. What I want to do here is to introduce some perhaps less familiar arguments and to show how geographers' understanding of concepts such as home, place, community and nation might be enhanced by exploring literatures in other disciplines that also place these spatial concepts at the centre of their analyses. That there is a reciprocal relationship goes almost without saying. Social and cultural theorists have also found much of value in geographical texts and papers. So, in terms of the authors and disciplines included here you might be forgiven for not immediately recognising the content of reader as geography. As McRobbie has suggested, the locus of the construction of identity has changed from traditional categories of class, work and community to 'other constellations of strong cultural meaning: the body, sexuality, or ethnicity for example; nationality, style, image, even subculture' (McRobbie, 1994 p. 6) and I wanted to be able to give at least a flavour of these new poles of identity and the way they are linked to a range of places. But even so, in so far as disciplinary boundaries matter at all, the reader is firmly addressed to a geographic audience. I hope it will not only encourage you to look at a range of books and journals that are probably less familiar than others but also that it will be a contribution towards rethinking the boundaries of social geography and the type of questions that are important at the end of the twentieth century.

If social geography, like anthropology, defines itself as 'the study of social relationships on the ground and the cultural values (including notions of community) that attended to such relationships' (Strathern, pp. 188–9), then there is little doubt that the nature of these social relationships and the ground or the territory on which they are played out, indeed constituted by, has shifted in the last half century. The connections and relations between people and their understandings of their place in the world have shifted fundamentally with economic restructuring, technological change and the transformation of knowledge. Thus what it means to be a man or a woman, to have a nationality and a sense of place has been transformed and reinvented as people's view of the world has changed. The assumption, for example, that the ideal individual was one who most nearly conformed to the rational western subject in a world that might be controlled through thought and nature has been undercut. Similarly, the idea that superior knowledge is a singular construction and is that which most nearly conforms to an exterior reality has also been challenged. The type of questions that we ask and the multiple ways in which we try to answer them means that geography as a discipline has also changed in the second half of the twentieth century. It is one of the most exciting subjects that it is possible to study as its focus on the links between identity and place are at the centre of the transformations of place and knowledge at the end of the millennium.

Further reading

Alexander, J. C. 1994: Modern, anti, post and neo. *New Left Review* **210**, 63–104.

Appadurai, A. 1990: Disjuncture and difference. *Theory, Culture and Society* **7**, 295–310.

Auge, M. 1996: *Non places: introduction to an anthropology of supermodernity.* London: Verso.

Barnes, T. and Gregory, D. (eds) 1997: *Reading human geography: the poetics and politics of inquiry.* London: Arnold.

Berman, M. 1991: Why modernism still matters. pp. 58–90 In Lash, S. and Friedman, J. (eds) *Modernity and identity.* Oxford: Blackwell, 58–90.

Barrett, M. 1992: *The politics of truth.* Cambridge: Polity.

Castells, M. 1989: *The information city.* Oxford: Blackwell.

Castells, M. 1997a: *The rise of the network society.* Oxford: Blackwell.

Castells, C. 1997b: *The power of identity.* Oxford: Blackwell.

Crook, S. et al. 1992: *Postmodernization: change in an advanced society.* London: Sage.

Featherstone, M. 1991: *Global culture.* London: Sage.

Giddens, A. 1979: *Central problems in social theory.* London: Macmillan.

Hall, S. and Jacques, M. (eds) 1990: *New times: the changing face of politics in the 1990s.* London: Lawrence and Wishart.

Hamnett, C. (ed.) 1996: *Social geography: a reader.* London: Arnold.

Harvey, D. 1989: *The condition of post modernity.* Oxford: Blackwell.

Jameson, F. 1991: Postmodernism, or the cultural logic of late capitalism. London: Verso.

McRobbie, A. 1994: *Postmodernism and popular culture.* London: Routledge.

Moore, H. 1994: *A passion for difference.* Cambridge: Polity.

Ross, K. 1995: *Fast cars, clean bodies: decolonisation and the reordering of French culture.* Cambridge, Mass.: MIT Press.

Sivanandan, A. 1990: *Communities of resistance.* London: Verso.

Smart, B. 1993: *Postmodernity.* London: Routledge.

Soja, E. 1989: *Postmodern geographies.* London: Verso.

Soja, E. 1996: *Thirdspace.* Oxford: Blackwell.

Strathern, M. 1992: *After nature.* Cambridge: Cambridge University Press.

Urry, I. 1995: *Consuming places.* London: Routledge.

Watts, M. 1992: Capitalisms, crises and cultures I: notes towards a totality of fragments. Introduction to Pred, A. and Watts, M. (eds) *Reworking modernity: capitalism and symbolic discontent.* Brunswick, New Jersey: Rutgers University Press.

Watts, M. 1991: Mapping meaning, denoting difference, imagining identity: dialectical images and postmodern geographies. *Geografiska Annaler* **73**, 7–16 (extracts reprinted in Barnes, T. and Gregory, D. (eds) 1997: *Reading human geography.* London: Arnold).

Young, R. 1995: *Colonial desire.* London: Routledge.

SECTION ONE

HOMEPLACE

Editorial introduction

In this part of the reader I want to consider questions about the relationship of home and identity, particularly its significance in the social construction of gender identities. I am interested in the power of the concept of 'home' and how this has changed over time as well as its key significance both in feminist writing and in the rhetoric of political parties of both the left and the right. The concept home is crucially tied to the idea of community and both terms are used at a variety of spatial scales. Thus we use home to refer to the house we live in, the local area, the town or city, even a region or country (for a student 'going home' may involve a journey to another part of the country or part of the world). We also talk about our home country or 'homeland', which with the exception of the specific use of the latter term in apartheid South Africa, tends to have connotations of belonging and a sense of national identity. Community or community ties, on the other hand, most commonly refer to a smaller spatial scale: that of a neighbourhood, although it is also used to refer to interest groups, a community of like-minded people perhaps. Despite these overlapping meanings, I want to distinguish the use of the terms home, community and homeland and, perhaps artificially, to examine each in turn in the context of a specific spatial scale. Thus in Section One, by home I mean the house or the dwelling itself. In Section Two I shall move up a spatial scale and examine communities defined as spatially fixed or grounded places. In Section Three I want to turn to fluid communities, non-place based associations of people defined by a particular characteristic or set of characteristics and united by their desire to escape the bounds of home and locality. In Section Four I shall turn to the meaning of home in the sense of 'homeland', the association between territories and national identities.

The term home is an evocative word, redolent with meaning and sets of associations. Ideally the term is associated with safety, with familiar and protective boundaries, with the family, the exclusion of unwanted others, with privacy, a haven in a heartless world. In UK and USA, this idealised view of the home has a particular set of associations: with white, patriarchal, Christian, middle class and liberal assumptions about the correct way to live. But the home, as feminist geographers and other feminist scholars have argued, is also a location and result of race, class and gender conflicts. (There is now a large feminist literature about the home, some of which is included in *Space, Gender, Knowledge*, a companion volume to this one that I have co-edited

(McDowell and Sharp, 1997) but see also Andrew and Moore-Milroy, 1988; Bowlby et al., 1989; Dowling and Pratt, 1993; McDowell, 1989; Mackenzie, 1989; Pratt 1990; Spain, 1992; Weisman, 1992.) For many women, it is argued, the home is a sphere of domination and oppression: if the Englishman's home is his castle, then his wife and children are his chattels (literally in Britain until the Married Women's Property Act was passed in 1883). While for many men the home is an arena of leisure, for women it is also the location of a great deal of work – the housework and childcare that keeps the family clean and fed, the house clean and the home functioning smoothly. Ideally it is a male bread-winner who, by venturing into the public world of work, maintains this division of labour at home, returning to his reward of privacy and relaxation in his home in the evenings and at weekends.

This particular division of space, into a private and domestic sphere associated with women and a masculine public arena of waged work and politics came to dominate capitalist societies, as an ideal if never a completely realised material reality, in the transition to industrial capitalism. Its establishment was an historically variable process and its greatest reach was among the middle classes. Working-class house-holds could seldom afford the privilege of a non-wage earning member and working-class women have long had to labour to contribute to their own and their families' well-being. The gendered assumptions also came to dominate theory, as the supposed shift from a work-based to a consumption-based identity in the transition from modernity to postmodernity may parallel many men's experience but the changing position of women tends to counter this generalisation.

The 1950s is the era which is most commonly regarded as the apotheosis of home-based family values in the west: the period in which it is commonly assumed that fathers went to work and women stayed at home nurturing their family, content to surrender the limited independence that they had gained during the Second World War. Indeed so pervasive was the ideology of domesticity that it is com-monly suggested that feminism as a social movement disappeared until the late 1960s. However, as Elizabeth Wilson has argued, in this period women were 'only halfway to paradise' and 'there was, some-where, still a knowledge and an understanding that women remained in many ways subordinate and oppressed' (Wilson, 1980, p. 2–3). But the ideology of familialism and domesticity was so strong that when Frieden came to write about the dissatisfaction of educated women whose main role in life was taking care of the home she could only term their discontent 'the sickness without a name' (Frieden, 1963). In the first reading in Section One, **Stephanie Coontz** examines the social conditions in 1950s USA and shows that the retrospective nostalgia for a more settled period is exactly that – nostalgia. In fact there was greater social diversity in household and family forms in the 1950s than is now allowed. In Great Britain, for example, the late 1990s have seen a moral crusade against alternative ways of living, especially against single parenthood and a demand for a return to the 'family values' of the 1950s.

But as Wally Seccombe (1993) has argued, in his historical assess-ment of the family, it has been the values of the marketplace that have influenced family far more than any supposed decline in moral values. During the fifties, with full male employment, strong trade unions and welfare provisions to ensure a basic minimum income the family became a significant consumption unit. New suburban homes began to be filled with consumer goods to support the privatised lifestyle of the expanding middle class and the relatively secure working class. Indeed, sociologist John Goldthorpe (1968) and his colleagues believed that they had identified a new privatised family form based on the embourgoisement of the working class. Home decorating, tinkering with the car and the garden not only kept a man busy at home but also relatively quiescent at work as he laboured in the car industry to earn the wages for this new consumption-based lifestyle. Unfortunately just after Goldthorpe *et al.* announced their new thesis, the Ford plant in Luton where they did their research was hit by industrial action and, more significantly, the collapse of the Fordist social compact that produced a family wage for the male aristocracy and was the foundation of the 1950s family. The male manual worker and job security have virtually vanished in an economy increasingly dependent on part-time, cheap, casual female and, in the USA more than in Britain, on minority labour working in exploited positions.

There is no doubt that the changing structure of the labour market has been the major challenge to the public/private divide and its association with gender. Rising unemployment among working-class men made them visible at home and in the urban landscape in the day in previously unknown ways. As the astute social commentator Beatrix Campbell (1984) noted of unemployed men in 1980s Great Britain:

> It's as if he is stranded in a place where he is an immigrant who doesn't speak the language – at home. Millions of men have fought for the right not to be there. They have spent most of their time not there, at work or out to play with other men. For them home is not where the heart is, it's where the wife is. (p. 170)

But growing numbers of wives are now leaving home every day (in the 1950s 15 per cent of all women worked for wages, now it's over half) and men have had to come to terms with cold dinners and unironed shirts, as well as with their roles as househusbands. Jane Wheelock (1990) has investigated the consequences of 'Husbands at Home' in a fine study of Geordie working-class men, once the epitome of the 'lads' who, as Campbell found in her later study *Goliath* (1993), now have to construct and reinforce their masculinity not in shared dangers in the workplace but in anti-social activities in the locality: in car thefts and chases for example. And for growing numbers of young working-class women in these communities, often peripheral council estates built in the more hopeful 1950s and 1960s, unemployed young men are now longer a 'catch' but a burden, yet another dependant in their lives. Young men without work, living on welfare benefits that anyway are now a smaller proportion of average incomes, and perhaps involved in anti-social activities, are a different prospect for women than the solid working men of the 1950s. Young women's choices to live alone or to raise children as single parents may be a rational response to current social conditions in the 1990s rather than the 'selfishness' or 'feminist self-assertion' too often reached for by popular commentators as an explanation for the breakdown of 'the family'.

The overall result, however, is a revaluation of the meaning of home and a challenge to the old patriarchal gender order on which its traditional meaning is based. Judith Stacey, in an interesting investigation into the changing meaning of the family and home in post-industrial California, while not necessarily mourning the old order, has some astute comments on the specificity and short-lived nature of that idealised form. She suggests that:

> One glimpses the ironies of class and gender history here (in the decline of the 'traditional' family). For decades industrial unions struggled heroically for a socially recognised male breadwinner wage that would allow the working class to participate in the modern gender order. These struggles, however, contributed to the cheapening of female labour that gradually helped to undermine the modern family regime. Escalating consumption standards, the expansion of mass collegiate coeducation, and the persistence of high divorce rates then gave more and more women ample cause to invest a portion of their identities in the 'instrumental' sphere of paid labor. Thus middle class women began to abandon their confinement in the modern family just as working class women were approaching its access ramps. The former did so, however, only after the wives of working class men had pioneered the twentieth century revolution in women's paid work. Entering employment during the catastrophic 1930s, participating in the defense industries in the 1940s, and raising their family incomes to middle-class standards by returning to the labor force rapidly after child rearing in the 1950s, working class women quietly modeled and normalized the post-modern standard of employment for women. Whereas in the 1950s the less a man earned, the more likely his wife was to be employed, by 1968 wives of middle income men were the most likely to be in the labor market.
>
> Thus the apotheosis of the modern family only temporarily concealed its immanent decline. (Stacey, 1992, p. 11)

But it was not only, or even at first mainly, women who abandoned the family and the modern gender order that underpinned it. As Barbara Ehrenreich (1983) has argued in her book *The Hearts of Men*, referred to in Chapter 1 by Coontz, the origins of changing family forms and what she terms the 'flight from commitment' are to be found in men's behaviour rather than in women's rejection of domesticity and their entry into waged labour. Like Seccombe and others she noted the rise of consumerism from the 1950s but instead of assessing the effects of women's spending patterns in their role as homemakers in the 1950s and 1960s, she has drawn attention to the importance of the spending power of single men. The 1950s and 1960s not only saw the rise of consumption based on a familial lifestyle but also the beginnings of new individualised consumption patterns. It was in the 1950s USA, for example, that Hugh Heffner started *Playboy*, recommending a hedonistic lifestyle based on a range of accoutrements such as fast cars, music centres, hair products, wine and good food for the single man who had the sense to evade the clutches of a wife and family. Figure 1 is a spoof ad from a 1963 article in *Playboy* advising men to evade female 'clutches'. *The Hearts of Men* is a light-hearted but serious

TIRED OF THE RAT RACE?
FED UP WITH JOB ROUTINE?
Well, then . . . how would you like to make $8,000, $20,000 – *as much as $50,000 and More* – working at Home in Your Spare Time? No selling! No commuting! No time clocks to punch!
BE YOUR OWN BOSS!!!
Yes, an Assured Lifetime Income can be yours *now*, in an easy, low-pressure, part-time job that will permit you to spend most of each and every day as *you please!* – relaxing, watching TV, playing cards, socializing with friends! . . .

Figure 1   A spoof ad for a wife. From a *Playboy* article entitled, 'Love, Death and the Hubby Image', published in 1963. (Quoted in Ehrenreich, B. 1983, p. 48.)

examination of the origins of that self-obsessed single man that now features so widely in current advertising campaigns. While Heffner's spoof advertisement shows that women's confinement in the home is open to various interpretations, a number of British sociologists and geographers also argued that the feminist critique of the ideology and actuality of the home was misguided. In a passionate piece, building on an earlier article with Peter Williams (Saunders and Williams, 1988) in which they set out an interesting agenda for geographical research on the home and its multiple meanings, Peter Saunders (1989) attempted to demolish feminists' assertions that housework was drudgery. Through a questionnaire survey he purported to demonstrate that the home was instead, for both his female and male respondents, an arena of love and affection, identified with leisure and relaxation. 'Women are just as enthusiastic as men in volunteering images of warmth, love and comfort when asked what the home means to them, and their answers should be respected as valid representations of their everyday experiences.' However, it is unlikely that through a questionnaire (and of course it also depends on what questions are asked) people will reveal the tensions that exist in their households. Further it is always dangerous to generalise on the basis of a sample that reflects conventional gender divisions. Women who leave home because of male violence and are either homeless or in a refuge and so not included in sample surveys like Saunders', might report different attitudes. What Saunders seems unable to accept is that the meaning of home is complex and contested. The home may be both a sphere of work and of relaxation – for men and for women. It may be a haven as well as a trap. But Saunders' work did play a valuable part in pushing feminist geographers to accept that their analyses might be overly deterministic or culturally specific. As the long quote from Stacey above reminds us, working-class women are being denied the home life that middle-class women rejected by changing economic circumstances. For middle-class women, waged work may be 'liberating' whereas for working class women, it is far more likely to be exploitative. In Chapter 2, **bell hooks** argues that white feminists' critique of the home misunderstood the significance of the home and family for black American women and men. For them the home has long been an important site of resistance against the intrusions of the state and, formerly, against slave owners.

The work of Coontz and hooks demonstrates the importance of understanding the variety of family forms, the historical and geographical specificity of the notion of home as well as the multiple ways in which consumption behaviour as well as participation in waged work has challenged traditional assumptions about gender divisions. The assumed privacy of the home and its separation from the 'public' world of politics has also had to be rethought as the growing penetration of television and then other forms of new communication technologies brought the world into the home for growing numbers from the 1950s onwards. **Lynn Spigel** in Chapter 3 examines the effects of television on the participation of different family members in the 'public' world of the local neighbourhood, its streets and associations, in the private arena of the home and in a placeless 'new community of values' (Spigel, p. 187 in original). Her argument thus parallels the work of the British sociologists mentioned earlier who also documented the withdrawal from the collective spheres of traditional communities. In Section Two the changing social bases of community in post-war Britain will be examined in more detail. Spigel also suggests that a profound reorganisation of social space and its gendered divisions was precipitated by television such that participation in activities that had previously taken place in public arenas – concerts, opera, watching sport – were relocated to the home. In Britain Jonathon Gershuny and Ray Pahl (1979) have also documented the move towards what they termed 'self-provisioning' as record players and then CD players allowed households to recreate the concert hall at home, as well as to use power tools and other equipment to improve their homes. Certainly a decline in attendance at cinemas, concert halls and in eating out in restaurants was evident throughout the 1960s and 1970s but it was reversed in the 1980s and 1990s, as new forms of niche-marketed performances – specialist small-screen cinemas in multiplex buildings for example and the rapid rise in new 'ethnic' and fast-food restaurants and chains – have drawn people back into the public arena. There is no doubt, however, that the advent and spread of television to virtually every home in Britain and the USA in the post-war decades had a remarkable impact on public consciousness and the organisation of spatial divisions. Not only were public and private divisions renegotiated but the organisation of rooms, of meals and control over technology in the home influenced the internal spaces of houses and the activities therein, as David Morley (1991) has demonstrated in his small-scale ethnographic work on who controls what equipment in contemporary British homes. In the final extract in Section One **Tim Putnam** ponders the significance of these changes and the importance of the fixed and parochial aspects of the home in a survey of the recent work on the home and the domestic arena. How is the home to be analysed as a 'terminal' in the vast networks of information that link places in the 'postmodern' world?

Geographical questions about the home are not exhausted by the articles included here but space, in the sense of words and pages, precludes more detailed assessment. These include a range of questions at two spatial scales: that of the relationships between the body and its positioning in space within the home and at the other end of the scale about the location of the home, or residential areas within the city. There are also important questions about people and households who are excluded by the conventional associations between the home and the nuclear family – single people, groups of unrelated household members or gay households for example. The significance of gay communities in the city will, however, be discussed in Section Two. There are also too many people in Britain and the USA who are permanently or temporarily homeless. Here work by scholars such as Sophie Watson (1986) and April Veness (1993) has challenged conventional definitions of homelessness and reminds us that it has (or should have) a wider meaning than being literally without a roof. Are children or the elderly who live 'in a home', as institutions are often and inaccurately termed, homeless or not? Other interesting work on the perspectives and lives of homeless people has been undertaken by a group of geographers at the University of Southern California (see for example Rowe and Wolch, 1990; Dear and Wolch, 1987).

Turning to the micro spatial scale there are questions about whose body is allowed in particular spaces, and whose is 'out of place' in a particular room or a building. There is widespread evidence of gendered divisions of internal and external space in a diverse range of societies and circumstances. In ancient Greece, for example, the Goddess Hestia was symbolised by the circular hearth, the enclosed privacy of the home whereas the God Hermes was the god of the threshold and the gate representing movement and social relations with strangers. The feminine and masculine associations are common to many societies. In Britain the domestic arena is that of the woman, the Victorian 'angel of the hearth' who parallels the ancient Greek goddess as women's role model. Here anthropologists, from the classic collection edited by Shirley Ardener (1981) *Women and Space* onwards, and geographers have added to our understanding of how sacred and profane spaces from houses to government and other public buildings are constructed and represented (see Spain 1992 for a recent comparative study of internal divisions of space). This work raises related questions about how much space *per se* a body should or does occupy and what are the class and gender differences here. Many commentators have noted that women tend to take up less space than men, having been taught to stand and sit neatly with their limbs close together. The science of proxemics has developed to study variations in the spatial arrangement of bodies, in for example public buildings, on different forms of public transport or in queues. Observation of cinema queues, for example, has revealed that when a 'sexy' film is showing queuers stand closer together than for westerns or other genres. Further, the body itself is often perceived as a portion of social space, as a territory, with its boundaries and frontiers. It might be invaded (by disease) or penetrated (by a lover or rapist). The inverse imagery, of territory as a body, is also common. The historian Richard

Sennett (1994), for example, has traced the associations between the city and bodily imagery from ancient Athens to the present in his book *Flesh and Stone: the body and the city in western civilisation.*

At the scale of the city itself, there are a range of questions that have long been the focus of geographical attention: about land uses and urban layout, of planning or zoning of land uses that results in western cities in the common separation of residential land uses from 'non-conforming' uses as well as the spatial separation of housing built for people of different social statuses and class positions. These perhaps more familiar issues of social and urban geography are well-represented in a companion volume in this series, *Social Geography: A Reader* edited by Chris Hamnett. I wanted here to introduce you to perhaps less familiar issues about the place of the home. In Section Two, however, I shall turn to the larger spatial scale to assess the changing place and social construction of communities in the shift from a modern to a postmodern world in late twentieth century Britain and the USA.

References and further reading

Andrew, C. and Moore-Milroy, B. 1988: *Life spaces: gender, household, employment.* Vancouver: University of British Columbia Press.

Ardener, S. (ed.) 1981: *Women and space: ground rules and social maps.* London: Croom Helm.

Bowlby, S., Lewis, J., McDowell, L. and Foord, J. 1989: The geography of gender. In Peet, R. and Thrift, N. (eds), *New models in geography*, vol. 1. London: Unwin Hyman, pp. 157–75.

Campbell, B. 1984: *Wigan Pier revisited.* London: Virago.

Campbell, B. 1993: *Goliath: Britain's dangerous places.* London: Virago.

Dear, M. and Wolch, J. 1987: *Landscapes of despair: deinstitutionalization and homelessness.* Princeton, New Jersey: Princeton University Press.

Dowling, R. and Pratt, G. 1993: Home truths: recent feminist reconstructions. *Urban Geography* **14**, 464–75.

Ehrenreich, B. 1983: *The hearts of men: American dreams and the flight from commitment.* London: Pluto.

Frieden, B. 1962: *The feminine mystique.* New York: Dell.

Gershuny, J. and Pahl, R. 1979: Work outside employment: some preliminary speculations. *New Universities Quarterly* **34**, 120–35.

Giddens, A. 1992: *The transformation of intimacy.* Cambridge: Polity.

Goldthorpe, J., Lockwood, D., Bechhofer, F. and Platt, J. 1968: *The affluent worker: political attitudes and behaviour.* Cambridge: Cambridge University Press.

Hamnett, C. (ed.) 1996: *Social geography: a reader.* London: Arnold.

Hayden, D. 1984: *Re-designing the American dream: the future of housing, work and family life.* New York and London: W. W. Norton.

Heron, L. 1993: *Streets of desire: women's writing on the city.* London: Virago.

Leslie, D. A. 1993: Feminity, postfordism and the new traditionalism. *Environment and Planning D: Society and Space* **11**.

Lessing, D. 1972: *The golden notebook.* London: Joseph.

Maitland, S. 1988: *Very heaven: Looking back at the 1950s.* London: Virago.

McDowell, L. 1986: Towards an understanding of the gender division of urban space. *Environment and Planning D: Society and Space* **1**.

McDowell, L. 1989: Women, gender and the organisation of space. In Gregory, D. and Walford, R. (eds), *Horizons in human geography*. London: Macmillan: pp. 131–51.

McDowell, L. and Sharp, J. (eds) 1997: *Space, gender, knowledge: a reader for feminist geographers*. London: Arnold.

Mackenzie, S. 1984: Women in the city. In Peet, R. and Thrift, N. (eds), *New models in geography*, vol. 1. London: Unwin Hyman, pp. 109–26.

Mackenzie, S. 1989: *Visible histories: women and environments in a post-war British city*. London: McGill-Queens University Press.

Morley, D. 1991: Where the global meets the local: notes from the sitting room. *Screen* **32**, 1–15.

Pratt, G. 1990: Feminist analyses of the restructuring of urban life. *Urban Geography* **11**, 594–605.

Roberts, M. 1991: *Living in a man-made world: gender assumptions in modern housing design*. London: Pluto.

Rowbotham, S. 1989: *The past is before us: feminism in action since the 1960s*. London: Pandora.

Rowe, S. and Wolch, J. 1990: Social networks in time and space: homeless women in skid row, Los Angeles. *Annals of the Association of American Geographers* **80**, 184–205.

Saunders, P. 1989: The meaning of 'home' in contemporary English culture. *Housing Studies* **4**, 177–92.

Saunders, P. and Williams, P. 1988 The constitution of the home: towards a research agenda. *Housing Studies* **3**, 81–93.

Seccombe, W. 1993: *Weathering the storm: working class families from the industrial revolution to the fertility decline*. London: Verso.

Sennett, R. 1994: *Flesh and stone: the body and the city in western civilisation*. London: Faber and Faber.

Spain, D. 1992: *Gendered spaces*. London: University of North Carolina Press.

Stacey, J. 1992: *Brave new families*. New York: Basic Books.

Veness, A. 1993: Neither homed nor homeless: contested definitions and the personal worlds of the poor. *Political Geography* **12**, 319–40.

Watson, S. 1986: *Women and homelessness*. London: Routledge.

Weisman, L. 1992: Discrimination by design: a feminist critique of the man-made environment. Chicago: University of Illinois Press.

Wheelock, J. 1990: *Husbands at home: the domestic economy in a post-industrial society*. London: Routledge.

Wilson, E. 1980: *Only half way to paradise: women in post-war Britain 1945–68*. London: Tavistock.

Young, I. M. 1990: *Throwing like a girl and other essays in feminist philosophy and social theory*. Bloomington: Indiana University Press.

1 Stephanie Coontz

'Leave It to Beaver' and 'Ozzie and Harriet': American Families in the 1950s

Excerpts from: *The way we never were: American families and the nostalgia trap.* New York, Basic Books (1992)

Our most powerful visions of traditional families derive from images that are still delivered to our homes in countless reruns of 1950s television sit-coms. When liberals and conservatives debate family policy, for example, the issue is often framed in terms of how many 'Ozzie and Harriet' families are left in America. Liberals compute the percentage of total households that contain a breadwinner father, a full-time homemaker mother, and dependent children proclaiming that fewer than 10 percent of American families meet the 'Ozzie and Harriet' or 'Leave It to Beaver' model. Conservatives counter that more than half of all mothers with preschool children either are not employed or are employed only part-time. They cite polls showing that most working mothers would like to spend more time with their children and periodically announce that the Nelsons are 'making a comeback', in popular opinion if not in real numbers.[1]

Since everyone admits that nontraditional families are now a majority, why this obsessive concern to establish a higher or a lower figure? Liberals seem to think that unless they can prove the 'Leave It to Beaver' family is on an irreversible slide toward extinction, they cannot justify introducing new family definitions and social policies. Conservatives believe that if they can demonstrate the traditional family is alive and well, although endangered by policies that reward two-earner families and single parents, they can pass measures to revive the seeming placidity and prosperity of the 1950s, associated in many people's minds with the relative stability of marriage, gender roles, and family life in that decade. If the 1950s family existed today, both sides seem to assume, we would not have the contemporary social dilemmas that cause such debate.

At first glance, the figures seem to justify this assumption. The 1950s was a profamily period if there ever was one. Rates of divorce and illegitimacy were half what they are today; marriage was almost universally praised; the family was everywhere hailed as the most basic institution in society; and a massive baby boom, among all classes and ethnic groups, made America a 'child-centered' society.[2]

In retrospect, the 1950s also seem a time of innocence and consensus: Gang warfare among youths did not lead to drive-by shootings; the crack epidemic had not yet hit; discipline problems in the schools were minor; no 'secular humanist' movement opposed the 1954 addition of the words *under God* to the Pledge of Allegiance; and 90 percent of all school levies were approved by voters. Introduction of the polio vaccine in 1954 was the most dramatic of many medical advances that improved the quality of life for children.

The profamily features of this decade were bolstered by impressive economic improvements for vast numbers of Americans. Between 1945 and 1960, the gross national product grew by almost 250 percent and per capita income by 35 percent. Housing starts exploded after the war, peaking at 1.65 million in 1955 and remaining above 1.5 million a year for the rest of the decade; the increase in single-family homeownership between 1946 and 1956 outstripped the increase during the entire preceding century and a half. By 1960, 62 percent of American families owned their own homes, in contrast to 43 percent in 1940. Eighty-five percent of the new homes were built in the suburbs, where the nuclear family found new possibilities for privacy and togetherness. While middle-class Americans were the prime beneficiaries of the building boom, substantial numbers of white working-class Americans moved out of the cities into affordable developments, such as Levittown.[3]

Many working-class families also moved into the middle class. The number of salaried workers increased by 61 percent between 1947 and 1957. By the mid-1950s, nearly 60 percent of the population had what was labeled a middle-class income level (between $3,000 and $10,000 in constant dollars), compared to only 31 percent in the 'prosperous twenties', before the Great Depression. By 1960, thirty-one million of the nation's forty-four million families owned their own home, 87 percent had a television, and 75 percent possessed a car. The number of people with discretionary income doubled during the 1950s.[4]

For most Americans, the most salient symbol and immediate beneficiary of their newfound prosperity was the nuclear family. The biggest boom in consumer spending, for example, was in household goods. Food spending rose by only 33 percent in the five years following the Second World War, and clothing expenditures rose by 20 percent, but purchases of household furnishings and appliances climbed 240 percent. 'Nearly the entire increase in the gross national product in the mid-1950s was due to increased spending on consumer durables and residential construction', most of it oriented toward the nuclear family.[5]

Putting their mouths where their money was, Americans consistently told pollsters that home and family were the wellsprings of their happiness and self-esteem. Cultural historian David Marc argues that prewar fantasies of sophisticated urban 'elegance', epitomized by the high-rise penthouse apartment, gave way in the 1950s to a more modest vision of utopia: a single-family house and a car. The emotional dimensions of utopia, however, were unbounded. When respondents to a 1955 marriage study 'were asked what they thought they had sacrificed by marrying and raising a family, an overwhelming majority of them replied, "Nothing".' Less than 10 percent of Americans believed that an unmarried person could be happy. As one popular advice book intoned: 'The family is the center of your living. If it isn't, you've gone far astray'.[6]

The novelty of the 1950s family

In fact, the 'traditional' family of the 1950s was a qualitatively new phenomenon. At the end of the 1940s, all the trends characterizing the rest of the twentieth century suddenly reversed themselves: For the first time in more than one hundred years, the age for marriage and motherhood fell, fertility increased, divorce rates declined, and women's degree of educational parity with men dropped sharply. In a period of less than ten years, the proportion of never-married persons declined by as much as it had during the entire previous half century.[7]

At the time, most people understood the 1950s family to be a new invention. The Great Depression and the Second World War had reinforced extended family ties, but in ways that were experienced by most people as stultifying and oppressive. As one child of the Depression later put it, 'The Waltons' television series of the 1970s did not show what family life in the 1930s was really like: 'It wasn't a big family sitting around a table radio and everybody saying goodnight while Bing Crosby crooned "Pennies from Heaven".' On top of Depression-era family tensions had come the painful family separations and housing shortages of the war years: By 1947, six million American families were sharing housing, and postwar family counselors warned of a widespread marital crisis caused by conflicts between the generations.

During the 1950s, films and television plays showed people working through conflicts between marital loyalties and older kin, peer group, or community ties; regretfully but decisively, these conflicts were almost invariably 'resolved in favor of the heterosexual couple rather than the claims of extended kinship networks, . . . homosociability and friendship.' Talcott Parsons and other sociologists argued that modern industrial society required the family to jettison traditional productive functions and wider kin ties in order to specialize in emotional nurturance, childrearing, and production of a modern personality. Social workers 'endorsed nuclear family separateness and looked suspiciously on active extended-family networks.'[8]

Popular commentators urged young families to adopt a 'modern' stance and strike out on their own, and with the return of prosperity, most did. By the early 1950s, newlyweds not only were establishing single-family homes at an earlier age and a more rapid rate than ever before but also were increasingly moving to the suburbs, away from the close scrutiny of the elder generation.

For the first time in American history, moreover, such average trends did not disguise sharp variations by class, race, and ethnic group. People married at a younger age, bore their children earlier and closer together, completed their families by the time they were in their late twenties, and experienced a longer period living together as a couple after their children left home. The traditional range of acceptable family behaviors – even the range in the acceptable number and timing of children – narrowed substantially.[9]

The values of 1950s families also were new. The emphasis on producing a whole world of satisfaction, amusement, and inventiveness within the nuclear family had no precedents. Historian Elaine Tyler May comments: 'The legendary family of the 1950s . . . was not, as common wisdom tells us, the last gasp of "traditional" family life with deep roots in the past. Rather, it was the first wholehearted effort to create a home that would fulfill virtually all its members' personal needs through an energized and expressive personal life.'[10]

Beneath a superficial revival of Victorian domesticity and gender distinctions, a novel rearrangement of family ideals and male-female relations was accomplished. For women, this involved a reduction in the moral aspect of domesticity and an expansion of its orientation toward personal service. Nineteenth-century middle-class women had cheerfully left housework to servants, yet 1950s women of all classes created makework in their homes and felt guilty when they did not do everything for themselves. The amount of time women spent doing housework actually *increased* during the 1950s, despite the advent of convenience foods and new, labor-saving appliances; child care absorbed more than twice as much time as it had in the 1920s. By the mid-1950s, advertisers' surveys reported on a growing tendency among women to find 'housework a medium of expression for . . . their femininity and individuality.'[11]

For the first time, men as well as women were encouraged to root their identity and self-image in familial and parental roles. The novelty of these family and gender values can be seen in the dramatic postwar transformation of movie themes. Historian Peter Biskind writes that almost every major male star who had played tough loners in the 1930s and 1940s 'took the roles with which he was synonymous and transformed them, in the fifties, into neurotics or psychotics'. In these films, 'men belonged at home, not on the streets or out on the prairie, . . . not alone or hanging out with other men'. The women who got men to settle down had to promise enough sex to compete with 'bad' women, but ultimately they provided it only in the marital bedroom and only in return for some help fixing up the house.[12]

The 'good life' in the 1950s, historian Clifford Clark points out, made the family 'the focus of fun and recreation.' The ranch house, architectural embodiment of this new ideal, discarded the older privacy of the kitchen, den, and sewing room (representative of separate spheres for men and women) but introduced new privacy and luxury into the master bedroom. There was an unprecedented 'glorification of self-indulgence' in family life. Formality was discarded in favor of 'livability', 'comfort', and 'convenience'. A contradiction in terms in earlier periods, 'the sexually charged, child-centered family took its place at the center of the postwar American dream.'[13]

On television, David Marc comment, all the 'normal' families moved to the suburbs during the 1950s. Popular culture turned such suburban families into capitalism's answer to the Communist threat. In his famous 'Kitchen debate' with Nikita Khrushchev in 1959, Richard Nixon asserted that the superiority of capitalism over communism was embodied not in ideology or military might but in the comforts of the suburban home, 'designed to make things easier for our women'.[14]

A complex reality: 1950s poverty, diversity, and social change

Even aside from the exceptional and ephemeral nature of the conditions that supported them, 1950s family strategies and values offer no solution to the discontents that underlie contemporary romanticization of the 'good old days'. The reality of these families was far more painful and complex than the situation-comedy reruns or the expurgated memories of the nostalgic would suggest.

In the first place, a full 25 percent of Americans, forty to fifty million people, were poor in the mid-1950s, and in the absence of food stamps and housing programs, this poverty was searing. Even at the end of the 1950s, a third of

American children were poor. Only half the population had savings in 1959; one-quarter of the population had no liquid assets at all. Even when we consider only native-born, white families, one-third could not get by on the income of the household head.[15]

In the second place, real life was not so white as it was on television. Television, comments historian Ella Taylor, increasingly ignored cultural diversity, adopting the motto "least objectionable programming", which gave rise to those least objectionable families, the Cleavers, the Nelsons and the Andersons'. Such families were so completely white and Anglo-Saxon that even the Hispanic gardener in 'Father Knows Best' went by the name of Frank Smith. But contrary to the all-white lineup on the television networks and the streets of suburbia, the 1950s saw a major transformation in the ethnic composition of America. More Mexican immigrants entered the United States in the two decades after the Second World War than in the entire previous one hundred years. Prior to the war, most blacks and Mexican-Americans lived in rural areas, and three-fourths of blacks lived in the South. By 1960, a majority of blacks resided in the North, and 80 percent of both blacks and Mexican-Americans lived in cities. Postwar Puerto Rican immigration was so massive that by 1960 more Puerto Ricans lived in New York than in San Juan.[16]

These minorities were almost entirely excluded from the gains and privileges accorded white middle-class families. The homemaker role was not available to the more than 40 percent of black women with small children who worked outside the home. Twenty-five percent of these women headed their own households, but even minorities who conformed to the dominant family form faced conditions quite unlike those portrayed on television. The poverty rate of two-parent black families was more than 50 percent, approximately the same as that of one-parent black ones. Migrant workers suffered 'near medieval' deprivations, while termination and relocation policies were employed against Native Americans to get them to give up treaty rights.[17]

African Americans in the South faced systematic, legally sanctioned segregation and pervasive brutality, and those in the North were excluded by restrictive covenants and redlining from many benefits of the economic expansion that their labor helped sustain. Whites resisted, with harassment and violence, the attempts of blacks to participate in the American family dream. When Harvey Clark tried to move into Cicero, Illinois, in 1951, a mob of 4,000 whites spent four days tearing his apartment apart while police stood by and joked with them. In 1953, the first black family moved into Chicago's Trumbull Park public housing project; neighbors 'hurled stones and tomatoes' and trashed stores that sold groceries to the new residents. In Detroit, *Life* magazine reported in 1957, '10,000 Negroes work at the Ford plant in nearby Dearborn, [but] not one Negro can live in Dearborn itself'.[18]

More complexities: repression, anxiety, unhappiness, and conflict

The happy, homogeneous families that we 'remember' from the 1950s were thus partly a result of the media's denial of diversity. But even among sectors of the population where the 'least objectionable' families did prevail, their values and behaviors were not entirely a spontaneous, joyful reaction to prosperity. If suburban ranch houses and family barbecues were the carrots offered to white middle-class families that adopted the new norms, there was also a stick.

Women's retreat to housewifery, for example, was in many cases not freely chosen. During the war, thousands of women had entered new jobs, gained new skills, joined unions, and fought against job discrimination. Although 95 percent of the new women employees had expected when they were first hired to quit work at the end of the war, by 1945 almost an equally overwhelming majority did not want to give up their independence, responsibility, and income, and expressed the desire to continue working.[19]

After the war, however, writes one recent student of postwar reconstruction, 'management went to extraordinary lengths to purge women workers from the auto plants,' as well as from other high-paying and nontraditional jobs. As it turned out, in most cases women were not permanently expelled from the labor force but were merely downgraded to lower-paid, 'female' jobs. Even at the end of the purge, there were more women working than before the war, and by 1952 there were two million more wives at work than at the peak of wartime production. The jobs available to these women, however, lacked the pay and the challenges that had made wartime work so satisfying, encouraging women to define themselves in terms of home and family even when they were working.[20]

Vehement attacks were launched against women who did not accept such self-definitions. As sociologist David Riesman noted, a woman's failure to bear children went from being 'a social disadvantage and sometimes a personal tragedy' in the nineteenth century to being a 'quasi-perversion' in the 1950s. The conflicting messages aimed at women seemed almost calculated to demoralize: At the same time as they labeled women 'unnatural' if they did not seek fulfillment in motherhood, psychologists and popular writers insisted that most modern social ills could be traced to domineering mothers who invested too much energy and emotion in their children.

Women who could not walk the fine line between nurturing motherhood and castrating 'momism', or who had trouble adjusting to 'creative homemaking,' were labeled neurotic, perverted, or schizophrenic. A recent study of hospitalized 'schizophrenic' women in the San Francisco Bay Area during the 1950s concludes that institutionalization and sometimes electric shock treatments were used to force women to accept their domestic roles and their husbands' dictates. Shock treatments also were recommended for women who sought abortion, on the assumption that failure to want a baby signified dangerous emotional disturbance.[21]

All women, even seemingly docile ones, were deeply mistrusted. They were frequently denied the right to serve on juries, convey property, make contracts, take out credit cards in their own name, or establish residence.

Men were also pressured into acceptable family roles, since lack of a suitable wife could mean the loss of a job or promotion for a middle-class man. Bachelors were categorized as 'immature,' 'infantile,' 'narcissistic,' 'deviant,' or even 'pathological.'

Even for people not directly coerced into conformity by racial, political, or personal repression, the turn toward families was in many cases more a defensive move than a purely affirmative act. Some men and women entered loveless marriages in order to forestall attacks about real or suspected homosexuality or lesbianism. Growing numbers of people saw the family, in the words of one husband, as the one 'group that in spite of many disagreements internally always will face its external enemies together'. Conservative families warned children to beware of communists who might masquerade as friendly neighbors; liberal children learned to confine their opinions to the family for fear that their father's job or reputation might be threatened.[22]

Americans were far more ambivalent about the 1950s than later retrospectives, such as 'Happy Days,' suggest. Plays by Tennessee Williams, Eugene O'Neill, and Arthur Miller explored the underside of family life. Movies such as *Rebel Without a Cause* (1955) expressed fears about youths whose parents had failed them. There was an almost obsessive concern with the idea that the mass media had broken down parental control, thus provoking an outburst of 'delinquency and youthful viciousness'.

Many families, of course, managed to hold such fears at bay – and it must be admitted that the suburbs and small towns of America were exceptionally good places for doing so. Shielded from the multiplying problems and growing diversity of the rest of society, residents of these areas could afford to be neighborly. Church attendance and membership in voluntary associations tended to be higher in the suburbs than in the cities, although contact with extended kin was less frequent. Children played in the neighborhoods and cul-de-sacs with only cursory warnings about strangers.[23]

For many other children, however, growing up in 1950s families was not so much a matter of being protected from the harsh realities of the outside world as preventing the outside world from learning the harsh realities of family life. Few would have guessed that radiant Marilyn Van Derbur, crowned Miss America in 1958, had been sexually violated by her wealthy, respectable father from the time she was five until she was eighteen, when she moved away to college.[24]

Beneath the polished facades of many 'ideal' families, suburban as well as urban, was violence, terror, or simply grinding misery that only occasionally came to light. Although Colorado researchers found 302 battered-child cases, including 33 deaths, in their state during one year alone, the major journal of American family sociology did not carry a single article on family violence between 1939 and 1969.

We will probably never know how prevalent incest and sexual abuse were in the 1950s, but we do know that when girls or women reported incidents of such abuse to therapists, they were frequently told that they were 'fantasizing' their unconscious oedipal desires. Although incest cases were common throughout the records of case-workers from 1880 to 1960, according to historian Linda Gordon's study of these documents, the problem was increasingly redefined as one of female 'sex delinquency'. By 1960, despite overwhelming evidence to the contrary, experts described incest as a 'one-in-a-million occurrence'. Not until the 1970s, heartened by a supportive women's movement, were many women able to speak out about the sexual abuse they had suffered in silent agony during the 1950s.

Less dramatic but more widespread was the existence of significant marital unhappiness. Between one-quarter and one-third of the marriages contracted in the 1950s eventually ended in divorce; during that decade two million legally married people lived apart from each other. Many more couples simply toughed it out.

A successful 1950s family, moreover, was often achieved at enormous cost to the wife, who was expected to subordinate her own needs and aspirations to those of both her husband and her children. In consequence, no sooner was the ideal of the postwar family accepted than observers began to comment perplexedly on how discontented women seemed in the very roles they supposedly desired most. In 1949, *Life* magazine reported that 'suddenly and for no plain reason' American women were 'seized with an eerie restlessness'. Under a 'mask of placidity' and an outwardly feminine appearance, one physician wrote in 1953, three was often 'an inwardly tense and emotionally unstable individual seething with hidden aggressiveness and resentment'.[25]

Although Betty Friedan's bestseller *The Feminine Mystique* did not appear until 1963, it was a product of the 1950s, originating in the discontented responses Friedan received in 1957 when she surveyed fellow college classmates from the class of 1942. The heartfelt identification of other 1950s women with 'the problem that has no name' is preserved in the letters Friedan received after her book was published.

Men tended to be more satisfied with marriage than were women, especially over time, but they, too, had their discontents. Even the most successful strivers after the American dream sometimes muttered about 'mindless conformity'. The titles of books such as *The Organization Man*, by William Whyte (1956), and *The Lonely Crowd*, by David Riesman (1958), summarized a widespread critique of 1950s culture. Male resentments against women were expressed in the only partly humorous diatribes of *Playboy* magazine (founded in 1955) against 'money-hungry' gold diggers or lazy 'parasites' trying to trap men into commitment.[26] [See spoof ad] (p. 17).

Contradictions of the 1950s family boom

Happy memories of 1950s family life are not all illusion, of course – there were good times for many families. But even the most positive aspects had another side. One reason that the 1950s family model was so fleeting was that it contained the seeds of its own destruction. It was during the 1950s, not the 1960s, that the youth market was first produced then institutionalized into the youth culture. It was also during this period that advertising and consumerism became saturated with sex.[27]

In the 1950s, family life was financed by economic practices that were to have unanticipated consequences in the 1970s. Wives and mothers first started to work in great numbers during the 1950s in order to supplement their families' purchasing power; expansion of household comforts came 'at the cost of an astronomical increase of indebtedness'. The labor-management accord of the

1950s helped erode the union movement's ability to oppose the takebacks and runaway shops that destroyed the 'family wage system' during the 1970s and 1980s.[28]

Family and gender strategies also contained some time bombs. Women who 'played dumb' to catch a man, as 40 percent of Barnard College women admitted to doing, sometimes despised their husbands for not living up to the fiction of male superiority they had worked so hard to promote. Commitment to improving the quality of family life by manipulating the timing and spacing of childbearing led to the social acceptability of family planning and the spread of birth-control techniques. Concentration of childbearing in early marriage meant that growing numbers of women had years to spare for paid work after the bulk of their child-care duties were finished. Finally, 1950s families fostered intense feelings and values that produced young people with a sharp eye for hypocrisy; many of the so-called rebels of the 1960s were simply acting on values that they had internalized in the bosom of their families.[29]

The problem of women in traditional families

People who romanticize the 1950s, or any model of the traditional family, are usually put in an uncomfortable position when they attempt to gain popular support. The legitimacy of women's rights is so widely accepted today that only a tiny minority of Americans seriously propose that women should go back to being full-time housewives or should be denied educational and job opportunities because of their family responsibilities. Yet when commentators lament the collapse of traditional family commitments and values, they almost invariably mean the uniquely female duties associated with the doctrine of separate spheres for men and women.

The crisis of commitment in America is usually seen as a problem associated with women's changing roles because women's family functions have histor-ically mediated the worst effects of competition and individualism in the larger society. Most people who talk about balancing private advancement and individ-ual rights with 'nurturance, mutual support, and long-term commitment' do not envision any serious rethinking of the individualistic, antisocial tendencies in our society, nor any ways of broadening our sources of nurturance and mutual assistance. Instead, they seek ways – sometimes through repression, sometimes through reform – of rebuilding a family in which women can continue to compensate for, rather than challenge, the individualism in our larger economy and polity.

Notes

1 *Boston Globe*, 11 April 1989; David Blankenhorn, 'Ozzie and Harriet, Alive and Well', *Washington Post*, 11 June 1989: 'Ozzie and Harriet Redux', *Fortune*, 25 March 1991; Richard Morin, 'Family Life Makes a Comeback: Maybe Ozzie and Harriet Had a Point', *Washington Post National Weekly Edition*, 25 November–1 December 1991.
2 William Chafe, *The American Woman: Her Changing Social, Economic, and Political Roles, 1920–1970* (New York: Oxford University Press, 1974), p. 217.

3 Joseph Mason, *History of Housing in the U.S., 1930–1980* (Houston: Gulf, 1982); Martin Mayer, *The Builders* (New York: Gulf, 1978), p. 132.

4 William Chafe, *The Unfinished Journey: America Since World War II* (New York: Oxford University Press, 1986), pp. 111–18; Stephen Mintz and Susan Kellogg, *Domestic Revolutions; A social History of American Family Life* (New York: Free Press, 1988), pp. 182–83; Elaine Tyler May, *Homeward Bound: American Families in the Cold War Era* (New York: Basic Books, 1988), p. 165.

5 May, *Homeward Bound*, p. 167; Clifford Clark, Jr., 'Ranch-House Suburbia: Ideals and Realities', in *Recasting America: Culture and Politics in the Age of Cold War*, ed. Lary May (Chicago: University of Chicago Press, 1989), p. 188.

6 David Marc, *Comic Visions: Television Comedy and American Culture* (Boston: Unwin Hyman, 1989), p. 50; May, *Homeward Bound*, p. 28; Mintz and Kellogg, *Domestic Revolutions*, p. 180.

7 Steven D. McLaughlin et al., *The Changing Lives of American Women* (Chapel Hill: University of North Carolina Press, 1988), p. 7; Donald Brogue, *The Population of the United States* (Glencoe, Ill.: Free Press, 1959).

8 Talcott Parsons and Robert Bales, *Family, Socialization, and Interaction Process* (Glencoe: Free Press, 1955); Judith E. Smith, 'The Marrying Kind: Working Class Courtship and Marriage in Postwar Popular Culture' (Paper presented at American Studies Association Conference, New Orleans, October 1990), p. 3; Linda Gordon, *Heroes of Their Own Lives: The Politics and History of Family Violence, 1880–1960* (New York: Viking, 1988), p. 161.

9 May, *Homeward Bound*, p. 137; Mary Ryan, *Womanhood in American from Colonial Times to the Present* (New York: Franklin Watts, 1983), pp. 271–72; Susan House-holder Van Horn, *Women, Work, and Fertility, 1900–1986* (New York: New York University Press, 1988); Landon Jones, *Great Expectations: America and the Baby Boom Generation* (New York: Ballantine, 1980), p. 34.

10 May, *Homeward Bound*, p. 11.

11 Glenna Mathews, *'Just a Housewife': The Rise and Fall of Domesticity in America* (New York: Oxford, University Press, 1987); Betty Friedan, *The Feminine Mystique* (New York: Dell, 1963), p. 204.

12 Peter Biskind, *Seeing Is Believing: How Hollywood Taught Us to Stop Worrying and Love the Fifties* (New York: Pantheon, 1983), pp. 252, 255.

13 Clifford Clark, *The American Family Home, 1800–1960* (Chapel Hill: University of North Carolina Press, 1986), pp. 209, 216; Clark, 'Ranch-House Suburbia', pp. 171, 182; May, *Homeward Bound*, p. 162.

14 Marc, *Comic Visions*, p. 81; May, *Homeward Bound*, p. 18.

15 James Patterson, *America Struggles Against Poverty, 1900–1985* (Cambridge: Harvard University Press, 1986), p. 13; Douglas Miller and Marion Nowak, *The Fifties: The Way We Really Were* (Garden City, N.Y.: Doubleday, 1977), p. 122; Michael Harrington, *The Other America: Poverty in the United States* (New York: Macmillan, 1962); *Social Security Bulletin*, July 1963, pp. 3–13; Chafe, *Unfinished Journey*, p. 143; Mark Stern, 'Poverty and the Life-Cycle, 1940–1960', *Journal of Social History* 24 (1991): 538.

16 Taylor, *Prime-Time Families*, p. 40; David Marc, 'The Sit-Com Sensibility,' *Washington Post*, 25 June 1989; Eric Barnouw, *Tube of Plenty: The Evolution of American Television* (New York: Oxford University Press, 1975); Richard Griswold del Castillo, *La Familla: Chicago Families in the Urban Southwest, 1848 to the Present* (Notre Dame: University of Notre Dame Press, 1984), pp. 113–14; Henretta et al., *America's History*, vol. 2, p. 845.

17 Glenda Riley, *Inventing the American Woman* (Arlington Heights, Va.: Harlan Davidson, 1987), p. 240; Harrington, *Other America*, p. 53; Edward R. Murrow, 'Harvest of Shame', *CBS Reports*, 25 November 1960; John Collier, 'Indian Take-away', *Nation*, 2 October 1954.

18 Herbert Shapiro, *White Violence and Black Response: From Reconstruction to Montgomery* (Amherst: University of Massachusetts Press, 1988); Michael Danielson, *The Politics of Exclusion* (New York: Columbia University Press, 1976); Miller and Nowak, *The Fifties*, pp. 199–201; *Life*, 9 November 1953, p. 151; 'The Negro and the North', *Life*, 11 March 1957, p. 163.

19 Joan Ellen Trey, 'Women in the World War II Economy', *Review of Radical Political Economics*, July 1972; Chafe, *American Woman*, pp. 178–79.

20 Ruth Milkman, *Gender at Work: The Dynamics of Job Segregation by Sex During World War II* (Urbana: University of Illinois Press, 1987), p. 102; Sheila Tobias and Lisa Anderson, 'What Really Happened to Rosie the Riveter', *MSS Modular Publications* 9 (1973); Steven D. McLaughlin et al., *The Changing Lives of American Women* (Chapel Hill: University of North Carolina, 1988), p. 24.

21 Carol Warren, *Madwives: Schizophrenic Women in the 1950s* (New Brunswick: Rutgers University Press, 1987); Hartmann, *Home Front*, p. 174.

22 May, *Homeward Bound*, p. 91.

23 For a defense of the suburbs, see Scott Donaldson, *The Suburban Myth* (New York: Columbia University Press, 1969). See also John Seeley, R. Alexander Sim, and E. W. Loosely, *Crestwood Heights: A Study of Culture in Suburban Life* (New York: Basic Books, 1956), and William H. Whyte, *The Organization Man* (New York: Simon & Schuster, 1956). Though Whyte criticized the lack of individualism in the suburbs he described, his description of boring group life might sound rather comforting to many alienated modern Americans.

24 Marilyn Van Derbur Atler, 'The Darkest Secret', *People*, 6 July 1991.

25 Mintz and Kellogg, *Domestic Revolutions*, p. 195; Miller and Nowak, *The Fifties*, p. 174. The physician reported that most of these women had fulfilled their wifely and motherly roles for years, in seemingly irreproachable ways, but were nevertheless unfulfilled. Unable to accept the logic of his own evidence, the doctor concluded that their problems were a result of their 'intense strivings for masculinity.'

26 Ehrenreich, *The Hearts of Men*.

27 Jones, *Great Expectations*, pp. 41–49; Friedan, *Feminine Mystique*, pp. 250–51.

28 Chafe, *Unfinished Journey*, p. 144.

29 Chafe, *Unfinished Journey*, p. 125; Eisler, *Private Lives*, p. 369; Chafe, *American Woman*, p. 218; Ryan, *Womanhood in America*, p. 277; May, *Homeward Bound*, pp. 149–52; Joseph Demartini, 'Change Agents and Generational Relationships: A Reevaluation of Mannheim's Problem of Generations', *Social Forces* 64 (1985).

2 bell hooks
Homeplace: A Site of Resistance

Excerpts from: *Yearning: Race, gender and cultural politics.*
Boston, South End Press (1991)

When I was a young girl the journey across town to my grandmother's house was one of the most intriguing experiences. It was a movement away from the segregated blackness of our community into a poor white neighborhood. I remember the fear, being scared to walk to Baba's (our grandmother's house) because we would have to pass that terrifying whiteness – those white faces on the porches staring us down with hate. Even when empty or vacant, those porches seemed to say 'danger', 'you do not belong here', 'you are not safe'.

Oh! that feeling of safety, of arrival, of homecoming when we finally reached the edges of her yard, when we could see the soot black face of our grandfather, Daddy Gus, sitting in his chair on the porch, smell his cigar, and rest on his lap. Such a contrast, that feeling of arrival, of homecoming, this sweetness and the bitterness of that journey, that constant reminder of white power and control.

I speak of this journey as leading to my grandmother's house, even though our grandfather lived there too. In our young minds houses belonged to women, were their special domain, not as property, but as places where all that truly mattered in life took place – the warmth and comfort of shelter, the feeding of our bodies, the nurturing of our souls. There we learned dignity, integrity of being; there we learned to have faith. The folks who made this life possible, who were our primary guides and teachers, were black women.

Their lives were not easy. Their lives were hard. They were black women who for the most part worked outside the home serving white folks, cleaning their houses, washing their clothes, tending their children – black women who worked in the fields or in the streets, whatever they could do to make ends meet, whatever was necessary. Then they returned to their homes to make life happen there. This tension between service outside one's home, family, and kin network, service provided to white folks which took time and energy, and the effort of black women to conserve enough of themselves to provide service (care and nurturance) within their own families and communities is one of the many factors that has historically distinguished the lot of black women in patriarchal white supremacist society from that of black men. Contemporary black struggle must honor this history of service just as it must critique the sexist definition of service as women's 'natural' role.

Since sexism delegates to females the task of creating and sustaining a home environment, it has been primarily the responsibility of black women to construct domestic households as spaces of care and nurturance in the face of the brutal harsh reality of racist oppression, sexist domination. Historically, African-American people believed that the construction of a homeplace, however fragile and tenuous (the slave hut, the wooden shack), had a radical political dimension.

Despite the brutal reality of racial apartheid, of domination, one's homeplace was the one site where one could freely confront the issue of humanization, where one could resist. Black women resisted by making homes where all black people could strive to be subjects, not objects, where we could be affirmed in our minds and hearts despite poverty, hardship, and deprivation, where we could restore to ourselves the dignity denied us on the outside in the public world.

This task of making homeplace was not simply a matter of black women providing service; it was about the construction of a safe place where black people could affirm one another and by so doing heal many of the wounds inflicted by racist domination. We could not learn to love or respect ourselves in the culture of white supremacy, on the outside; it was there on the inside, in that 'homeplace,' most often created and kept by black women, that we had the opportunity to grow and develop, to nurture our spirits. This task of making a homeplace, of making home a community of resistance, has been shared by black women globally, especially black women in white supremacist societies.

I want to remember these black women today. The act of remembrance is a conscious gesture honoring their struggle, their effort to keep something for their own. I want us to respect and understand that this effort has been and continues to be a radically subversive political gesture. For those who dominate and oppress us benefit most when we have nothing to give our own, when they have so taken from us our dignity, our humanness that we have nothing left, no 'homeplace' where we can recover ourselves. I want us to remember these black women today, both past and present. I want to speak about the importance of homeplace in the midst of oppression and domination, of homeplace as a site of resistance and liberation struggle. Writing about 'resistance', particularly resistance to the Vietnam war, Vietnamese Buddhist monk Thich Nhat Hahn says:

> . . . resistance, at root, must mean more than resistance against war. It is a resistance against all kinds of things that are like war . . . So perhaps, resistance means opposition to being invaded, occupied, assaulted and destroyed by the system. The purpose of resistance, here, is to seek the healing of yourself in order to be able to see clearly . . . I think that communities of resistance should be places where people can return to themselves more easily, where the conditions are such that they can heal themselves and recover their wholeness.

Historically, black women have resisted white supremacist domination by working to establish homeplace. It does not matter that sexism assigned them this role. It is more important that they took this conventional role and expanded it to include caring for one another, for children, for black men, in ways that elevated our spirits, that kept us from despair, that taught some of us to be revolutionaires able to struggle for freedom. In his famous 1845 slave narrative, Frederick Douglass tells the story of his birth, of his enslaved black mother who was hired out a considerable distance from his place of residence. Describing their relationship, he writes:

> I never saw my mother, to know her as such more than four or five times in my life; and each of these times was very short in duration, and at night. She was hired by Mr Stewart, who lived about twelve miles from my house. She made her journeys to see

me in the night, traveling the whole distance on foot, after the performance of her day's work. She was a field hand, and a whipping is the penalty of not being in the field at sunrise . . . I do not recollect of ever seeing my mother by the light of day. She was with me in the night. She would lie down with me and get me to sleep, but long before I waked she was gone.

After sharing this information, Douglass later says that he never enjoyed a mother's 'soothing presence, her tender and watchful care' so that he received the 'tidings of her death with much the same emotions I should have probably felt at the death of a stranger'. Douglass surely intended to impress upon the consciousness of white readers the cruelty of that system of racial domination which separated black families, black mothers from their children. Yet he does so by devaluing black womanhood, by not even registering the quality of care that made his black mother travel those twelve miles to hold him in her arms. In the midst of a brutal racist system, which did not value black life, she valued the life of her child enough to resist that system, to come to him in the night, just to hold him.

Now I cannot agree with Douglass that he never knew a mother's care. I want to suggest that this mother, who dared to hold him in the night, gave him at birth a sense of value that provided a groundwork, however fragile, for the person he later became. Holding him in her arms, Douglass' mother provided, if only for a short time, a space where this black child was not the subject of dehumanizing scorn and devaluation but was the recipient of a quality of care that should have enabled the adult Douglass to look back and reflect on the political choices of this black mother who resisted slave codes, risking her life, to care for her son. I want to suggest that devaluation of the role his mother played in his life is a dangerous oversight. Though Douglass is only one example, we are currently in danger of forgetting the powerful role black women have played in constructing for us homeplaces that are the site for resistance. This forgetfulness undermines our solidarity and the future of black liberation struggle.

Douglass's work is important, for he is historically identified as sympathetic to the struggle for women's rights. All too often his critique of male domination, such as it was, did not include recognition of the particular circumstances of black women in relation to black men and families. To me one of the most important chapters in my first book, *Ain't I A Woman: Black Women and Feminism*, is one that calls attention to 'Continued Devaluation of Black Womanhood'. Overall devaluation of the role black women have played in constructing for us homeplaces that are the site for resistance undermines our efforts to resist racism and the colonizing mentality which promotes internalized self-hatred. Sexist thinking about the nature of domesticity has determined the way black women's experience in the home is perceived. In African-American culture there is a long tradition of 'mother worship'. Black autobiographies, fiction, and poetry praise the virtues of the self-sacrificing black mother. Unfortunately, though positively motivated, black mother worship extols the virtues of self-sacrifice while simultaneously implying that such a gesture is not reflective of choice and will, rather the perfect embodiment of a woman's 'natural' role. The assumption then is that the black woman who works hard to be a responsible caretaker is only doing what she should be doing. Failure to recognize the realm of choice, and the remarkable re-visioning of both woman's role and the idea of 'home' that black women consciously exercised in practice, obscures the political commitment to racial uplift, to eradicating racism, which was the philosophical core of dedication to community and home.

Though black women did not self-consciously articulate in written discourse the theoretical principles of decolonization, this does not detract from the importance of their actions. They understood intellectually and intuitively the meaning of homeplace in the midst of an oppressive and dominating social reality, of homeplace as site of resistance and liberation struggle. I know of what I speak. I would not be writing this essay if my mother, Rosa Bell, daughter to Sarah Oldham, granddaughter to Bell Hooks, had not created homeplace in just this liberatory way, despite the contradictions of poverty and sexism.

In our family, I remember the immense anxiety we felt as children when mama would leave our house, our segregated community, to work as a maid in the homes of white folks. I believe that she sensed our fear, our concern that she might not return to us safe, that we could not find her (even though she always left phone numbers, they did not ease our worry). When she returned home after working long hours, she did not complain. She made an effort to rejoice with us that her work was done, that she was home, making it seem as though there was nothing about the experience of working as a maid in a white household, in that space of Otherness, which stripped her of dignity and personal power.

Looking back as an adult woman, I think of the effort it must have taken for her to transcend her own tiredness (and who knows what assaults or wounds to her spirit had to be put aside so that she could give something to her own). Given the contemporary notions of 'good parenting' this may seem like a small gesture, yet in many post-slavery black families, it was a gesture parents were often too weary, too beaten down to make. Those of us who were fortunate enough to receive such care understood its value. Politically, our young mother, Rosa Bell, did not allow the white supremacist culture of domination to completely shape and control her psyche and her familial relationships. Working to create a homeplace that affirmed our beings, our blackness, our love for one another was necessary resistance. We learned degrees of critical consciousness from her. Our lives were not without contradictions, so it is not my intent to create a romanticized portrait. Yet any attempts to critically assess the role of black women in liberation struggle must examine the way political concern about the impact of racism shaped black women's thinking, their sense of home, and their modes of parenting.

An effective means of white subjugation of black people globally has been the perpetual construction of economic and social structures that deprive many folks of the means to make homeplace. Remembering this should enable us to understand the political value of black women's resistance in the home. It should provide a framework where we can discuss the development of black female political consciousness, acknowledging the political importance of resistance effort that took place in homes. It is no accident that the South African apartheid regime systematically attack[ed] and destroy[ed] black efforts to construct homeplace, however tenuous, that small private reality where black women and men can renew their spirits and recover themselves. It is no accident that this homeplace, as fragile and as transitional as it may be, a makeshift shed, a small bit of earth where one rests, is always subject to violation and destruction. For when a people no longer have the space to construct homeplace, we cannot build a meaningful community of resistance.

Throughout our history, African-Americans have recognized the subversive value of homeplace, of having access to private space where we do not directly encounter white racist aggression. Whatever the shape and direction of black liberation struggle (civil rights reform or black power movement), domestic space has been a crucial site for organizing, for forming political solidarity. Homeplace has been a site of resistance. Its structure was defined less by whether or not black women and men were conforming to sexist behavior norms and more by our struggle to uplift ourselves as a people, our struggle to resist racist domination and oppression.

That liberatory struggle has been seriously undermined by contemporary efforts to change that subversive homeplace into a site of patriarchal domination of black women by black men, where we abuse one another for not conforming to sexist norms. This shift in perspective, where homeplace is not viewed as a political site, has had negative impact on the construction of black female identity and political consciousness. Masses of black women, many of whom were not formally educated, had in the past been able to play a vital role in black liberation struggle. In the contemporary situation, as the paradigms for domesticity in black life mirrored white bourgeois norms (where home is conceptualized as politically neutral space), black people began to overlook and devalue the importance of black female labor in teaching critical consciousness in domestic space. Many black women, irrespective of class status, have responded to this crisis of meaning by imitating leisure-class sexist notions of women's role, focusing their lives on meaningless compulsive consumerism.

Identifying this syndrome as 'the crisis of black womanhood' in her essay, 'Considering Feminism as a Model for Social Change', Sheila Radford-Hill points to the mid-sixties as that historical moment when the primacy of black woman's role in liberation struggle began to be questioned as a threat to black manhood and was deemed unimportant. Radford-Hill asserts:

> Without the power to influence the purpose and the direction of our collective experience, without the power to influence our culture from within, we are increasingly immobilized, unable to integrate self and role identities, unable to resist the cultural imperialism of the dominant culture which assures our continued oppression by destroying us from within. Thus, the crisis manifests itself as social dysfunction in the black community – as genocide, fratricide, homicide, and suicide. It is also manifested by the abdication of personal responsibility by black women for themselves and for each other ... The crisis of black womanhood is a form of cultural aggression: a form of exploitation so vicious, so insidious that it is currently destroying an entire generation of black women and their families.

This contemporary crisis of black womanhood might have been avoided had black women collectively sustained attempts to develop the latent feminism expressed by their willingness to work equally alongside black men in black liberation struggle. Contemporary equation of black liberation struggle with the subordination of black women has damaged collective black solidarity. It has served the interests of white supremacy to promote the assumption that the wounds of racist domination would be less severe were black women conforming to sexist role patterns.

We are daily witnessing the disintegration of African-American family life that is grounded in a recognition of the political value of constructing homeplace as a site of resistance; black people daily perpetuate sexist norms that threaten our survival as a people. We can no longer act as though sexism in black communities does not threaten our solidarity; any force which estranges and alienates us from one another serves the interests of racist domination.

Black women and men must create a revolutionary vision of black liberation that has a feminist dimension, one which is formed in consideration of our specific needs and concerns. Drawing on past legacies, contemporary black women can begin to reconceptualize ideas of homeplace, once again considering the primacy of domesticity as a site for subversion and resistance. When we renew our concern with homeplace, we can address political issues that most affect our daily lives. Calling attention to the skills and resources of black women who may have begun to feel that they have no meaningful contribution to make, women who may or may not be formally educated but who have essential wisdom to share, who have practical experience that is the breeding ground for all useful theory, we may begin to bond with one another in ways that renew our solidarity.

When black women renew our political commitment to homeplace, we can address the needs and concerns of young black women who are groping for structures of meaning that will further their growth, young women who are struggling for self-definition. Together, black women can renew our commitment to black liberation struggle, sharing insights and awareness, sharing feminist thinking and feminist vision, building solidarity.

With this foundation, we can regain lost perspective, give life new meaning. We can make homeplace that space where we return for renewal and self-recovery, where we can heal our wounds and become whole.

References

Douglass, Frederick. *Narrative of the Life of Frederick Douglass*. Cambridge, MA: Belknap Press, 1960.

Hahn, Thich Nhat. *The Raft Is Not the Shore*. Boston: Beacon Press, 1975.

Radford-Hill, Sheila. 'Considering Feminism as a Model for Social Change,' *Feminist Studies, Critical Studies*, ed. Teresa de Lauretis. University of Indiana, 1986.

3 Lynn Spigel
The Suburban Home Companion: Television and the Neighborhood Ideal in Postwar America

Excerpts from: *Sexuality and space.* Princeton, NJ: Princeton Architectural Press (1992)

It is a truism among cultural historians and media scholars that television's growth after World War II was part of a general return to family values. Less attention has been devoted to the question of another, at times contradictory, ideal in postwar ideology – that of neighborhood bonding and community participation. During the 1950s, millions of Americans – particularly young white couples of the middle class – responded to a severe housing shortage in the cities by fleeing to new mass-produced suburbs. In both scholarly studies and popular literature from the period suburbia emerges as a conformist-oriented society where belonging to the neighborhood network was just as important as the return to family life. Indeed, the new domesticity was not simply experienced as a retreat from the public sphere; it also gave people a sense of belonging to the community. By purchasing their detached suburban homes, the young couples of the middle class participated in the construction of a new community of values; in magazines, in films, and on the airwaves they became the cultural representatives of the 'good life'. Furthermore, the rapid growth of family-based community organizations like the PTA suggests that these neosuburbanites did not barricade their doors, nor did they simply 'drop out'. Instead, these people secured a position of meaning in the *public* sphere through their new-found social identities as *private* landowners.

In this sense, the fascination with family life was not merely a nostalgic return to the Victorian cult of domesticity. Rather, the central preoccupation in the new suburban culture was the construction of a particular *discursive space* through which the family could mediate the contradictory impulses for a private haven on the one hand, and community participation on the other. By lining up individual housing units on connecting plots of land, the suburban tract was itself the ideal articulation of this discursive space; the dual goals of separation from and integration into the larger community was the basis of tract design. Moreover, the domestic architecture of the period mediated the twin goals of separation from and integration into the outside world.[1] Applying principles of modernist architecture to the mass-produced housing of middle-class America, housing experts of the period agreed that the modern home should blur distinctions between inside and outside spaces. The central design element used to create an illusion of the outside world was the picture window or 'window wall' (what we now call sliding glass doors), which became increasingly popular in the postwar period. As Daniel Boorstin has argued, the widespread dissemination of large plate-glass windows for both domestic and commercial use 'leveled the environment' by encouraging the 'removal of sharp distinctions between indoors and outdoors' and thus created an 'ambiguity' between public and private space.[2] This kind of spatial ambiguity was a reigning aesthetic in postwar home magazines which repeatedly suggested that windows and window walls would establish a continuity of interior and exterior worlds.

Given its ability to merge private with public spaces, television was the ideal companion for these suburban homes. In 1946, Thomas H. Hutchinson, an early experimenter in television programming, published a popular book designed to introduce television to the general public, *Here is Television, Your Window on the World*.[3] Commentators in the popular press used this window metaphor over and over again, claiming that television would let people imaginatively travel to distant places while remaining in the comfort of their homes.[4]

Indeed, the integration of television into postwar culture both precipitated and was symptomatic of a profound reorganization of social space. Leisure time was significantly altered as spectator amusements – including movies, sports, and concert attendance – were increasingly incorporated into the home. While in 1950 only 9 percent of all American homes had a television set, by the end of the decade that figure rose to nearly 90 percent and the average American watched about five hours of television per day.[5] Television was caught in a contradictory movement between private and public worlds, and it often became a rhetorical figure for that contradiction. In the following pages, I examine the way postwar culture balanced these contradictory ideals of privatization and community involvement through its fascination with the new electrical space that television provided.

Postwar America witnessed a significant shift in traditional notions of neighbourhood. Mass-produced suburbs like Levittown and Park Forest replaced previous forms of public space with a newly defined aesthetic of prefabrication. At the center of suburban space was the young, upwardly mobile middle-class family; the suburban community was, in its spatial articulations, designed to correspond with and reproduce patterns of nuclear family life. Playgrounds, yards, schools, churches, and synagogues provided town centers for community involvement based on discrete stages of family development. Older people, gay and lesbian people, homeless people, unmarried people and people of color were simply written out of these community spaces – relegated to the cities.

Although the attempt to zone out 'undesirables' was never totally successful, this antiseptic model of space was the reigning aesthetic at the heart of the postwar suburb. Not coincidentally, it had also been central to utopian ideals for electrical communication since the mid-1800s. As James Carey and John Quirk have shown, American intellectuals of the nineteenth century foresaw an 'electrical revolution' in which the grime and noise of industrialization would be purified through electrical power.[6] Electricity, it was assumed, would replace the pollution caused by factory machines with a new, cleaner environment. Through their ability to merge remote spaces, electrical communications like the telephone and telegraph would add to this sanitized environment by allowing people to occupy faraway places while remaining in the familiar and safe locales of the office or the home. Ultimately, this new electrical environment was linked to larger concerns about social decadence in the cities. In both intellectual and popular culture, electricity became a rhetorical figure through which people imagined ways to cleanse urban space of social pollutants; immigrants and class conflict might vanish through the magical powers of electricity.

In the postwar era, the fantasy of antiseptic electrical space was transposed onto television. Numerous commentators extolled the virtues of television's antiseptic spaces, showing how the medium would allow people to travel from their homes while remaining untouched by the actual social contexts to which they imaginatively ventured. Television was particularly hailed for its ability to keep youngsters out of sinful public spaces, away from the countless contaminations of everyday life. At a time when juvenile delinquency was considered a number one social disease, audience research showed that parents believed television would keep their children off the streets.[7]

But television technology promised more than just familial bliss and 'wholesome' heterosexuality. Like its predecessors, it offered the possibility of an intellectual neighborhood, purified of social unrest and human misunderstanding. As NBC's president Sylvester 'Pat' Weaver declared, television would make the 'entire world into a small town, instantly available, with the leading actors on the world stage known on sight or by voice to all within it'. Television, in Weaver's view, would encourage world peace by presenting diverse people with homogeneous forms of knowledge and modes of experience. Television, he argued, created 'a situation new in human history in that children can no longer be raised within a family or group belief that narrows the horizons of the child to any belief pattern. There can no longer be a We-Group, They Group under this condition. Children cannot be brought up to laugh at strangers, to hate foreigners, to live as man has always lived before.' But for Weaver, this democratic utopian world was in fact a very small town, a place where different cultural practices were homogenized and channeled through a medium whose messages were truly American.

But more than just offering family fun, these new home theaters provided postwar Americans with a way to mediate relations between public and private spheres. By turning one's home into a theater, it was possible to make outside spaces part of a safe and predictable experience. In other words, the theatricalization of the home allowed people to draw a line between the public and the private sphere – or, in more theatrical terms, a line between the proscenium space where the spectacle took place and the reception space in which the audience observed the scene.

According to the popular wisdom, television had to recreate the sense of social proximity that the public theater offered; it had to make the viewer feel as if he or she were taking part in a public event. At the same time, however, it had to maintain the necessary distance between the public sphere and private individual upon which middle-class ideals of reception were based.

The impossibility of maintaining these competing ideals gave rise to a series of debates as people weighed the ultimate merits of bringing theatrical experiences indoors. Even if television promised the fantastic possibility of social interconnection through electrical means, this new form of social life wasn't always seen as an improvement over real community experiences. The inclusion of public spectacles in domestic space always carried with it the unpleasant possibility that the social ills of the outside world would invade the private home.

The more that the home included aspects of the public sphere, the more it was seen as subject to unwelcome intrusions.

This was especially true in the early years of innovation when the purchase of a television set quite literally decreased privacy in the home. Numerous social scientific studies showed that people who owned television receivers were inundated with guests who came to watch the new set.[8] But this increased social life was not always seen as a positive effect by the families surveyed. As one woman in a Southern California study complained, 'Sometimes I get tired of the house being used as a semiprivate theater. I have almost turned the set off when some people visit us'.[9] Popular media were also critical of the new 'TV parties'. In 1953, *Esquire* published a cartoon that highlighted the problem entailed by making one's home into a TV theater. The sketch pictures a living room with chairs lined up in front of a television set in movie theater fashion. The residents of this home theater, dressed in pajamas and bathrobes with hair uncombed and feet unshod, are taken by surprise when the neighbors drop in – a bit too soon – to watch a wrestling match on television. Speaking in the voice of the intruders, the caption reads, 'We decided to come over early and make sure we get good seats for tonight's fight'.

Such popular anxieties are better understood when we recognize the changing structure of social relationships encountered by the new suburban middle class. These people often left their families and life-long friends in the city to find instant neighborhoods in preplanned communities. Blocks composed of total strangers represented friendships only at the abstract level of demographic similarities in age, income, family size, and occupation. This homogeneity quickly became a central cause for anxiety in the suburban nightmares described by sociologists and popular critics. In *The Organization Man* (1957), William H. Whyte argued that a sense of community was especially important for the newcomers who experienced a feeling of 'rootlessness' when they left their old neighborhoods for new suburban homes. As Whyte showed, the developers of the mass-produced suburbs tried to smooth the tensions caused by this sense of rootlessness by promising increased community life in their advertisements. But when newcomers arrived in their suburban communities, they were likely to find something different from the ideal that the magazines and advertisements suggested. Tiny homes were typically sandwiched together so that the Smiths' picture window looked not onto rambling green acres, but rather into the Jones' living room – a dilemma commonly referred to as the 'goldfish bowl' effect. In addition to this sense of claustrophobia, the neighborhood ideal brought with it an enormous amount of pressure to conform to the group. As Harry Henderson suggested in his study of Levittown (1953), the residents of this mass-produced suburb were under constant 'pressure to "keep up with the Joneses" ', a situation that led to a 'kind of superconformity' in which everyone desired the same luxury goods and consumer lifestyles.

These nightmarish visions of the preplanned community served as an impetus for the arrival of a surrogate community on television. Television provided an illusion of the ideal neighborhood – the way it was supposed to be. Just when people had left their life-long companions in the city, television sitcoms pictured romanticized versions of neighbor and family bonding.

The burgeoning television culture extended these metaphors of neighborhood bonding by consistently blurring the lines between electrical and real space. Television families were typically presented as 'real families' who just happened to live their lives on TV. Ricky and Lucy, Ozzie and Harriet, Jane and Goodman Ace, George and Gracie, and a host of others crossed the boundaries between fiction and reality on a weekly basis. Promotional and critical discourses further encouraged audiences to think that television characters lived the life of the stars who played them.

These televised neighbors seemed to suture the 'crack' in the picture window. They helped ease what must have been for many Americans a painful transition from the city to the suburb. But more than simply supplying a tonic for displaced suburbanites, television promised something better: it promised modes of spectator pleasure premised upon the sense of an illusory – rather than a real – community of friends. It held out a new possibility for being alone in the home, away from the troublesome busybody neighbors in the next house. But it also maintained ideals of community togetherness and social interconnection by placing the community *at a fictional distance*. Television allowed people to enter into an imaginary social life, one which was shared not in the neighborhood networks of bridge clubs and mahjong gatherings, but on the national networks of CBS, NBC, and ABC.

Indeed, television – at its most ideal – promised to bring to audiences not merely an illusion of reality as in the cinema, but a sense of 'being there', a kind of *hyperrealism*. Television producers and executives devised schemes by which to merge public and private worlds into a new electrical neighbourhood.

Television's promise of social interconnection has provided numerous postwar intellectuals – from Marshall McLuhan to Joshua Meyrowitz – with their own utopian fantasies. Meyrowitz is particularly interesting in this context because he has claimed that television helped foster women's liberation in the 1960s by bringing traditionally male spaces into the home, thus allowing women to 'observe and experience the larger world, including all male interactions and behaviours.' 'Television's first and strongest impact', he concludes, 'is on the perception that women have of the public male world and the place, or lack of place, they have in it. Television is an especially potent force for integrating women because television brings the public domain to women. . . .'[10] But Meyrowitz bases this claim on an essentialist notion of space. In other words, he assumes that public space is male and private space is female. However, public spaces like the office or the theater are not simply male; they are organized according to categories of sexual difference. In these spaces certain social positions and subjectivities are produced according to the placement of furniture, the organization of entrances and exits, the separation of washrooms, the construction of partial walls, and so forth. Thus, television's incorporation of the public sphere into the home did not bring 'male' space into female space; instead it transposed one system of sexually organized space onto another.

Not surprisingly in this regard, postwar media often suggested that television would increase women's social isolation from public life by reinforcing spatial hierarchies that had already defined their everyday experiences in patriarchal cultures. The new family theaters were typically shown to limit opportunities for social encounters that women traditionally had at movie theaters and other forms of public entertainment.

Social scientific studies from the period show that the anxieties expressed in popular representations were also voiced by women of the period. One woman in a Southern California study confessed that all her husband 'wants to do is to sit and watch television – I would like to go out more often'. Another woman complained, 'I would like to go for a drive in the evening, but my husband has been out all day and would prefer to watch a wrestling match on television'.[11]

If television was considered to be a source of problems for women, it also became a central trope for the crisis of masculinity in postwar culture. According to the popular wisdom, television threatened to contaminate masculinity, to make men sick with the 'disease' of femininity. As other scholars have observed, this fear of feminization has characterized the debates on mass culture since the nineteenth century. Culture critics have continually paired mass culture with patriarchal assumptions about femininity. Mass amusements are typically thought to encourage passivity, and they have often been represented in terms of penetration, consumption, and escape. As Andreas Huyssen has argued, this link between women and mass culture has, since the nineteenth century, served to valorize the dichotomy between 'low' and 'high' art (or modernism). Mass culture, Huyssen claims, 'is somehow associated with women while real, authentic culture remains the prerogative of men.'[12] The case of broadcasting is especially interesting because the threat of feminization was particularly aimed at men. Broadcasting quite literally was shown to disrupt the normative structures of patriarchal (high) culture and to turn 'real men' into passive homebodies.

As popular media often suggested, television threatened to rob men of their powers, to usurp their authority over the image, and to turn them into passive spectators. This threat materialized in numerous representations that showed women controlling their husbands through television. Here, television's blurring of private and public space became a powerful tool in the hands of housewives who could use the technology to invert the sexist hierarchies at the heart of the separation of spheres. In this topsy turvy world, women policed men's access to the public sphere and confined them to the home through the clever manipulation of television technology.

In contemporary culture, the dream of social interconnection through antiseptic electrical space is still a potent fantasy. In 1989, in an issue entitled 'The Future and You', *Life* magazine considered the new electronic space that the home laser holographic movie might offer in the twenty-first century. Not coincidentally, this holographic space was defined by male desire. As Marilyn Monroe emerged from the screen in her costume from *The Seven Year Itch*, a male spectator watched her materialize in the room. With his remote control aimed at the set, he policed her image from his futuristic La-Z-Boy Lounger. Although the scene was clearly coded as a science-fiction fantasy, this form of home entertainment was just the latest version of the older wish to control and purify public space. Sexual desire, transported to the home from the Hollywood cinema, was made possible by transfiguring the celluloid image into an electrical space where aggressive and sadistic forms of cinematic pleasure were now sanitized and made into 'passive' home entertainment. The aggression entailed in watching Monroe was clearly marked as passive aggression, as a form of desire that could be contained within domestic space. But just in case the desire for this electronic fantasy woman could not be properly contained, the article warned readers to 'fasten the seatbelt on your La-Z-Boy.'[13]

As this example shows, the utopian dreams of space-binding and social sanitation that characterized television's introduction in the fifties is still a dominant cultural ideal. Electronic communications offer an extension of those plans as private and public spaces become increasingly intertwined through such media as home computers, fax machines, message units, and car phones. Before considering these social changes as a necessary part of an impending 'electronic revolution' or 'information age', we need to remember the racist and sexist principles upon which these electrical utopias have often depended. The loss of neighborhood networks and the rise of electronic networks is a complex social phenomenon based on a series of contradictions that plague postwar life. Perhaps being nostalgic for an older, more 'real' form of community is itself a historical fantasy. But the dreams of a world united by telecommunications seem dangerous enough to warrant closer examination. The global village, after all, is the fantasy of the colonizer, not the colonized.

Notes

1 See my article 'Installing the Television Set: Popular Discourses on Television and Domestic Space, 1948–55', *Camera Obscura* 16 (March 1988): 11–47; and my dissertation, 'Installing the Television Set: The Social Construction of Television's Place in the American Home' (University of California-Los Angeles, 1988).

2 Daniel J. Boorstin, *The Americans: The Democratic Experience* (New York: Vintage Books, 1973), pp. 336–345. Boorstin sees this 'leveling of place' as part of a wider 'ambiguity' symptomatic of the democratic experience.

3 Thomas H. Hutchinson, *Here is Television, Your Window on the World* (1946; New York: Hastings House, 1948), p. ix.

4 For more on this, see my article 'Installing the Television Set: Popular Discourses on Television and Domestic Space, 1948–55' and my dissertation, 'Installing the Television Set: The Social Construction of Television's Place in the American Home'.

5 The data on installation rates vary slightly from one source to another. These estimations are based on Cobbett S. Steinberg, *TV Facts* (New York: Facts on File, 1980), p. 142; 'Sales of Home Appliances', and 'Dwelling Units', *Statistical Abstract of the United States* (Washington, D.C., 1951–56); Lawrence W. Litchy and Malachi C. Topping, *American Broadcasting: A Source Book on the History of Radio and Television* (New York: Hastings House, 1975), pp. 521–522. Note, too that there were significant regional differences in installation rates. Television was installed most rapidly in the Northeast; next were the central and western states, which had relatively similar installation rates; the South and southwest mountain areas were considerably behind the rest of the country. See 'Communications', in *Statistical Abstract of the United States* (Washington, D.C., 1959); U.S. *Bureau of the Census, Housing and Construction Reports*, Series H-121, nos. I–5 (Washington, D.C., 1955–58). Average hours of television watched is based on a 1957 estimate from the A. C. Nielsen Company printed in Leo Bogart, *The Age of Television: A Study of Viewing Habits and the Impact of Television on American Life* (1956; New York: Frederick Unger, 1958), p. 70.

6 James W. Carey and John J. Quirk, 'The Mythos of the Electronic Revolution', in *Communication as Culture*, ed. James W. Carey (Boston: Unwin Hyman, 1989), pp. 113–141. For related issues, see Leo Marx, *The Machine in the Garden: Technology and the Pastoral Ideal in America* (New York: Oxford University Press, 1964); John F. Kasson, *Civilizing the Machine: Technology and Republican Values in America, 1776–1900* (New York: Penguin, 1977); Wolfgang Schivelbusch, *Disenchanted Night: The Industrialization of Light in the Nineteenth Century*, trans. Angela Davies (Berkeley: University of California Press, 1988).

7 For a detailed study of the widespread concern about juvenile deliquency, see James Gilbert, *A Cycle of Outrage: American's Reaction to the Juvenile Delinquent in the 1950s* (New York: Oxford University Press, 1986).

8 After reviewing numerous studies from the fifties, Bogart claims in *The Age of Television*, 'In the early days, "guest viewing" was a common practice' (p. 102). For a summary of the actual studies, see Bogart, pp. 101–107. For additional studies that show the importance of guest viewing in the early period, see John W. Riley et al., 'Some Observations on the Social Effects of Television', *Public Opinion Quarterly* 13, no. 2 (Summer 1949): 233 (this article was an early report of the CBS-Rutgers University studies begun in the summer of 1948); McDonagh et al., 'Television and the Family', p. 116; 'When TV Moves In', *Televiser 7*, no. 8 (October 1950): 17 (a summary of the University of Oklahoma surveys of Oklahoma City and Norman, Oklahoma); Philip F. Frank, 'The Facts of the Medium', *Televiser* (April 1951): 14; and 'TV Bonus Audience in the New York Area', *Televiser* (November 1950): 24–25.

9 McDonagh et al., 'Television and the Family', p. 116.

10 Joshua Meyrowitz, *No Sense of Place: The Impact of Electronic Media on Social Behaviour* (New York: Oxford University Press, 1985), pp. 223–224.

11 McDonagh et al., 'Television and the Family', pp. 117, 119.

12 Andreas Huyssen, *After the Great Divide: Modernism, Mass Culture, Postmodernism* (Bloomington: Indiana University Press, 1986) p. 47.

13 *Life* (February 1989): 67.

4 Tim Putnam

Beyond the Modern Home: Shifting the Parameters of Residence

Excerpts from: J. Bird, B. Curtis, T. Putnam, G. Robertson and L. Tickner (eds), *Mapping the futures: Local cultures, global change*, pp. 150–165. London: Routledge (1993)

Where – and what – is home in a postmodern geography? What do contemporary concerns with problems of 'identity', 'situation' and 'consumption' have to do with the process of making a home, and the work done by constructs of 'the home'? Reading across the dislocations in the discourses that impinge on the domestic, is it possible to characterize a shift in the parameters of residence?

Although the making of houses into homes is a paradigmatic form of emplacement, there may be a temptation to avoid enquiring too deeply into residence when addressing orientation in a global context. From their earliest existence, the discourses of political, economic and ideological determination surveyed their respective terrain with this 'private' sphere firmly behind them. Now that the world can no longer be represented as a federation of families, it has become commonplace to consider the domestic as dominated and decentred, a territory of 'consumption' and 'reproduction' rather than signifying or consequential action. The social sciences which have engaged with 'home' and 'family' have been bent on their regulation and reconstruction. Even in critical social and cultural studies, more accustomed to eliciting difference, subtle barriers exist to recognizing what transpires in this backstage sphere, associated with the dominated gender. In times favourable to 'local narratives', the domestic still connotes parochial interests, trivialized commitments, unacknowledged groundings. Those who would nevertheless speak about emplacement in 'the home' must steer their discourse between past reifications in a vortex of infinitely varied modes of living. Far easier, in conducting a discussion of 'being at home' in the contemporary world, to evade the vagaries of the domestic altogether.

However, it is impossible in a discussion of emplacement to neglect the principal site where material culture is appropriated in mutual relationships. In a postmodern context, the agency exercised in home-making becomes less trivial – and its qualities less readily apparent – than accounts of mass production and mass media would allow. In recent years, investigators of several kinds have become fascinated by the relations between these micro-mysteries and large-scale processes. England, France, Germany, Sweden and the US have seen major exhibitions in which 'domestic creation' is held up for contemplation (Putnam and Newton 1990; Segalen 1990; Pallowski 1988; Lofgren 1990; Galassi 1991). While this interest reflects the greater care in treating the relations between local narratives and global schemata of all kinds, it has also been fed by an awareness that both the ends and means of home-making have altered in our life time. We can now see that from the moment the dream of 'the modern home' began to be realized, domestic consumption practice departed in new directions, pointing, as Lefebvre suggested at the time, to a new epoch (Lefebvre 1971).

In one respect, this new home-making may be characterized as a problematic of discretionary consumption, where a stimulated individual agency contemplates an enhanced field of choice. The traditional autonomy of householders to establish common meanings through collaboration in fashioning and maintaining a shared environment, refracted by commoditization and eclipsed by the mass diffusion of the apparatus of modern living, appeared to take on renewed significance. Boundless bricolage laid to rest any notion of mass consumption as a passive relation, and forced attention on creative autonomy, even resistance and subversion, in the 'everyday' and the 'banal' (Goodall 1983; Saunders 1984; Forrest and Murie 1987; Tomlinson 1990; Putnam 1991; cf. Duncan 1981). But the symbiotic relation between these initiatives and the extensive new promotions of domestic design might still be judged as culturally arbitrary, not only lacking the complex local 'groundedness' of pre-industrial tradition (cf. Heidegger) but, like other phenomena of postmodern consumption, any secure referent whatsoever (cf. Baudrillard 1981). The cultural agency of householders might be considered as confined within a problematic of 'distinction' (cf. Bourdieu 1984). The adequacy of such perspectives has not, however, been tested against the diversity of lived relationships which produce contemporary homes and rely on them as supports. Those who have pondered dislocations in material culture have only recently come to recognize that they must deal with those who encounter, enact and envisage 'the home', and that the domestic sphere has witnessed such extensive renegotiations of generation and gender relations that the viability of this concept as the goal of a joint project has been brought into question.

Domestic happenings, then, have had enough vitality to escape the compass of discourses. The professions and industries that service and support the domestic, increasingly unsure of their object, have called for new lines of research. The initiatives of householders have surpassed and often surprised the calculations of architects and those who package domestic design for living. Feminism and critical self-awareness in the human sciences have exposed the bulk of what passes as 'knowledge' about actual homes as quasi-instrumental, constituted for the reconstruction and regulation of such objects as 'housing', 'the family', 'hygiene', 'leisure' and 'consumption'. The authority of these reifications, reinforced by their conjuncture in the all-enveloping project to install the 'modern home', has waned with the achievement of that project and the disintegration of its constituent parts.

The current state of uncertainty can be read in this recent call for European collaboration in defining housing quality:

> Twenty years ago it was not very difficult to define housing quality. At that time it was a matter of dwelling size and sanitary standard. Later, the awareness grew that housing quality also included such aspects as architectural and landscaping qualities and maybe also the degree of public and commercial service facilities in the neighbourhood.
>
> Today it is much more difficult to come to grips with what constitutes housing quality. Instead of the quantitative aspects, should it rather be described in qualitative terms like community, participation, belonging and the home? Or still, is housing

quality a mixture of the quantitative and qualitative aspects? In that case; how can housing quality be measured? And how can housing quality be compared between different segments of the population, and indeed, between different countries? (Gaunt 1989)

A recent attempt to reformulate a discourse about the domestic was made in *Home: a Short History of an Idea* (1986), by Witold Rybczynski. An architect reared on modernism, Rybczynski deplores the 'postmodern' packaging of lifestyle decor, but accepts the inadequacy of functionalism to give an adequate definition of home. He is wary in particular of functionalism's associations with a technological definition of domestic comforts (Brindley 1989). Like others emerging, willingly or not, into a postmodern awareness, Rybczynski realizes that 'home' embraces imaginative, social and material orders. He perceives that its effectiveness as a cultural category depends on local interpretation, but elevates personalization to the status of a universal principle. Looking out from his study across the centuries, he only manages to recognize the domestic arrangements and ideals that conform to the petty-bourgeois model in which domestic privacy is celebrated as a separate sphere for the creative achievements of the protected female.

The artful way in which Rybczynski historicizes his own sensibility as a resolution of the dilemmas of modernism and postmodernism has had some resonance, particularly in North America, and may be taken as a 'sign of the times'. It also turns on a question which is embedded in many recent approaches to the domestic: authenticity. Rybczynski juxtaposes the discriminating detritus of a busy life with lifestyle packaging. This effects a double disdain: first, manifestly, of the pretension of the commodity to set not only the standard, but tone of living, and second, implicitly, of the home-making lack in the consumer that this would fill. Like weightier cultural theorists, the author of *Home* is personally quite confident in his own aura of authenticity yet dubious as a spectator on a macro-scale.

This problem of authenticity has been around long enough to be regarded as a condition of modernity (Berman 1988). To the extent that we live in a world that permits diverse interaction and self-redefinition, that escapes the degree of closure accorded to the society of tradition, recirculates goods and reattaches signifiers, polyvalence of meaning becomes the norm (Appadurai 1986; Miller 1987). Extreme consequences have been extrapolated, for example, by Heidegger and Baudrillard. However, one has to ask for whom is authenticity what kind of problem? While contemplating the movement of images and objects in the abstract may pose epistemological dilemmas (Baudrillard 1981; Hebdige 1988), the problems of practical reason for those involved in making a home appear to be of a different order. Home-making establishes proximate relations between aspects of spatial and artefactual order, social practice, discourse and the imagination. Although the resulting order can be very sophisticated, it does not exhibit a high level of consistency either internally or in relation to external referents (Kaufman 1991). Households use commodities and general cultural resources in a fundamentally opportunistic way to carry out internal strategies which evolve relatively slowly. While the household's parameters must be secured to a tolerable degree, its object-world does not need to be grounded in the transcendental sense of which Heidegger speaks; nor is it undermined by the lack of ultimate referents (de Certeau 1984).

As Rybczynski recognizes, home-making in producing its own field of value generates an effective authenticity. And while the pursuit of external guarantees or tokens of authenticity for domestic consumption is bound to prove quixotic, this is beside the point so long as it serves to generate a range of satisfactions – hardly exhausted by Bourdieu's construction of social identity through discrimination. Those who would maintain that such emplacement is illusionistic, inauthentic, should 'take a holiday' (as Chambers suggests) at home. The work of framing 'the home' in a macro order is not the same as that of constituting meaning on a domestic scale. As we know little about how these local narratives serve to place people in the world and how their meanings are communicated and accumulated on a wider social scale in a variety of ways, the sovereignty of the discourse from which such 'talk' is assessed is not beyond question.

Attempts to recapture the domestic in discourse have attempted to produce a degree of convergence on a common complex object of study. The two main lines of policy research relevant to the domestic sphere – that which informs housing design and provision and that which monitors the family as a unit of social reproduction – customarily disregarded each other's domains, sometimes with comic or catastrophic results. However, domestic design research, which long ago invoked psychology in a search for universally valid solutions, has been forced to recognize social process and cultural difference. Family studies, having previously acknowledged the importance of external contexts, are beginning to take a greater interest in the immediate home environment. Policy myopia and established differences in disciplinary focal length have meant that the most productive encounters have often been interdisciplinary: where geographers' work on emplacement can cross the threshold and meet accounts of family processes, or recent attempts in psychology to deal with sociocultural difference (Altman and Low 1991; Bernard 1990).

The scope and texture of home-making agency and its relation to macro-processes has been most readily accommodated in that branch of ethnography focused on metropolitan milieux (Lofgren 1990; Miller 1987, 1990; Segalen 1990; Silverstone 1992). In an ethnographic perspective, the significance of 'home' as a cultural category depends primarily on how it is used by subjects to place themselves in both an elemental social nexus ('household', 'family') and a secure and malleable environment ('habitat', 'house'), comprising a spatial and artefactual order. The ethnographic ability to identify the immanence of cultural value in everyday spatial and temporal order, and spot homologies and supports between the orders of discourse, social process and material culture, has characterized the most notable recent research on the home. Although the articulation of the multiple relations between personal, social and material dimensions of home as locality has yet to be adequately explored, and its bearing on the inflection of domestic ideologies remains unspecified, it is now widely acknowledged that all work on the domestic must attempt to situate itself in this context. The ambiguity of the English word 'home' can no longer go unremarked: does the usage indicate an environment encountered, relationships enacted, an ideal envisaged, or an articulation of all three?

To remind ourselves of what is at play in home-making does not in itself give an account of the parameters of residence; the quality of the integration sought or achieved by home-makers must then be related to differentiated accounts of how the externally produced elements and supports of home-making are mediated. Here the ethnographic paradigm is at an apparent disadvantage by comparison with those disciplines that have focused on the macro-structures and processes that govern the transformation of the built environment, employment, the production and mediation of goods and representations, the formation of social fractions and the instruments and legitimation of power. Little work has been done to reintegrate the pertinent findings from these relatively well-developed fields with the articulation of home as a cultural form (but see Putnam and Newton 1990).

Thus, quite different emphases can arise from privileging particular parameters of residence. It has been possible to assert, on the one hand, that the global connectedness brought by electronic technology has so eroded the boundaries of home as to attenuate any sense of own place, and, on the other, that increasing investment of time or elaboration of commodities produced for consumption in the domestic sphere constitutes a new privatism. Such apparently contradictory assertions may depict complementary, even interdependent, trends if brought to bear on each other. As part of such an encounter, it is incumbent upon each discourse to acknowledge its own history and, in particular, its implication in the project to create the modern home. Each needs to disentangle itself from the aftermath of this project and assess the extensive effects of its disintegration – not least on the legitimation of the public realm, of technocracy, of mobility and the aspirations of classes, genders, nations and generations.

Deconstructing 'the modern home'

In speaking of a modern home, we are talking about more than technologized comforts. The modern home is inconceivable except as a terminal, affording the benefits of but also providing legitimating support to a vast infrastructure facilitating flows of energy, goods, people and messages. The near-completion today of great infrastructures of modern material culture: of hygiene, energy, transport and communication, has dramatically transformed what is possible and desirable in homes. The most obvious aspect has been a qualitative transformation of the technical specification of houses and their redefinition as terminals of networks. This infrastructure-building has had a strong public character even though, in most countries, it was largely carried by private investments of science-based industry, and it was linked to the diffusion of a new 'modern' standard of living through the cultivation of popular aspiration. Widening access to this standard and the means of its achievement was an important dimension of the new democratic politics. Proffered improvements to the material environment of the home figured in the agendas of all tendencies that aspired to mobilize support on a broad basis.

There is much to be done to disentangle the ideological aspirations and consequences of the extension of these networks, and their impact on the household as a social nexus. Some work has been done on the programmes, proclamations and institutional politics of hygiene, energy promotion, and public housing provision, but relatively little on the reception of these initiatives or their significance in a wider political context (aspects have been problematized – see Madigan and Munro 1990; Matrix 1984; Cowan 1989). It has been shown both that promotion of new infrastructure dependent appliances as labour-saving was strong, and that their adoption has led to a greater change in the quality than in the quantity of housework. The feminist critique of 'liberating consumption' has demonstrated the non-equivalence of technological possibility, recommended consumption, social practice and householder's ideals, and leaves open many questions about the relations between them. How, for example, should we understand the abandonment of labour-saving as a proffered rationale for appliance consumption since the 1950s in favour of rhetorics of leisure, pleasure and higher standards of consumption?

The rhetorics of appliance promotion remind us of the symbiotic relationship of the public and private sectors during this period of the consolidation of infrastructures and mass legitimation, and also of the close links with family policy, the politics of gender and generation. Here, connections could be made between existing research about the demobilization of women after the Second World War, infrastructural development, and housing policy and design. In some countries, Britain included, the qualitative improvement of the housing stock, in public discussion, has often been overshadowed by a politics of housing quantity. The outlines of postwar public housing programmes are relatively well known and certain aspects – comprehensive redevelopment, system building – have been the subject of extensive discussion. It is important not to overlook the fact that much of the demand for new housing units came from those who would otherwise have had to share with family or strangers, and that housing provision is thus linked with a change in the norm for the number of people per housing unit and especially with a decline in the number of generations sharing family accommodation (in Italy, this norm is currently in process, see Rullo 1990).

The aspiration of couples to have their 'own house' extended beyond public sector tenants, but public guarantees and incentives, backed by provision in the public sector, fundamentally changed the horizon of what it was legitimate to expect. This movement towards generational autonomy was consonant with the reduced role of intergenerational family contacts in transmitting occupational opportunity and life chances which followed industrial concentration and reorganization and the increasing importance of formal education as mediator of cultural resources and opportunities. The disruptive subordination of local communities to the new networks has been much discussed, as has the unreality of various attempts to synthesize substitutes. The neighbourhood aspect of residence has been transformed to the extent that the shared experience of work and home life within a local geographical scale has become a sign of social and economic marginality. This is not simply an effect of mobility, but of changes in the means by which sociality and identity are established and cultural and economic resources can be accessed, which have altered the parameters within which meaningful lives can be made.

In the advanced countries, the installation of the infrastructures of modernity is fundamentally over, although restructuring and relocation continues. For the great majority, the modern home is no longer a dream but an unspoken premise – the conjuncture which linked diverse elements of a political programme has been broken by its own fulfilment and the exposure of its internal contradictions, perhaps crystallized in the universal 'happy housewife' of 1950s advertising. One can sense the transmutation of inspiration into structure in the shifting rhetoric of advertising and propaganda in the 1960s – as well as in the independent courses struck both by privileged social groups and youth, rediscovering the charms of the antique and rustic, and or a style self-consciously postmodern, disenchanted with the new housing estates, even as many were still struggling to realize the essential items of modern domestic consumption.

The disintegration of the project of the modern home brings to the fore some long-established features of discretionary, differentiating consumption, display and confirmation of status. In the present period, these have been fed by a conjunctural attenuation of egalitarian ideals, restructuring and increasing disparities of income. Nevertheless, the appearance of massive new domestic investment is misleading; there has been a shift in the typical objects of investment rather than an increase in the proportion of income invested in domestic consumption.

The infrastructures of modernity are, however, presumed in a permanently changed domestic environment. They afford vastly more choice in the use of the home, but the quality of life which can thus be achieved, accessed through the network in the broadest sense, depends on the natural resources and modes of operation of the inhabitants. Otherwise the installed comforts are as bleak as the eviscerated neighbourhoods. Perceptions of residential satisfaction are very strongly related to confidence in accessibility of external support and opportunity (Bonvalet 1989; Feldman 1989; Franklin 1990). Communication and transport may vie for importance, but the resource to be accessed is crucial: family, friends, shopping, healthcare and leisure facilities, schooling and employment opportunities. So situation signifies first in this broader sense. The Kantian astheticized contemplation of situation depends on the satisfaction of this range of needs' wider agenda.

Home as a project: a 'new conjugality'

Having been relocated culturally, socially and economically by participation in an external educational and employment system, more couples now enjoy and, in a sense, must create, living environments representing their distinctive trajectories – rising from personal and familial histories to self-projection in social space. Greater confidence permits diversity and experiment. The question of the interrelation of projects within the household has gained a new importance. Several recent studies have argued for a relation between a 'new conjugality' and involvement in home improvements. As a sense of joint project is compatible with the maintenance of strong gender definitions in both space and activity patterns in the household, it is interesting to find that this 'new conjugality' has been linked with changes in the relation between generations, sense of neighbourhood and community, new patterns of alteration and redefinition of gender and space in the home (Almquist 1989; Franklin 1990; Miller 1990). There are reasons to treat such findings with some scepticism.

First, the conjugal home project occurs principally as a story told by women (Segalen 1990; see also accounts by Almquist 1989, Lofgren 1990 and Morley 1990b). There is the possibility of a certain irony in these accounts, in that the hidden subject of the account is the woman's specific role both in creating the design of the home and, in a larger sense, taking responsibility for the relationship which is being celebrated (Swales 1988). As well as embodying a displacement, such accounts are also an idealization, for the self and for the visitor. In one sense, the ability to produce such accounts amounts to a kind of examination on social norms; in another sense it represents the necessity of a pole of orientation.

Second, the celebration of conjugality in the domestic interior is hardly new, and a number of 'new domesticity' revolutions have been discovered by historians: the shift of aristocratic family focus from lineage to couple in the eighteenth century (Trumbach 1977); the domestication of the early nineteenth-century bourgeoisie, associated with 'separate spheres' (Davidoff and Hall 1987); companionate marriage in the early twentieth century (Franklin 1989) and its extension in 'joint marriage' (Bott 1957; cf. Partington 1989 and Morley 1990a). One needs to disentangle the specificity of each from a tradition–modernity polarity.

The 'jointness' in current British discussions of conjugality refers principally to shared leisure time, both within and outside the home. In comparison with marriages of separate spheres, such partnerships typically involve paid employment for women outside the home, as well as more time spent by the man within it. However, as Franklin (1990) points out, the new conjugality may involve a strong complementarily of roles – the extent to which this may have been attenuated in the territory of housework has been the subject of some dispute (see Cowan 1989) – and there may also be a greater and more obvious domination of shared time and space by the male than in a segregated marriage.

Third, although contemporary conjugality is not one end of a tradition–modernity polarity, the long-term processes of centralization of economic life and formalization of knowledge which have instigated the installation of the modern home, have a differential impact both within and between households. Those forces that appear to enhance the autonomy of the couple may also challenge the previously defined unity of the conjugal household (Allen and Crow 1989). The home environment is not only able but expected to cater for individualized activities. One sign of this is the demand for dedicated personal space, and appropriate technologies, for all household members (Morley and Silverstone 1990).

Recent studies reveal two patterns in the new conjugality in home-making in Britain. The first pattern involves a general blurring of functionally defined room boundaries and a weakening of the association of gender with particular spaces in the house. Instead there is a focus on particular pursuits – work or leisure – and their supports, which often involve new technologies. Where rooms are dedicated, this is more likely to be on a personal rather than a functional basis. The home is a welter of improvisation and a base for activities. Received gender roles in the household have been transgressed to some extent. The socioeconomic position of this household is likely to be in that part of the intermediate strata where cultural rather than economic capital is predominant.

The second pattern involves extensive DIY or bricolage which has not, however, blurred the functional boundaries of rooms (Almquist's 'carpentry culture'). In this case, there appears to be a link between renovation work and gender complementarity (Miller's 'neo-traditionalism'). This redefined complementarity pervades domestic roles, although the man spends more time at home and the woman less than in a segregated marriage. The place of technologies in the home is likely to be constrained within these patterns. Overall, the home is a project to be realized as well as a centre for leisure and recuperation. The socioeconomic position of this household is likely to be of well-paid manual or clerical workers.

The two groups of couples have a few things in common. Both are likely to be house-owners, although each pattern can be sustained in rented accommodation and in flats. Both kinds of couple believe they occupy a social place and direction of movement somewhat different from that of their parents. Both possess a diverse network of chosen social contacts relating to work or leisure interests. In both kinds of household, domestic labour and responsibility is typically compulsory for women, and optional for men.

The statistical importance of each pattern in contemporary Britain is not known.[1] While the studies surveyed are neither fully comparable nor, taken together, comprehensive, the two patterns identified are sufficiently different to cast doubt on any generalizations which might be made about the temper of a 'new conjugality'. The first is oriented around the relationship between mutuality and individuality and moves towards the deconstruction of its own presuppositions. The second is oriented around the relationship between reciprocity and identity, and works towards the reconstruction of tradition.

These emphases are related to distinctive preoccupations in domestic design. The classic problem for the first group is the redefinition of the comforts and opportunities afforded by a large old house (Swales 1988). This provides endless opportunities for personal exploration and collective realignment; the adaptations of the house are a metaphor for the social process of the family group. The classic problem for the second group is room 'improvement or modernization' – especially of the kitchen, as in Miller (1990). Whereas a previous generation would have occupied separate spheres, now design and labour were being exchanged to demonstrate the complementarity of the genders within the household. Where, as in Miller's black households, gender division was not connected with room use, men took their own decorative initiatives in the kitchen.

The two groups produce their greatest impact on the home environment at different points in the life stage. The first pattern involves the redefinition of space and accumulation of objects as family, careers and avocational interests develop. The second pattern is an extension of the practice of 'setting-up home' before rearing children. The obligation of both partners to work to acquire the requisites of contemporary comfort reinforces jointness in relaxation and in project-planning at home. The importance of role redefinition is dramatized by the juxtaposition of shared experience and internalized gender norms, reinforced by peer and family expectations. When children arrive, both resources for modification and symbolic tension are reduced, to re-emerge at a later stage when the children leave and the couple must redefine its internal dynamic, personal and conjugal goals and social identity.

These distinctive orientations towards home investment and domestic consumption characterize distinct social fractions in the process of formation. The first group crystallizes around educational investment and culture as a disciplined pleasure. The improvised quality of much of its home arrangement reflects both cultural capital and the capacity for further development, with the latter aspect receiving particular emphasis. The second group creates the home as a refuge and a mark of achievement, and its consumption reflects not so much investment in the socioeconomic system as compensation for effort. Social goods are displayed as marks of achievement rather than as developmental means. Although this group's consumption goods may be more varied, its reference more current and 'ambitious' than among traditional segregated marriages, traditions persist in areas that are important to family role definition, such as the three-piece suite (Morley 1990a).

Role expectations for mothers in these studies carry the marks of class-differentiation in segregated marriages. Mother as educator is very different from mother as servant (Hunt 1989). In contemporary conjugality, both of these roles are reinforced by the kind of work which could be performed by the women concerned 'outside' the home (Franklin 1990). Child-rearing forces a prioritization which clarifies the extent and basis of jointness in practice. The ways in which gender categories are redefined also, however, leave their mark on the home environment.

These studies indicate something of how social relations are refashioned in the home-making matrix. Class-distinctive modes of consumption are evident beneath an apparent accession to a standard of living with global access. This difference in social practice implies differential modes of emplacement, appropriation of objects, and social encounter. Such differences revolve around the extent to which subjects see the world as transactable, with perhaps a greater emphasis on compensatory autonomy on the one hand and control of change on the other. These are hardly new themes in the sociology of class formation (Bernstein 1971; Bourdieu 1984); they underline the fact that the dream of the modern home has been realized to the extent that people have been able to make full use of resources of the network.

Much further work is needed on the contemporary home as a cultural site, especially on the dialectic of transmission between generations in a household. Modern infrastructures of education and communication, and an enhanced standard of accommodation would appear to have given greater autonomy to the young in relation to parental cultural practices while increasing their interaction with outside cultural influences. In this sense, 'the home' as an ideological category may come to have a less definite and fully specified signification, which is not to say that the area it designates has become less important in emplacement.

Note

1 Swales's (1988) survey in Birmingham's 'pink triangle' turned up only variants on the first pattern. Both groups appeared in Johnston's (1980) survey of Guildford, among which owner-occupiers predominated, along with a group influenced by feminism to aspire to the first pattern but who were unable to move their partners in this direction Miller's 1984 (1990) survey of Islington council tenants and Wallman's (1984) survey in Battersea produced the second pattern in a sample which also included more segregated marriages. Segregated marriages were dominant in the older London working-class samples interviewed by Morley (1990b) and Brain-Tyrrel (1990). Hunt's (1989) interviews with five middle-class and five working-class families in Stoke did not include an example of either pattern. (This discussion draws on these studies, together with those of Devine 1989, Segalen 1990 and Almquist 1989.)

References

Allen, G. and Crow, G. (1989) *Home and Family*, London: Macmillan.

Almquist, A. (1989) 'Who wants to live in collective housing?', unpublished paper, Gavle, Swedish Institute for Building Research.

Altman, I. and Low, S. (eds) (1991) *Place Attachment*, New York: Plenum.

Appadurai; A. (ed.) (1986) *The Social Life of Things*, Cambridge: Cambridge University Press.

Baudrillard, J. (1981) *For a Critique of the Political Economy of the Sign*, London: Telos.

Berman, M. (1988) *All That is Solid Melts into Air*, New York: Penguin.

Bernard, Y. (1990) 'Life-style evolution and home use', unpublished paper, Paris, Laboratoire Psycologie D'Environnement.

Bernstein, B. (1971–7) *Class, Codes and Control* (3 vols), London: Routledge.

Bonvalet, C. (1989) 'Housing and the life cycle', unpublished paper, Paris, Institution Nationale d'Études Démographiques.

Bott, E. (1957) *Family and Social Network*, London: Tavistock.

Bourdieu, P. (1984) *Distinction*, London: Routledge.

Brain-Tyrrell, A. (1990) 'Objects of necessity', MA dissertation in the History of Design, Middlesex University.

Brindley, T. (1989) 'Social theory and architectural innovation in housing design', unpublished paper, Leicester Polytechnic.

Cowan, R. S. (1989) *More Work for Mother*, London: Free Associations.

Davidoff, L. and Hall, C. (1987) *Family Fortunes*, London: Hutchinson.

de Certeau, M. (1984) *The Practices of Everyday Life*, Berkeley: California University Press.

Deem, R. (1989) *All Work and No Play?*, Milton Keynes: Open University Press.

Devine, F. (1989) 'Privatised families and their homes', in G. Allen and G. Crow (eds) *Home and Family*, London: Macmillan.

Duncan, J. (ed.) (1981) *Housing and Identity*, London: Croom Helm.

Feldman, R. (1989) 'Psychological bonds with home places in a mobile society', unpublished paper, University of Illinois.

Forrest, R. and Murie, A. (1987) 'The affluent homeowner', in N. Thrift and P. Williams (eds) *Class and Space*, London: Routledge.

Franklin, A. (1989) 'Working class privatism: an historical case study of Bedminster, Bristol', *Society and Space* 7 (1).
—— (1990) 'Variations in marital relations and the implications for women's experience of the home', in T. Putnam and C. Newton (eds) *Household Choices*, London: Futures.
Galassi, P. (1991) *Pleasures and Terrors of Domestic Comfort*, New York: Museum of Modern Art.
Gaunt, L. N. (1983) 'Housing Quality', Lund: National Board for Housing, Building and Planning.
Goodall, P. (1983) 'Gender, design and the home', *BLOCK* 8.
Harris, C. (ed.) (1979) *The Sociology of the Family*, Keele: British Sociological Association.
Hebdige, R. (1988) *Hiding in the Light*, London: Comedia.
Heidegger, M. (1975) *Poetry, Language, Thought*, New York: Harper.
Hunt, P. (1989) 'Gender and the construction of home life', in Allen and Crow (eds) *Home and Family*, London: Macmillan.
Johnston, D. (1980) 'Women's attitudes towards the kitchen', unpublished PhD thesis, University of Surrey.
Kaufman, J. C. (1991) 'Les habitudes domestiques', in F. de Singly (ed.) *La Famille*, Paris: La Découverte.
Lefebvre, H. (1971) *Everyday Life in the Modern World*, London: Allen Lane.
Lofgren, O. (1990) 'Consuming interests', *Culture and History* 7.
Madigan, R. and Munro, M. (1990) 'Ideal homes: gender and domestic architecture', in T. Putnam and C. Newton (eds) *Household Choices*, London: Futures.
Marshall, G., Newby, H., Rose, D. and Vogler, C. (1988) *Social Class in Modern Britain*, London: Hutchinson.
Mass Observation (1944) *People's Homes*, London: Mass Observation.
Matrix (1984) *Making Space: Women and the Manmade Environment*, London: Pluto.
Miller. D. (1987) *Material Culture and Mass Consumption*, Oxford: Basil Blackwell.
—— (1990) 'Approaching the state on the council estate', in T. Putnam and C. Newton (eds) *Household Choices*, London: Futures.
Morley, C. (1990a) 'The three piece suite', MA dissertation in History of Design, Middlesex University.
—— (1990b) 'Homemakers and design advice in the postwar period', in T. Putnam and C. Newton (eds) *Household Choices*, London: Futures.
Morley, D. and Silverstone, R. (1990) 'Families and their technologies', in T. Putnam and C. Newton (eds) *Household Choices*, London: Futures.
Pahl, R. (1984) *Divisions of Labour*, Oxford: Basil Blackwell.
Pallowski, K. (1988) 'Die Schwierigkeit der Designer mit dem Geschmack der Leute', *Designactuel* 1: 26–43.
Partington, A. (1989) 'The designer housewife in the 1950s', in J. Attfield and P. Kirkham (eds), *A View From the Interior*, London: Women's Press.
Putnam, T. (1991) 'The aesthetics of the living room', *Issue* 7, London: Design Museum.
—— (1992) 'Regimes of closure and the representation of cultural process in the making of home', in R. Silverstone (ed.) *Consuming Technologies*, London: Routledge.
Putnam, T. and Newton, C. (eds) (1990) *Household Choices*. London: Futures.
Rullo, G. (1990) 'Experience of the home among young adults', unpublished paper, Rome, Instituto di Psicologia.
Rybczynski, W. (1986) *Home: A Short History of an Idea*, New York: Viking.

Saunders, P. (1984) 'Beyond housing classes', *International Journal of Urban and Regional Research*, pp. 201–27.

Saunders, P. and Williams, P. (1988) 'The constitution of the home: towards a research agenda', *Housing Studies* 3 (2).

Segalen, M. (1990) *Être bien dans ses meubles*, research report, Paris: Centre d'Ethnologie Française.

Silverstone, R. (ed.) (1992) *Consuming Technologies*, London: Routledge.

Swales, V. (1988) 'Personal space and room use in Moseley', MA research paper in History of Design, Middlesex University.

Tomlinson, A. (1990) *Consumption, Identity and Style*, London: Routledge.

Trumbach, R. (1977) *The Rise of the Egalitarian Family*, London: Academic Press.

Wallman, S. (1984) *Eight London Households*, London: Routledge.

SECTION TWO

IN PLACE: PLACE AND COMMUNITY

Editorial introduction

In Section Two, we move up the spatial scale to examine the changing relationships between identity and the neighbourhood or locality, as a fixed or bounded geographical area. This is a spatial scale that is more familiar to social geographers than the very local level of the home and the body. Indeed traditional social geography texts have long had a predominant focus on this spatial scale (see for example Knox, 1987; Ley, 1983 and also the companion reader in this series by Hamnett (1996)), with their concern to address issues about the spatial differentiation of communities and their social bases in class and ethnic divisions. This is the focus here too, but the specific aim is to unite a temporal and geographic focus and look at the new bases for community that have become increasingly important in western cities in the post-war period.

Neighbourhoods, the locality, residential areas (or whatever we choose to call them) may be a source of security and the basis of a supportive network for many people whose lives are relatively restricted in an everyday sense to a small area, although they might equally well be a source of irritation, danger, and even despair depending on their location and social characteristics. Further as Marcuse (1989) has argued, the locality is 'also a source of identity, a definition of who a person is and where that person belongs in society as a whole'. An alternative view, however, was expressed by Putnam (Chapter 4), where he suggested that 'the neighbourhood aspect of residence has been transformed to the extent that the shared experience of work and home life within a local geographical scale has become a sign of social and economic marginalisation' (p. 52). Whereas the affluent are 'free in space', able to move literally and metaphorically across the globe, it is the poor and deprived who are trapped 'in place', without the resources to surf the net or fly between global cities. But even the most mobile of the newly mobile, the most global of global actors – the international financier is the quintessential example – is rooted, grounded in the place where he (or less likely she) lives, even if temporarily. And the workplace is also an 'embedded' location as well. Global financiers may spend time on the telephone or on the net but the culture of the City of London, of New York or Frankfurt makes a difference to their everyday lives. Issues of workplace culture are now beginning to permeate the literature of economic geography although they are equally appropriately considered by social/cultural geographers. In this set of readings, however, it is the residential rather than the working environment that is the focus.

In urban sociology and geography there has been a long tradition of 'community studies', examining the dense networks of everyday life in predominantly working-class and often single industry areas, either in the inner areas of larger cities or in industrial villages in areas of the country dominated by heavy industry. Thus in the USA, the Chicago school of urban sociologists dominated by the 'great men' Park, Burgess, and Wirth, concentrated on the dangerous and the deviant but also analysed the social structure of solidly working-class communities in the inner areas of Chicago. As well as the gangs, the hobos and the drifters who peopled their monographs, street life or corner society of working-class Italian men (and the specificity of men is important as I shall argue in a moment) or of the Polish migrants working in the slaughter houses and meat packing plants of the great mid-western metropolis in the interwar period were examined. In Britain, the key influences in the post-war development of community studies were the social commentators Michael Young and Peter Willmott who studied the lives of working-class families in Bethnal Green (Young and Willmott, 1957), as well as the impact of rehousing policies that evacuated these families to the unfamiliar fields at the edge of London (Willmott and Young, 1960). Life in Greenleigh lacked the dense networks of solidarity built up in the face of common hardships in the inner city. A whole range of other studies of rural and urban areas in Britain in the 1950s and 1960s were published, of places where social relationships were also based on solidaristic communal values arising from a shared tradition of material deprivation. These included villages where men worked in the mines, in the fishing industry or in agriculture, 'villages' in the conurbations and middle-class communities in small towns or the suburbs (for an early sample of the genre see Frankenberg, 1966). This work opened a window on a world that was just about to vanish in the face not only of the tide of consumer goods and secure work in the new factories of the south, but also as economic restructuring drew women into the labour market to replace their menfolk. In the 'older' communities, it was argued, people were united through their common working situation and familiarity with each other: 'the people of such communities are simultaneously fellow-workers, husbands and wives of fellow workers, neighbours, interlinked extended families' (Williams, 1983). The nostalgia for this way of life, and the implications for the decline of left-wing politics are explored in Andre Gorz's *Farewell to the Working Class* (1982).

But, of course, the working class had not vanished. It had merely changed its gender. Both Gorz and the community studies school may be criticised for their overly romanticised view of this vanishing world of working-class hardship. Social divisions within the community and within the household were glossed over. For example, the domestic work of women that kept the family going, and was a crucial part of the community cement, as Young and Willmott implicitly recognised in their focus on the social networks among women in their classic study of *Family and Kinship in East London* (1957), tended to be recognised only in that stereotypical figure, 'Our Mam'. She is familiar in academic texts, and also in boyhood memories of working-class men made good (Hoggart, 1957; Jackson and Marsden, 1986; Seabrook 1982) and in plays and films, tied to the sink or the cooker in the 'back kitchen', sacrificing her ambitions for her menfolk. But as Caroline Steedman revealed in her analysis of her mother's life, born and raised in a Lancashire working-class community in the 1930, and her own life (she was a child in the 1950s), her mother was utterly different. She yearned for 'nice things' and longed to escape the stifling bounds of 'community'. The 'old left' in Britain has long had an anti-consumerism bias, and clearly regrets the (mis)remembered warmth of an earlier era, as well as its 'discipline' and certainty. Steedman's memoir reveals how this form of socialism had no appeal to women like her mother. The sentimental, nostalgic view of the 1950s apparent in the urban sociology and geography of the 1960s, and it might be argued mirrored in the return to 'locality studies' by geographers in the mid-1980s (see Cooke, 1988; Jackson, 1991; Massey, 1991 and Chapter 22 in this reader), as well as in 1960s left-wing politics, has recently been resurrected by both the political left and the right. It lies behind the 1990s moral crusade with its call to return to earlier moral certainties supposedly being undermined by the siren song of postmodern relativism, with its valorisation of difference, and according to critics, inability to choose between different positions. The recent book by Richard Hoggart (1995) *The Way We Live Now* is an excellent illustration of this position and is particularly interesting as his earlier memoir *The Uses of Literacy* (1957), along with Jeremy Seabrook's (1982) *Working Class Childhood*, is one of the nostalgic versions of working-class life of which Steedman is so critical.

The first extract I have included in Section Two, by **Huw Benyon, Ray Hudson, Jim Lewis, David Sadler and Alan Townsend**, is from the locality projects funded by the British Economic and Social Research Council between 1984 and 1987. This is a clear-sighted description and assessment of the impact of national and international economic changes on a predominantly heavy industry/single class area that was solidly labour voting and exemplified the old 'community' bonds referred to by Raymond Williams and others. Here a team of researchers reflect on the results of their work in Middlesborough and the prospects for the town in the future, which are bleak and have not significantly improved in the last decade. But despite the regret for the old class-based communities of the inner city and the 'north' and, crucially, political anger at the waste of local energy and resources in towns like this, the story of spatially bounded communities like this one is not entirely negative. Local and international migration, new ways of living and a new spirit of openness in Britain (in the USA with its different history of settlement, ethnic diversity had long been a basis of residential segregation), have resulted in new forms of community, as well as the greater acceptance of difference and diversity within localities. The long-running British soap Coronation Street is a good litmus test. In its early episodes in the 1960s, it was based on a stereotype of family life 'up North' but over the years a more varied cast of characters has appeared and questions about 'race', ethnicity, divorce and sexuality have appeared. Rob Shields' discussion of Coronation Street in his book *Places on the Margin* (Shields, 1991) is a good introduction for readers who are interested in pursuing this representation of solidaristic communities (and I shall return to imagined places in Section Five).

During the 1960s and 1970s the old inner area communities that were studied by geographers began to change. Many of them were torn down and replaced by new state-built housing which was often high rise both when built on the site of the older terraced housing or in suburban estates. While the white British working class moved into this new accommodation, new households and families replaced them in the inner city. In both Britain and North America a twofold movement occurred that has resulted in greater social polarisation in the inner areas of many cities. In the USA, ethnicity, 'race' and skin colour had long been a significant basis of residential concentration and segregation. The inner area neighbourhoods of cities are marked by the movement of a wide range of in-migrants to the United States in the twentieth century and by the interwar and post-war movement north to industrial jobs of the black population. In Britain a similar migration, but from British colonies and ex-colonies, first in the Caribbean and then from the Indian sub-continent, rather than from within the nation, brought a visible minority to Britain from the 1950s onwards, also to fill vacancies in the expanding labour market that could not be filled by the 'natives'. In Britain, London Transport and British Rail actively recruited in a number of Caribbean islands, including Barbados and Jamaica, and the initial migrants, who were mainly men but were later followed by their partners and family, were faced with shortages and racism in the London housing market. Particular small areas, or locales, especially where privately rented accommodation was available, became associated with different groups of in-migrants. Barbadians tended to settle in Notting Hill in West London, for example, whereas Jamaicans were more likely to look for housing in Brixton in South London. In the paper by **John Western**, included as Chapter 6, the reasons for the development of patterns of 'racial' concentration are first examined, before the author assesses his assumption that these areas might have developed as places that are important for communal organisation, with a particular symbolic meaning for different minority groups.

As Western documents, Notting Hill was a scene of social unrest from the 1950s onwards and conflicts between the newcomers and the low-income and deprived white population in Notting Dale flared into rioting in 1958. Other 'deprived' areas, characterised by an ethnic mixture in their populations, in Britain and in the USA, have also witnessed rioting. The names of the predominantly black areas of Los Angeles: Watts and Compton are virtually synonymous with unrest. In Britain Brixton in London, Toxteth in Liverpool, Small Heath in Birmingham have all periodically been in the news between the 1950s and 1990s. But riots, despite their overwhelming presentation as such by the popular media, are not solely a 'race'-based phenomenon. White youths have also been involved and, in Britain at least, in the riots of 1991, outer council estates in Oxford, as well as Newcastle, where materially deprived families were penned up out of sight of policy makers and politicians, were also the site of riots. In Chapter 7 **Beatrix Campbell** examines one such deprived estate – Meadowell in Newcastle – and explains what pushes young men to turn on their own neighbours in her examination of the links between power and space and the inadequacies of Neighbourhood Watch schemes to deal with conflicts within urban communities.

Inner-city areas have also been the location of a new form of residential change in the post-war period: gentrification or the re-emergence of middle-class areas of housing in inner areas. The phenomenon began to take off in the mid to late 1970s as building societies altered their mortgage allocation policies and began to lend on inner-area properties in locations that had previously been 'redlined' (Williams, 1976). In the elegant squares and streets of late Georgian and Victorian property in British cities and the brown-stone apartments of East Coast US cities, the middle class purchased and returned to single family residence property that had previously been subdivided and rented. Western concludes his article with an assessment of the impact of gentrification in Notting Hill. For the 'new' urban middle-class inner-area residence confirmed their 'difference' in positive ways. Middle-class families who chose inner area living tended to see themselves as pioneers of a back to the city movement. Indeed Smith (1996) deliberately compares them with the frontiersmen and women of the Wild West. These pioneers emphasise their difference from 'conformist' suburbanites, their participation in the cultural activities of the surrounding area and the city centre and their tolerant delight in 'cultural difference', a euphemism for the social mix in these areas. The middle-class residents who are there by choice live in close spatial proximity to the urban poor who are trapped in these areas.

There are numerous interesting articles on middle-class gentrification in Britain and the USA (Abu-Lughod, 1994; Allen and McDowell, 1989; Hamnett, 1991; Hamnett and Williams, 1980; Mills, 1988; Smith and Williams, 1986). Neil Smith's book *The Revanchist City* (1996) is a fascinating collection of this geographer's work on gentrification in New York over the past two decades and is well worth reading. Here, however, I have chosen to include an extract from a collection of essays by **Patrick Wright** about the significance of the national past in contemporary Britain. In his analysis of the reasons for and impact of a middle-class movement into Stoke Newington, an area of nineteenth-century working-class housing in the north of the London Borough of Hackney, which is one of the most deprived areas in the nation (see also Wright's 1991 book – *A Journey through the Ruins*), Wright reveals how the new middle-class and older working-class residents of the area have constructed a different memory of the place. While the working-class residents hark back to the sort of 1950s community discussed above, the newer, more affluent gentrifiers have recreated an older 'village' past as their sense of community.

Inner city areas have long been places where people who do not conform to the dominant heterosexual familial way of living have been able to find a sufficient degree of tolerance or anonymity to live, as well as places for strangers and migrants to begin their new lives. A number

of geographers and other urban scholars have documented the importance of inner-city living for gay men and for lesbians. Like the 'new'
middle class, these households have often acted as gentrifiers, buying
into declining neighbourhoods while the property is relatively inex-
pensive and also developing businesses, shops and leisure facilities in
the same areas. The work of Larrie Knopp in the USA, originally with
Micky Lauria, and later alone, has been particularly influential in
uncovering what they (Lauria and Knopp, 1985) termed the role of gay
men in the 'urban renaissance'. For gay men, it has been argued, a
definable and visible territorial basis is a crucial element in the definition of their identity. The statement by Harry Britt, a gay community
leader in San Francisco, that 'when gays are spatially scattered, they
are not gay because they are invisible' is now well-known and widely
quoted, by, for example, Castells (1983) in his mapping of gay spaces in
San Francisco itself and more recently by Paul Hindle (1994), in his
exploration of the development of 'gay spaces' in Manchester. In the
paper included here, by **Benjamin Forest**, the significance of place in
the assertion of a gay identity is explored in relationship to West
Hollywood in Los Angeles. Forest takes a different position: that the
spatial 'containment' of gays has a marginalising effect similar to that
in 'ethnic enclaves' in cities.

The extent to which the visible association with a space or a location
in the city is equally important for lesbians is a disputed issue (see
Castells, 1983 and Adler and Brenner, 1992 for alternative views). But as
women as a group generally earn less than men, and are more often
single parents, they are seldom as affluent as gay men. It has also been
argued that women's fear of violence makes lesbians more cautious in
openly displaying their sexual preferences (Valentine, 1993) although
gay men are also often subjected to street violence in 'gay areas' of
cities such as Sydney, San Francisco, London and Manchester. Women
are, however, visible in the social and cultural lives of 'gay villages' in
these cities (Adler and Brenner, 1992; Winchester and White, 1988). The
increasing importance of sexual identities as a basis for residential
location and of sexual politics has also blurred that divide between the
public and private spheres examined in Section Two by organising
around issues that are traditionally regarded as 'private' where AIDS is
the classic example (see Geltmaker, 1992 and Bell and Valentine,
1995).

Class and social status, life cycle stage, sexuality and ethnicity do not
exhaust the social bases of residential segregation although I am not
able to consider additional divisions here. A range of other geographically fixed communities are also evident in contemporary western societies. Some of these are less familiar than others: small-scale
communities based on religion perhaps, from monastic settlements to
New Age encampments, as well as the temporally fluid groups of road
protesters and the long-lasting women's peace camp at Greenham
Common. One of the most evident social changes, however, is the
increasing longevity of the population of Britain and the USA, and new
forms of communities based on the age of their residents are being
developed – from 'homes' for the elderly, often located in converted

country houses or large urban properties for an older monied bourgeoisie, to purpose-built settlements for senior citizens. In a fascinating book *Cities on a Hill*, Frances Fitzgerald (1987) has examined a number of extreme examples of exclusive communities, including Sun City where all residents must be over 50 and without younger dependants, and a religious community founded by Bhagwan Shree Rajneesh.

In all the examples I have included here, community, whether based on associations of class, 'race', sexuality, or age, is a spatially bounded arena, in which the members live in proximity and from which 'others' are excluded. However, as Wright argues, the members of the same locality may have a different memory of the place and a different sense of its present. Communities are both real and imagined places, constructed by social relations and images, meanings and symbols. While this section has concentrated mainly on the social construction of communities, in Section Five we shall examine places in which the imagined construction of place plays a greater role in their definition.

Wright's paper also makes clear that a diverse group of people may live in the same locality. The boundaries of neighbourhoods, in the sense of small areas, are porous to greater or lesser degrees. In single-class communities the poor may be excluded by the cost of properties in that area, whereas in other areas – outer council estates for example – the more affluent are excluded both by choice and by local housing authorities' allocation criteria. In high-status residential suburbs or estates areas, the affluent may be considered a community of exclusion, in that they have the economic power to exclude others. Communities of the poor, and some based on ethnicity, on the other hand are what Parkin (1974), drawing on Weberian concepts, has defined as communities of solidarism, in which social closure or the use of power to exclude others is based on mechanisms of usurpation, rather than the exclusionary mechanisms of the already powerful. This exemplifies the fact that all territorial associations, communities, or places, are constructed by, based on and riven with power relations. The extract from Campbell's book *Goliath* in which she considers the internal relations of power within a place, is perhaps the one of all the extracts here that is most explicit about power, but while reading all the pieces it is salutary to keep in mind questions about how communities and localities are defined and maintained and, crucially, in whose interest. What all these authors do make clear is that territorially based individual and group identities are a crucial part of everyday life in the contemporary west. The traditional, or modern, basis of community, that of patriarchal class relations may have been undercut by the social and economic changes of the second half of the twentieth century, but it is clear that the significance of place has been reconstituted rather than undermined.

References and further reading

Abu-Lughod, J. (ed.) 1994: *From urban village to east village*. Oxford: Blackwell.

Adler, S. and Brenner, J. 1992: Gender and space: lesbians and gay men in the city. *International Journal of Urban and Regional Research* **16**, 24–34.

Allen, J. and McDowell, L. 1989: *Landlords and property.* Cambridge: Cambridge University Press.

Bell, D. and Valentine, G. 1995: *Mapping desire.* London: Routledge.

Castells, M. 1983: *The city and the grassroots.* London: Arnold.

Campbell, B. 1993: *Goliath: Britain's dangerous places.* London: Methuen.

Cooke, P. (ed.) 1988: *Localities: the changing face of urban Britain.* London: Unwin Hyman.

Dennis, N. Henriques, F. and Slaughter, C. 1969: *Coal is our life.* London: Tavistock.

Fitzgerald, F. 1987: *Cities on a hill: a journey through contemporary American cultures.* London: Picador.

Frankenberg, R. 1996: *Communities in Britain.* Harmondsworth: Penguin.

Frankenburg, R. 1976: In the production of their own lives, man(?) . . . sex and gender in British community studies. In Barker, D. and Allen, S. *Sexual divisions and society.* London: Tavistock.

Geltmaker, T. 1992: The queer nation acts up: health care, politics and sexual diversity in the County of Angels. *Environment and Planning D: Society and Space* **10**, 609–50.

Gorz, A. 1982: *Farewell to the working class.* London: Pluto.

Hamnett, C. 1991: The blind man and the elephant: explanations of gentrification. *Transactions, Institute of British Geographers* **16**, 173–89.

Hamnett, C. and Williams, P. 1980: Social change in London: a study of gentrification. *Urban Affairs Quarterly* **15**, 469–87.

Hamnett, C. (ed.) 1996: *Social geography: a reader.* London: Arnold.

Hindle, P. 1994: Gay communities and gay space in the city. In Whittle, S. (ed.) *The margins of the city: gay men's urban lives.* London: Arena, pp. 7–25.

Hoggart, R. 1957: *The uses of literacy.* London: Chatto and Windus

Hoggart, R. 1995: *The way we live now.* London: Pimlico.

Jackson, B. and Marsden, D. 1986: *Education and the working class.* London: Ark Paperbooks.

Jackson, P. 1991: Mapping meanings: a cultural critique of locality studies. *Environment and Planning A* **23**, 215–38.

Knopp, L. 1987: Social theory, social movements and public policy: recent accomplishments of the gay and lesbian movements in Minneapolis, Minnesota. *International Journal of Urban and Regional Research* **11**, 243–61.

Knopp, L. 1990: Some theoretical implications of gay involvement in an urban land market. *Political Geography Quarterly* **9**, 337–52.

Knopp, L. 1992: Sexuality and the spatial dynamics of capitalism. *Environment and Planning D: Society and Space* **10**.

Knox, P. 1987: *Urban social geography.* London: Longman, 2nd edition.

Lauria, M. and Knopp, L. 1985: Towards an analysis of the role of gay communities in the urban renaissance. *Urban Geography* **6**, 152–69.

Ley, D. 1983: *A social geography of the city.* London: Harper and Row.

Marcuse, P. 1989: Dual city: a muddy metaphor for a quartered city. *International Journal of Urban and Regional Research* **13**, 697–708.

Massey, D. 1991: The political place of locality studies. *Environment and Planning A* **23**, 267–81.

Mills, C. 1988: Life on the upslope: the postmodern landscape of gentrification. *Environment and Planning D: Society and Space* **6**, 169–89.

Parkin, F. 1974: *The social analysis of class structure.* London: Tavistock.

Seabrook, J. 1982: *Working class childhood.* London: Gollancz.

Shields, R. 1991: *Places on the margin*. London: Routledge.

Smith, N. 1987: Of yuppies and housing: gentrification, social restructuring and the urban dream. *Environment and Planning D: Society and Space* **5**, 151–72.

Smith, N. 1996: *The new urban frontier: gentrification and the revanchist city*. London: Routledge.

Smith, N. and Williams, P. (eds) 1986: *Gentrification of the city*. London: Allen and Unwin.

Steedman, C. 1986: *Landscape for a good woman*. London: Virago.

Valentine, G. 1993: (Hetero) sexing space: lesbian perceptions and experiences of everyday spaces. *Environment and Planning D: Society and Space* **11**, 395–413 (an edited extract is also included in the companion reader in this series: McDowell L. and Sharp, J. (eds) 1997: *Space, gender, knowledge: a reader for feminist geographers*. London: Arnold).

Williams, P. 1976: The role of institutions in the inner London housing market: the case of Islington. In *Transactions of the Institute of British Geographers* **3**, 23–34.

Williams, R. 1983: *Keywords*. London: Fontana.

Willmott, P. and Young, M. 1960: *Family and class in a London suburb*. London: Routledge and Kegan Paul.

Winchester, H. and White, P. 1988: The location of marginalised groups in the city. *Environment and Planning D: Society and Space* **6**, 37–54.

Wright, P. 1991: *A journey through the ruins: the last days of London*. London: Radius.

Young, M. and Willmott, P. 1957: *Family and kinship in East London*. London: Routledge.

Zukin, S. 1987: Gentrification, culture and capital in the urban core. *Annual Review of Sociology* **13**, 127–47.

Zukin, S. 1988: *Loft living: capital and culture*. London: Radius.

Zukin, S. 1995: *The cultures of cities*. Oxford: Blackwell.

5 Huw Benyon[1], Ray Hudson, Jim Lewis, David Sadler and Alan Townsend

'It's All Falling Apart Here': Coming to Terms with the Future in Teeside

Excerpts from: P. Cooke (ed.), *Localities: the changing face of urban Britain*, pp. 267–95. London: Unwin Hyman (1989)

Introduction

In the autumn of 1933, as part of his celebrated journey through England, the author J. B. Priestley visited Stockton-on-Tees. From the bridge over the river, he saw:

> the shipyards and slips, the sheds that are beginning to tumble down, the big chimneys that have stopped smoking, the unmoving cranes, and one small ship where once there were dozens. The other men who are standing on this bridge – they have just shuffled up from the Labour Exchange – used to work in those yards and sheds, as riveters and platers and fitters, used to be good men of their hands, but are now, as you can quickly see, not good men of their hands any longer, but are depressed and defeated fellows, sagging and slouching and going grey even in their very cheeks. (Priestley, 1934, p. 343)

What is particularly significant about his account is the comparison which can be made not just with the environmental dereliction and mass unemployment of large parts of Teeside today, but with the causes of this decline. For he was adamant that the demise of shipbuilding at Stockton-on-Tees was one, undesirable consequence of an *international* crisis.

This meant that

> For such a place as Stockton, the game was up. Such new industries as we have had went south. Stockton and the rest, miles from London, and with soaring rates, were useless as centres for new enterprises. They were left to rot. And that would perhaps not have mattered very much, for the bricks and mortar of these towns are not sacred, if it were not for one fact. These places left to rot have people living in them. Some of these people are rotting too. (p. 345)

Is 'the game up' for Teeside today? And what do the 'rules' of this 'game' mean for the ways in which people can and do live their lives? Such questions inform this chapter. It explores how international currents of production and trade (not just in shipbuilding, but in other industries like coal, steel, and chemicals, and more recently even in some service sector activities) have flowed into and out of Teeside. These processes have both shaped and been shaped by the changing social, economic and political character of a changing locality.

Changing times in Teesside

Adventurous, almost intrepid, as Priestley was in his journey through an England ravaged by depression, he did not have it in him to visit Middlesbrough. His comments on this place were short and to the point. It was he said, 'a product of the new Iron Age'. Its growth was fuelled at first by coal, as the town's harbour despatched the output of the Durham coalfield. Later the discovery of iron ore in the adjacent Cleveland hills, together with supplies of coking coal from Durham, left Teesside ideally placed to meet the growing demands for iron and steel of a country experiencing industrial revolution. Sustained growth between the middle and the end of the nineteenth century saw its population rise to over one hundred thousand, where previously there had been only hundreds. And the area came of age, so to speak, with developments in another, newer industry. Across the river at Billingham, Brunner-Mond from 1918, and ICI from 1926 developed a vast chemicals complex, and another company new town sprang up. By 1939 ICI had built 2300 houses as labour migrated into the town. Whereas the steel companies had sought coking coal from the adjoining Durham coalfield, ICI wanted labour power.

After 1945 ICI developed a second major chemicals complex at Wilton on the south bank of the river. This expanded under very different conditions of economic development and labour supply. In a national environment of postwar growth, and relatively full employment, the first concern of planning authorities was to ensure at Wilton an adequate supply, not just of raw materials, but most especially of labour. The concerns of the 1930s expressed by Priestley or the flood of immigrants to Billingham must have seemed a million miles away. Certainly this was how it appeared to consultants Pepler and Macfarlane, who reported in 1949 in their Interim Outline plan for the North-east Development Area prepared for the Ministry of Town and Country Planning. 'The development of such a large unit as Wilton on Teesside,' they argued, 'the one area where male labour is in short supply, must entail some influx of population if its labour requirements are to be fully met' (p. 100). Given the high *national* priority accorded to the expanding chemical industry, 'it would be unwise to prejudice the redeployment of Cleveland labour in heavy chemicals ... by offering it alternative male employment' (p. 76).

Alongside the heightened importance of securing now relatively scarce labour (even if it mean discouraging the introduction of industries which might compete with ICI in its phases of expansion during the 1950s), another feature of the post-1945 period was the increasing attention paid to *planning* for growth. Most significant of such plans was the response to what then seemed a cyclical decline in 1963, typified in Lord Hailsham's report on regional development in the North East (HMSO, 1963). In this, Teesside was to act as a vital part of a regional growth zone capable of generating and attracting new employment to compensate for continuing job losses in the surrounding area, especially the Durham coalfield reproducing the links established in previous years. Teesside was earmarked for industrial expansion supported by infrastructural investment. In 1965, a land use and transportation strategy was commissioned to provide a framework for this planned expansion. Its optimistic rhetoric captures the spirit of the times:

Teesside, born in the Industrial Revolution, offers to the second half of the twentieth century both a tremendous challenge and an almost unique opportunity. The challenge lies in the legacy of nineteenth century obsolescence; the opportunity is to make it one of the most productive, efficient and beautiful regions in Britain; a region in which future generations will be able to work in clean and healthy conditions, live in dignity and content and enjoy their leisure in invigorating surroundings. For Teesside already possesses in abundant measure those fundamental characteristics which provide the foundations for a full life. In few places does one find such modern industries, providing for man's economic prosperity, in such close proximity to a beautiful and spacious countryside, which can be the means of satisfying his recreational and spiritual needs. (HMSO, 1969, p. 3)

Teesplan exemplifies a reappraisal of the region's potential which took place in the mid-1960s. It represented a clear expression of contemporary optimism over the prospect of managed expansion in new light industry to compensate for (and indeed far exceed) gentle decline in the traditional industries. The structure of the statutory planning framework meant that in practice, though, there was little opportunity to safeguard the environment. There was no shortage of problems in this direction:

The air along the Tees, full of smoke from its belching factories, forced downwards by cold tidal breezes, must be as badly polluted an anywhere in Britain. (Gladstone, 1976, p. 44)

But there was little protest on these matters in Teesside.

For a period during the late 1960s and early 1970s questions to do with the objectives of planning and planners could be pushed quietly into the background. It looked as if, for once, things were going according to plan on Teesside. The area boomed with new investment, substantially underwritten by the British state. An almost breathless article from the *Sunday Times* in 1976 enthused over this new growth:

If only the spectators could see this. So said Henri Simonet, Vice-President of the European Commission, when he visited Teesside ten days ago. More than a billion pounds is being invested there in steel and chemical plant, nuclear power and oil installations, and the area can fairly claim to be Europe's most dynamic industrial site.

Here then was a new prospect for this part of the north-east: high fixed capital investment and outward looking capital meant that the area seemed to have conquered the 'British disease' and turned the corner towards economic prosperity. But even in 1976, at the high point of this 'boom', problems were visible. The clearest was the above average level of unemployment, especially amongst school-leavers.

By 1981, these difficulties had become very apparent, as the *Financial Times* explained:

A new beginning with new industries was thought to be the answer. Teesside eventually got that new industry but today it looks back and realises that capital intensive companies are not necessarily the answer. Cleveland has become a model of the new industrial Britain and it still has unemployment problems as serious as almost anywhere in the country.

One explanation for the area's economic problems was couched in terms of its lack of diversity. The *Sunday Times* had argued that way in 1976:

Teessiders admit reluctantly that their area lacks the tradition of independent small business that made so many towns in the south prosperous.

The lack of variety was seen as part of the area's heritage, leaving aside any consideration of how and why such a tradition could reproduce itself. Yet, despite this, by all conventional wisdom Cleveland's economic base should have been dynamic:

If ever a place should have boomed, it was Cleveland. It was a test-tube for regional development policies, for the orthodox wisdom of bribing private investors with fat grants. (*Guardian*, 1983)

In truth, of course, behind the 'tradition' lay the problem of the dominance of big companies like ICI and the British Steel Corporation (BSC) in the local labour market.

Teesside's record levels of unemployment have increasingly become the focus of outside attention. In 1984 the *Guardian* portrayed Colin Armstrong staring from his third-storey window across to BSC's Lackenby works, which had made him redundant two years previously.

'If only Mrs Thatcher could see us,' he said. He lives on an estate where 91 per cent of heads of households are unemployed.

By this time the county has the highest rate of unemployment in recession-hit Britain and the echoes of Priestley's *English Journey* were all too loud. Media attention increasingly focused on the 'north–south divide'. Reporters, visibly shaken, expressed surprise at what they saw as stoicism in the face of adversity.

A sense of injustice is widespread. Yet it has produced surprisingly little political radicalism ... These are people who are too proud to show how much it hurts. (*Financial Times*, 1985)

Proud or not, the media showed them to the nation. In May 1986 the national BBC TV current affairs programme Panorama reported on 'Hard Times' in Middlesbrough. Presenter Gavin Hewitt spoke of the 'growing number of women for whom separation from their husbands is the only answer to unemployment'; of a 'new jobless society', a 'Giro cheque economy'; and of 'the airport that was intended to entice business to Middlesbrough, which now serves instead the departure of its skilled workers'.

Similar processes can be observed in Teesside's second major manufacturing employer, the chemicals giant ICI (see Beynon et al., 1986b). The very name –

Imperial Chemical Industries – conjures up images of an industry developing from the outset to serve an international market. As the protected outlets of the old Empire disappeared in the postwar period, the company expanded instead into other international markets. The interwar growth of a major inorganic chemicals complex at Billingham, and postwar development of the organic complex at Wilton, played important roles in this global strategy (see Pettigrew, 1985; Reader 1970, 1975). ICI has invested heavily in the most up-to-date technologies to meet an overall growth in demand for chemicals and to protect its position against competition. From 1962 onwards, for example, major replacement investment took place in ammonia production at Billingham, replacing coal and coke as a basic feedstock first with naphtha, then later with natural gas. Larger production units and cheaper feedstock dramatically reduced costs; in addition, labour requirements were slashed and several thousand jobs were shed. Continuing employment growth in new plant across the river at Wilton, though, meant that in this period most of the job losses could be accommodated through transfers.

In the 1970s, as demand for bulk chemicals slumped, ICI found itself facing a new, intensified set of international competitive pressures. Overcapacity in ethylene production, the basic building block of most plastics such as those produced at the Wilton site, was increasingly in evidence throughout Europe, the USA and Japan. In response, ICI initiated a series of plant closures at Wilton, especially after 1980 when the company recorded its first-ever net loss. In a letter to all MPs in 1982, the company even threatened to close the entire Wilton site, claiming it was suffering unfairly from the tax concessions granted to its UK competitors over their supplies of ethane, in contrast of the tax arrangements for its naphtha-based plant at Wilton. Four years later ICI won a prolonged court battle over this issue and, with crude oil prices tumbling, petrochemical profits increased again. Nonetheless, intensive competitive pressures remain from companies and countries with access to still cheaper feedstocks, most especially in the Middle East. Billingham's main product, agricultural fertilizers, is also a market area under great strain in the UK, with demand falling and a strong competitor emerging since 1982 in the Norwegian conglomerate Norsk Hydro.

In response to such pressures in the UK market, ICI has increasingly located production overseas. The Wilton works, for example, now has a parallel production facility at Wilhelmshaven in West Germany. ICI employment in the UK has fallen from three quarters of global company employment at the start of the 1970s to less than a half. On Teesside this has meant a reduction from 31,500 jobs at ICI in 1965 to 14,500 by 1985. In addition, priority has shifted away from the so-called 'bulk' or commodity chemicals of plastics, petrochemicals and fibres, and into speciality chemicals where the emphasis is on high value-added, low-volume production. Employment in the manufacture of these new chemicals is relatively low and it is a high-risk business. The most significant such investment on Teesside, the 'Pruteen' plant, was opened in 1980 but has only operated intermittently – and when it does, it employs a handful of people to produce just 150 tonnes of artificial protein daily.

ICI's labour policies have also evolved in a sophisticated fashion over the years. From the earliest days, trade union activity was incorporated into ICI consultative councils as a deliberate technique of the company's personnel department. This was especially marked at Billingham in the interwar period, where the expanding company town was largely serviced by the ICI Works Council, not the town council. Wilton grew under very different conditions of labour market supply and was characterized, for example, by more marked resistance to ICI's major work measurement scheme, the 1969 Weekly Staff Agreement (see Beynon et al., 1986b; Roeber, 1975). Wilton workers were also more actively involved in a loose ICI combine committee, eschewing the more collaborative form of trade unionism.

The use of contractors' labour has escalated significantly, especially for maintenance. In 1984 ICI proposed that process workers should do some maintenance work. Whilst this met with strong union opposition, a more portentous move was announced in 1986 and implemented over the heads of trade unions the following year – the hiving off from ICI of the commodity chemicals side to a new subsidiary company, ICI Chemicals and Polymers, in which redundancy and demarcation agreements are completely open to renegotiation. The future prospects for ICI employment on Teesside are grim. Large-scale, continuing job losses for at least the next five years have already been publicly forecast by senior ICI management.

ICI brings a high level of political awareness to this process of managing decline in the UK. 'Resettlement' schemes were established at Wilton from 1981 to encourage workers to find alternative employment or (mostly) retire. The promotional agency Saatchi and Saatchi has been appointed with a £10 million budget to highlight ICI's strengths. In 1987 a science park was established alongside ICI Billingham, with the intention of attracting alternative jobs into small factory units (and deflecting criticism of the company's labour shedding). Throughout its history, and especially recently, ICI has shown itself keenly aware of the need for large employers to maintain a political presence both locally and nationally.

Like ICI, other sectors of the Teesside economy display both continuity and change in decline. Not all sectors of employment in the area are in such drastic decline, of course. *Teesplan* was strident in its insistence that there was a need to diversify the employment structure through expansion as a sub-regional service centre (HMSO, 1969). To some extent this happened, though on nothing like the scale necessary to mop up job loses elsewhere in the local economy. Many of the spaces reserved for office development now stand as vast 'temporary' car parks on the fringes of Middlesbrough town centre, a silent testimony to unfulfilled expectations. A net decline of 20,000 manufacturing jobs in the period 1971–1981 far exceeded a net gain of 4000 service jobs – with most of this latter increase as female part-time employment. Full-time service sector employment actually fell. By 1981 chemicals and steel still accounted for 60 per cent of all manufacturing jobs but over a half of Teesside's jobs were in the service sector. The rise of the service sector, if sometimes exaggerated and somewhat more complex than it is often presented, is nevertheless significant, especially in bringing large numbers of (married) women back into the labour market or into employment for the first time.

Much of this service sector growth was in financial services such as banking and insurance, which increased employment by 80 per cent between 1971 and 1981 (see Lewis, 1987). This has been vulnerable to technological changes, particularly the increased use of networked computers. Many insurance and finance companies have recently reorganized their operations within the north-east, typically concentrating on Newcastle and closing offices in Middlesbrough. As credit boomed via plastic money, a significant component of financial services employment in Teesside has arisen in Barclaycard, which opened offices at Middlesbrough and Stockton in 1973 and 1974 respectively. In 1985 these employed 700 people, 85 per cent of them women. The main tasks are processing of remittances and sales vouchers – entailing the continuous use of a VDU – and handling customer enquiries.

Service employment in the public sector is typified by the National Health Service. Here too the workforce is predominantly female, but the private sector has not dominated development as it has in financial services. There are no private hospitals in the area and there is a strong commitment to the public provision of health care through the NHS. The industrial and urban legacy of the area is apparent in its relatively large numbers of hospitals and hospital beds, with a standardized mortality rate 22 per cent above the national average (see Townsend, A., 1986; Townsend, P. *et al.*, 1986). In certain areas close to the river, mortality rates are even higher, one impact of long-term environmental pollution. The main planning issue for the South Tees Health Authority concerns the fragmentation among a large number of aging hospitals. Recent years have seen centralization of investment on the new South Cleveland hospital in Middlesbrough and a continuing tension over patient access to facilities in east Cleveland. Centralization in south Cleveland has not been opposed by the trade unions but they have been strong opponents of a more recent development, the planned contracting out, via privatization, of tasks such as catering, cleaning and domestic work.

These changes in steel and chemical production, and in financial and health service provision, represent a series of portraits of the dominant characteristics of change in the economic structure of Teesside. The two major manufacturing industries have acted in a changing and increasingly competitive national and international environment to attempt to secure continuing production, with a considerable degree of state support in the form of investment subsidies and, in the case of steel, nationalization. Regional Development Grant payments and the debt financing of BSC served to underpin investment in new technologies but did not, indeed could not, generate employment gains as in the initial period of absolute expansion of production. Both steel and chemicals industries currently rest on a precarious toehold in Teesside, subject to overseas competition in export markets and import penetration of the UK market. While service sector activities have become, almost by default, of greater significance than manufacturing in terms of the number of jobs, these are often of a qualitatively different character. Service sector growth has been dominated by unskilled and poorly paid part-time female employment, vulnerable to renewed technological change in the case of financial services, and to government-imposed financing limits in the case of the health service.

Such developments in these four very different industries mesh together in the character of the Teesside labour market. The dominant feature of this is the shortage of jobs. Cleveland County now has the highest rate of unemployment in Britain. In such a climate, changes in labour practices including the use of subcontractors or the spread of 'flexibility', both in and out of work and within work, are more easily imposed. Pressures in this direction, as we have seen, are evident in both manufacturing and service sectors, and indeed are increasingly apparent nationally (see NEDO, 1986). On Teesside there is a further emergent trend, one rooted in the earlier prevalence of skilled yet temporary employment in building the new chemicals and steel plants – that of migration of skilled construction workers overseas on lucrative, short-term contract work, to areas such as the Middle East. The irony could not be sharper. As Teesside's industries stagnate or die, Teesside's workers – at least those with the necessary skills acquired in an earlier era – are forced to find employment in precisely those countries where competing industries are emerging. Britain's role within an evolving international division of labour could not be clearer.

Experiences of life in Teesside

High unemployment does not directly affect all households to the same degree or in the same way, but the depressed labour market does increasingly have a general effect on those in work or seeking work. One household affected in this way is that of the Cowdreys, a couple in their thirties who live in Acklam with their two young sons. Martin held a variety of jobs, later joining British Steel. In 1982 his part of the works went out to private contractors and he was laid off. It took him two years to find another job. 'I thought I'd probably get another job straight away, but it wasn't to be.' Now he works as a driver on the buses. After two years back at work, he was still struggling to pay off the bills, even though he had invested his redundancy payment wisely. His wife's earnings helped as well, but without the redundancy money 'there's no way we could have stayed here with the mortgage – we'd have had to think about moving into the town'. He was relatively lucky – he'd just heard from one of the men laid off from British Steel at the same time, who still hadn't found a job. Carol worked in a number of office jobs, most recently at Barclaycard. She regularly works overtime at weekends, to earn a bit extra for the holidays. She can't see their sons getting a job locally.

> A few years ago you could pick and choose your own jobs, but if you've got a job now you're better off sticking with it. My first job I left because I didn't like the bloke I worked with, but now I wouldn't do that.

In many ways the Morecambes are a similar household. Richard and Lucy have been buying their council house at Coulby Newham for several years. Richard works shifts at ICI so that they have to negotiate over who has the car because the estate is quite a way out of the town. Richard started his early working life in the steelworks before moving around the industrial plant in the area. 'It's a funny thing, that, I always stayed in one job for about two years,' he remarked. In the early 1970s they thought about moving away, but then he got a job with ICI. Some weeks before we first met he was told that his part of the works was soon to be closed, and he was in the process of evaluating his options once he was laid off. He's convinced the chemicals industry is dying on Teesside. 'When I first went to Wilton it took you half an hour to get into the site because of all the cars queuing to get in. Now you drive from one end to the other in a matter of minutes.' At night, the evidence of decline is even more apparent. 'There's plant that should have lights on, and there's nothing. They've closed down – vast open spaces in the middle of the works.' If Lucy didn't work, Richard would work overtime. As it is they've got by, although they're obviously concerned for the future. Lucy said she'll encourage their two young daughters to travel and see the world – 'because I don't think there's going to be anything for them here'.

Even at service sector employers like Barclaycard, the pressures in work are increasing. Irene and Tom Oldham live in Acklam with their two daughters. Irene wants to work and almost always has. 'I have to work, it would drive me mad not to work, just being a housewife. I want something that I can do, a bit of independence. But we were all a lot happier three or four years ago when there was no pressure on.' She recognizes something else: that at Barclaycard, 'few women have husbands who are unemployed. It seems people who work in places like that don't come across unemployment so much. There's different sets of people. Tom has always maintained that if he lost his job he'd find another one.' This he has done successfully on numerous occasions in the past, and both their daughters now also have office jobs.

In their various ways, these households are illustrative of a range of people in Teesside who have evolved ways of coping with the exigencies of its changing labour market. What they indicate too is a growing polarization of society within Teesside between those who are given, or take, the opportunity to adjust to a lifetime of insecure, shifting or intensified employment, and those who do not even get that chance (see also Foord et al, 1985, 1987).

In a work-oriented society, unemployment frequently brings domestic tensions. One fifteen-year-old recalled how his father was out of work for more than a year.

> He got very depressed; he took up golf, but couldn't stick it. He just started getting depressed and niggly. You couldn't blame him really. You could understand. He was used not so much to a lot of excitement, but pressure. I think my parents nearly did split up at that time.

Becoming unemployed has been an everyday experience for some time now in Teesside. It is an area which has experienced the effects of recession before, in the 1920s and 1930s (see Nicholas, 1986). One man laid off from the engineering industry put this process into words. He described it as 'just a small item on page seven – you know, the industrial news – and you're out of a job'. Finding another one, or a first one, is not so easy. In the Middlesbrough Job Centre a twenty-three year old butcher, unemployed for two years, commented:

> I've lost count how many jobs I've applied for. I'd say about fifty. Now I've got to force myself to go and look at the cards on the board. A lot of people have given up – just sit at home saying there's no jobs so there's no point looking.

One reason for giving up lies in a perception of what employers are looking for. Mary, an unemployed catering worker, described one job she had applied for:

> At 53 I was too old. The woman behind me in the queue was too young at 20! If you want a job these days you have to be between 25 and 35. If you're old you've no choice. But there doesn't seem to be any jobs for younger ones either.

Increasingly, finding a job is as much about who you know as what you know. There is widespread, tacit acknowledgement of the significance of informal networks in finding work. 'Though a friend of a friend' is frequently heard in this context. Sometimes these connections extend into the greyer areas of the labour market, although there is more than a suspicion that the extent of the so-called 'black economy' is overstated and the benefits to potential (often waged) consumers under-emphasized, as living 'on the fiddle' is fraught with dangers. One unemployed man put it this way:

> There's a lot of people on the fiddle – but then again a lot are jealous of the money they earn. There's a lot of anonymous phone calls to the DHSS. I wouldn't tell on anybody myself, but I wouldn't take part either for fear of being caught.

Moving away in search of work is one escape sometimes seen as an inevitable process. A local businessman, for example, felt that 'nobody goes from the South to the North, it's against the natural routes'. Not that there is anything necessarily 'natural' about a journey in the other direction. It has its complications, and can be very traumatic. A school-leaver explained how his father worked in the Middle East on contract work.

> A year ago he came back from Oman. It's expensive to take the rest of the family over there so he has to go out for three months at a time and when he comes back, he's only back for a week and a half or something, so you don't see much of him. You get used to it – he's always away. Now, he's not away, he's working in Watford, down south. He works weekdays and he's back at the weekend.

Some concluding comments

The main features of Teesside's recent economic history are clear and easily understood – a transition from boom town to slump city; an area now of high unemployment where capital intensive investment created few new jobs, where service sector growth was unable to compensate for manufacturing decline, leaving migration in search of work the only realistic option for a substantial minority. Yet the story is also a complex and intricate one, because its plot was written by so many different and competing interests. Diversity in decline is a marked characteristic of Teesside, and other areas of the north, today. In this tension lies a source of polarization and fragmentation.

The evolution of Teesside has been a social process, entailing political decisions. Its current situation poses acute questions, also of a political nature, which are significant both nationally and locally. Different phases of Teesside's development have coincided with different dominant conceptions of the appropriate relationship between national and local government. In the boom years of the 1960s and early 1970s, the area was a highly attractive location for some forms of capital investment in manufacturing. Local authorities in that era were engaged not so much in planning for growth, as responding to growth. Planning was orchestrated nationally whilst, as Cleveland County Council's leader, Bryan Hanson, recalls, local authorities 'were moving like hell to keep up with everything'. He reflects today that 'the amazing thing is that the boom was for such a short period'. The question now is whether such conditions will recur, or whether alternative futures might emerge in and from an area which has seen enough of the social and environmental costs of industrial growth and decline.

Note

1 Huw Beynon is in the Department of Sociology, University of Manchester; Ray Hudson, Jim Lewis and Alan Townsend in the Department of Geography, University of Durham; and David Sadler in the Department of Geography, St. David's University College, Lampeter.

References

Benn, C. and Fairley, J. (1986) *Challenging the MSC on Jobs, Training and Education*, London: Pluto.

Beynon, H., Hudson, R. and Sadler, D. (1986a) 'Nationalised industry policies and the destruction of communities: some evidence from north-east England', *Capital and Class* 29, pp. 27–57.

Beynon, H., Hudson, R. and Sadler, D. (1986b) 'The growth and internationalisation of Teesside's chemicals industry', Middlesborough Locality Study, Working Paper No. 3.

Boswell, J. S. (1983) *Business Policies in the Making: Three Steel Companies Compared*, London: Allen and Unwin.

Carney, J., Lewis, J. and Hudson, R. (1977) 'Coal combines and interregional uneven development in the U.K.', in D. Massey and P. Batey (eds), *Alternative Framework for Analysis*, London: Pion.

Cleveland County Council (1983) *The Economic and Social Significance of the British Steel Corporation to Cleveland*, Middlesbrough.

Fevre, R. (1986) 'Contract work in the recession', in H. Purcell, et al. (eds), *The Changing Experience of Employment: Restructuring and Recession*, London: Macmillan.

Financial Times (1981) 'A capital intensive cul de sac', 3 June.

Financial Times (1985) 'A town too proud to let the anger show', 19 March.

Foord, J., Robinson, F. and Sadler, D. (1985) *The Quiet Revolution: Social and Economic Change on Teesside, 1965–1985*, Newcastle upon Tyne: BBC (NE).

Foord, J., Robinson, F. and Sadler, D. (1987) 'Living with economic decline: Teesside in crisis', *Northern Economic Review*, 14, pp. 33–48.

Globe and Mail, Toronto, 7 February 1987, 'England's great divide: the south prospers while the industries of the north crumble'.

Guardian (1983) 21 March.

Guardian (1984) 'Community without hope where the loan sharks prosper', 24 December.

HMSO (1963) *The North-east: A Programme for Regional Development and Growth*, London: HMSO, Cmnd. 2206.

HMSO (1985) *National Plan*, London: HMSO.

HMSO (1969) *Teesside Survey and Plan*, London: HMSO.

HMSO (1973) *British Steel Corporation: Ten Year Development Strategy*, London: HMSO, Cmnd. 5236.

HMSO (1978) *British Steel Corporation: The Road to Viability*, London: HMSO, Cmnd. 7149.

Hudson, R. and Sadler, D. (1984) *British Steel Builds the New Teesside? The Implications of BSC Policy for Cleveland*, Middlesbrough: Cleveland County Council.

Hudson, R. and Sadler, D. (1985) 'The development of Middlesbrough's iron and steel industry, 1841–1985, Middlesbrough Locality Study, Working Paper No. 2.

Labour Research Department (1939) *Coal Combines in Durham*, London: Farleigh.

Lewis, J. (1987) 'Employment trends in the financial services sector in Middlesbrough', Middlesbrough Locality Study, Working Paper No. 7.

Lock, M. (1946) *Middlesbrough Survey and Plan*.

NEDO (1986) *Changing Working Patterns: How Companies Achieve Flexibility to Meet New Needs*, London.

Nicholas, K. (1986) *The Social Effects of Unemployment in Teesside, 1919–39*, Manchester: Manchester University Press.

Pepler, G. and Macfarlane, P. W. (1949) *North-east Development Area Outline Plan*, Ministry of Town and Country Planning, interim confidential edition.

Pettigrew, A. (1985) *The Awakening Giant: Continuity and Change in ICI*, Oxford: Blackwell.

Priestley, J. B. (1934) *English Journey*, London: Heinemann.

Reader, W. J. (1970, 1975) *Imperial Chemical Industries: A History* (2 vols), London: Oxford University Press.

Roeber, J. (1975) *Social Change at Work: The I.C.I. Weekly Staff Agreement*, London: Duckworth.

Sadler, D. (1986) 'The impacts of offshore fabrication in Teesside', Middlesbrough Locality Study, Working Paper No. 5.

Sunday Times (1976) 'If only the speculators could see this'.

Townsend, A. (1986) 'Rationalisation and change in Teesside's health service', Middlesbrough Locality Study, Working Paper No. 4.

Townsend, P., Phillimore, P. and Beattie, A. (1986) *Inequalities in Health in the Northern Region*, Newcastle on Tyne: Northern RHA.

6 John Western
Ambivalent Attachments to Place in London: Twelve Barbadian Families

Excerpts from: *Environment and Planning D: Society and Space* **11**, 147–70 (1993)

Introduction

As a social geographer eagerly looking for symbolism in 'place', I came to live in a multiracial sector of London in 1987–88 hoping to uncover some redolently 'black' locales. I had assumed that talking with *any* Afro-Carribean[1] Londoners, whoever they might be, would soon point me to particular London districts. I must admit that I already had my mind made up about certain places in advance. 'Common sense', gained from being raised in Britain until 1968, from reading the academic literature, and from attending to years of media reports, predisposed me for an encounter with, among others, the place names 'Brixton' and 'Notting Hill'.

There was, however, a central feature of the research project that, I soon found, did not mesh at all well with my concern for symbolic 'black' places. That is, I had already committed myself to concentrating on persons from just one West Indian island, Barbados. It became evident to me that my thirty-four Barbadian-origin informants were not necessarily representative of all people of West Indian origin in London. Many of the Barbadians were convinced of their own particularity (and for some, of their own superiority) among persons from the Commonwealth Caribbean. That these Barbadians would point me the way, then, to *typical* 'black' London locales or to particular foci of some general 'black' sentiment was certainly not an outcome of which I could be certain. In addition, most of my informants had achieved a level of relative economic success ... or, at the least stability. Having migrated from an economically depressed Barbados in the 1956–64 period, all of them had assiduously maintained continuous employment, and with varying degrees of frugality had saved enough to become owners (the last in 1986) of London property. With the appreciation of real estate in the metropolis over the previous decade or more, they were potentially quite well off. Were they to decide to sell up and return to Barbados, they could indeed do so in style, visibly successful – thereby achieving what for many had been their aim when they left their home island thirty years earlier.

It was indeed the interviewees' particularity which in large degree accounted for my unmet expectations about significant black London places. Neither arbitrary nor serendipitous, the interviewees' lack of attachment to a Brixton or a Notting Hill had a logical basis, given their specific life experiences and situations. Thus was I compelled to come to appreciate the truth of Eyles's observation that it is indeed 'a tortuous business [to learn] to see the world of individuals or groups as they see it' (1988, page 2). Evidently, from the predominant, white English Londoners' viewpoint, a 'black minority' does exist in London, membership ascribed unproblematically by these persons' very appearance of Africanity. This 'black minority', most of whose ancestral pasts include a forced detour via the West Indies slave plantations, is assumed to have 'its' Brixton, for example. But 'from the inside looking out', although over half of my respondents had once lived in either Notting Hill or Brixton, none did so now; none seemed to feel in any way that they somehow should; and very few were convinced that any locales existed in London that could be specifically deemed important for either Barbadians or black people in general. Even when I directly asked my informants about what is arguably the most visible single element of the partial Caribbeanization of London life, the enormous end-of-summer Notting Hill street carnival, responses were on the whole rather tepid.

That I held the incorrect expectations I did was not necessarily the result of simple ignorance on my part or pure misrepresentation on the part of either the media or previous scholarship. I would point to at least three very plausible reasons why Notting Hill might be important to Afro-Caribbean Londoners. First, it was, with Brixton, the earliest zone of settlement for blacks in the modern period, and saw antiblack riots (and black resistance thereto) in 1958. Second, there is the history of police harassment of the Mangrove Restaurant, a focus of black politics, from its founding in 1969 to its demise in 1991. And third, there is the annual Notting Hill carnival. We shall consider each of these three strands of explanation below.

Caribbean colonization of London

The literature agreed, first, that in the modern, postwar period of large-scale black settlement in London, it was the two neighborhoods of Brixton and Notting Hill that represented the first putting down of roots in an alien land. Given the postblitz housing shortage, the lack of means of the West Indian immigrants, and widespread racial discrimination by white natives (Burney, 1967; McIntosh and Smith, 1974; Lee, 1977), there were only limited areas of London where Caribbean people found it possible to settle in any numbers. Mostly, these were inner-city locales offering downmarket furnished rooms near to less-skilled job opportunities. A first such settlement focus appeared adjacent to the Labour Exchange[2] on Somerleyton Road, Brixton, after 1948. Before long a second focus of black settlement emerged: the sector west and northwest from Paddington Railway Station. Glass found that 'migrants who come from the same territory in the Caribbean tend to live fairly near to one another in London. This is true especially of people from Barbados, from British Guiana and from Trinidad, the majority of whom are concentrated in the main areas of West Indian settlement north of the river . . . Over 40 per cent of the migrants from Barbados . . . live in the West area [Notting Hill and vicinity] – The Jamaicans are in general more widely distributed than the migrants from other territories; but there is one cluster of them in South London, in the South-West area [Brixton–Stockwell]' (Glass, 1960, page 40). These patterns have persisted somewhat (Peach, 1984).

The post-War in-migrants have made Britain into a consciously multiracial society.[3] As agents of this social change – a change surely unwilled by the majority of the white British – they often met covert or overt hostility. Paging through old copies of the *Kensington News* (Notting Hill is loosely synonymous with northern Kensington), one with ease finds 'to let' advertisements such as these:

> A large bed-sitting room. 2 divans. Newly decorated, Small cooker. H&C. Europeans only. References required. £3.10s weekly including cleaning. FLA 4785 (5 February 1958).

> Double room for two Englishmen. Breakfast and 6 pm dinner £2.18s.6d each (September 1954).

Or 'no colour', 'no Irish', or, revealingly, 'Coloured, respectable businesspeople. References essential' (5 September 1958). The experience of Tony and Sandra Gill is thus in no way surprising:

> We would look in the papers and in the shops. We had some difficulties. I remember going to one house after work, the lights were on, and as soon as I rang the bell they went off. They were looking at us. No one ever answered the door. Funny, isn't it?

It was perfectly legal at that time for landlords to discriminate on racial or other grounds. The incoming West Indians had few choices, and had to pay what, for their limited means and for the quality of accommodation received, were very high prices. It was a landlord's market . . . and not surprisingly, it attracted some who were less than ethical as they sensed easy profit. The most infamous of the slumlords was Perec 'Peter' Rachman, the chicanery of whose complex property deals and whose innovatively brutal strong-arm enforcement tactics gave rise to a new word in the English lexicon: 'Rachmanism'.

Rachman started buying up properties in and around St Stephen's Gardens in the Colville area of Notting Hill in 1955. (All street names mentioned may be found in figure 6.1.) The typically three-storied or four-storied row houses he acquired usually had 'statutory tenants' in them, who had relative security of tenure along with a measure of rent control. Rachman of course wanted them out. Two ploys were found to be particularly effective in ejecting the long-standing tenants: the introduction by Rachman of prostitutes to adjacent apartments; and the introduction of black neighbors.

That the two should be so unthinkingly equated as threats by so many statutory tenants is quite an indictment of British racism, as indeed is the very phraseology of an official government report when commenting on the matter:

> Coloured people were [for, Rachman] welcome. Cheerful people, and given to much singing, to playing radiograms and to holding parties, they were not always appreciated as neighbours by the remaining statutory tenants in Rachman's houses. These started to move out, and what perhaps began naturally, Rachman began to exploit, seeing, perhaps, no point in paying controlled tenants to go if they could be persuaded to do so by other means (Milner Holland, 1965, page 252).

In most of these rooming areas, conditions generally ranged from the dingy, down to the squalid and indeed unhealthy. Official statistics revealed the Colville

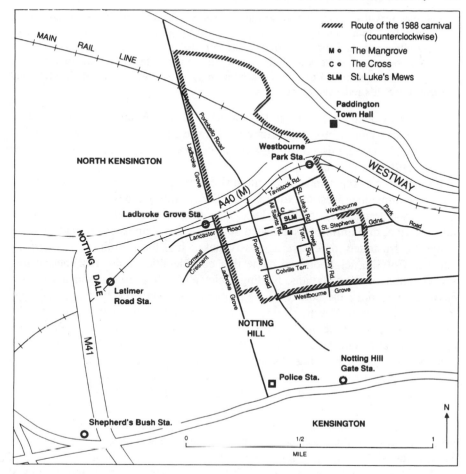

Figure 6.1 Notting Hill landmarks, 1948–88.

area experienced the greatest overcrowding in Britain, Glasgow excepted. Cohen has gone so far as to call it 'the worst housing situation the country had ever seen' (1982, page 27).

According to a recent report by a public-interest organization (Police Foundation, 1987, page 6.6), in Powis Terrace in the late 1950s sixteen houses were divided into 140 lettings, occupied by 300 people, most of whom were black. Yet, Rachman was one of the relative few, after all, to offer any accommodation (on his own terms) to West Indian people. This leads to a certain ambiguity, a certain reluctance on the part of West Indians to condemn him out of hand. As Frank Springer said to me: 'I mean, we couldn't get accommodation, and he provided it'. Audley Simmons stayed in a Rachman-owned property in Powis Terrace for five months in 1958–59. That he moved on so soon implies that he may not have found it ideal, but today he can recall a number of the neighborhood's positive features which were emerging as supports for the growing local West Indian community:

One plus was that you could go to the pub for a drink [that is, it welcomed blacks. Many pubs – quite legally then – refused to serve blacks]. It was at the corner of Portobello Road. And shopping, oh yes, the Portobello market. You could get some of the Caribbean food; there was a little shop on the corner that was open on Sundays. The Calypso Club was round the corner in a basement in Ledbury Road. We'd drink and dance there. It's redeveloped now.

There was evidently a liveliness to Notting Hill. For Frank Springer, distance nowadays can lend a little enchantment: 'Life in Ladbroke Grove was wonderfully sleazy', he reminisced. Such was the context in which the 1958 antiblack Notting Hill riots flared.

It seems that a major impetus for these riots emerged from Notting Dale, which lay to the immediate west of Notting Hill. Today largely redeveloped (much of it a freeway interchange), Notting Dale was then a poor white neighborhood: introspective, disdainful of established authority, and demeaned by outsiders – a recipe for frustrated, defensive aggressiveness.[4] Into the rooming houses of the adjacent Colville neighborhood came this considerable, rapid, and inescapably *visible* influx of West Indians. Notting Dale felt itself at risk – and it was was easy to misidentify the source of its problems as being the black immigrants. Young males from Notting Dale – a number in the distinctive antiauthority 'Teddy Boy' garb of the day[5] – were the precipitators of the violence. Pilkington asserts that, although there is a lack of *hard* evidence, 'it appears that the rioters were acting with the blessing of the majority of Notting Dale's white community . . . The vast majority . . . maintained a complicit silence' (1988, page 116).

The first major incident occurred on Saturday night, 23 August 1958, when, after some drinks in a pub, a group of nine white youths – armed with chains, a car starting-handle, and the like – crammed into a car and went 'nigger hunting' around Notting Hill and adjacent Shepherd's Bush. They picked on any solitary West Indian they could see; three West Indians ended up in hospital for several weeks. Tensions in the area increased. By the following weekend large groups of young whites roamed about looking for black people to beat up; and later they attacked the police as well. On Monday, 1 September, Pilkington states flatly that 'Notting Hill experienced some of the worst rioting that Britain has seen this century' (1988, page 115). The following two nights saw serious disturbances too.

The Mangrove and The Cross

If the streets of Notting Hill encapsulate the history of black travail and black resistance in 1958, it became clear to me soon after moving onto St Lukes Road in September 1987 that another focus of resistance stood at the far end of our block. This, the second strand of meaning, was the Mangrove Restaurant on All Saints Road. A brief London florescence of the Black Power movement – there were indeed a few Black Panther cells – focused on the Mangrove, established by Trinidadian Frank Crichlow in 1969. The police responded with continual raids. Demonstrations against these raids were organized. At the first, the police arrested a number of persons, including Crichlow, for 'conspiracy to riot'. The trial of the 'Mangrove Nine' (as they were inevitably dubbed) dragged on, becoming a cause célèbre. Glitterati such as writer Colin MacInnes (who lived in and wrote about this section of Notting Hill),[6] actor Vanessa Redgrave, and barrister Lord Gifford rose to the defence of the Nine at the trial. Twenty-two charges were dismissed; only seven minor counts (for example, operating a restaurant after hours) were proven. Drugs were implicated in this affair, as well as politics. It was common knowledge that All Saints Road was (and is) a major center in London for obtaining cannibis.[7] The Apollo pub, fifty yards along All Saints Road from the Mangrove, saw much drug dealing. (There were apparently some illegal drinking premises on All Saints Road too.) In the late 1970s Crichlow was again arrested, along with five others, on a drugs-supply charge (the 'Mangrove Six'); again he was acquitted.

Then, in the summer of 1981, widespread civil unrest broke out in numerous, mainly inner-city locales throughout England. Many expected Notting Hill to see a major riot, whose most predictable geographical flashpoint was likely to be the vicinity of All Saints Road.

All Saints Road had clearly become, in a term coined with not a little self-conscious braggadocio, a Front Line to activist blacks. Keith (1986, page 222) reports that Chief Superintendent Archer[8] was set upon 'taking away the symbolism of All Saints Road'; that from April 1982 onward new police tactics introduced high-technology targeting and surveillance; and that a story was intentionally leaked to a London Sunday tabloid about the 'festering sore' of the Apollo pub, whose brewery promptly shut it down. For six months in 1981–82, a police photographer took pictures of everyone using the Mangrove. Then, from August 1982 carnival onward, six officers were to be stationed in the road's immediate vicinity twenty-four hours a day.

Police attention continued. Throughout 1988 casual observation revealed that cars seemed to be invariably ticketed outside the Mangrove, whereas elsewhere in the immediate vicinity such zeal was not evident. There was a sense of harassment. Then, on the night of 20 May, a contingent of forty-eight police (some in paramilitary riot gear) raided the Mangrove for drugs and arrested Frank Crichlow and others. He was held in Wormwood Scrubs prison on remand for six weeks and then permitted bail on condition he did not go near his restaurant business. Also, his assets were frozen.

When Frank Crichlow came to trial in June 1989, Judge Lloyd dismissed the case, despite the testimony of thirty-six police officers. It transpired that the drugs – heroin and cannabis – had been planted: 'there was a conspiracy from Constable through the ranks of Inspector to Chief Inspector to Superintendent Pearman [then in charge at Notting Hill Police Station]' (Bailey, 1989, page 68). But the police did at least in part achieve their aim, because many of the clientele were scared away, and in February 1991 a bankrupt Mangrove Restaurant was auctioned off.

The Notting Hill Carnival

The third strand, another focus of black resistance – or as Cohen (1982, page 23) has termed it, a 'ritual of rebellion' – is to be found in the Notting Hill Carnival. Trinidad-born journalist Claudia Jones first had the notion in 1959 of setting up an annual pre-Lenten carnival reminiscent of the celebrated bacchanal in her native land (Ramdin, 1987; Bacchra, 1988). In choosing for its locus the inner west zone, she was choosing from among London's two major early foci of Caribbean immigration that which was more associated with Trinidadians than 'Jamaican' Brixton. This Carnival was, however, a limited affair, held indoors at nearby Paddington Town Hall because of London's miserable late-winter Shrove Tuesday weather. Claudia Jones died in 1964, but as Cohen (1980) has detailed, the Carnival was reformulated in 1966 by a remarkable white community worker, Rhaunee Laslett, was further developed by Trinidadian Leslie Palmer, and became by the early 1970s a predominantly black, ever-growing, end-of-summer outdoor festival. In 1975 the Carnival received a great commercial fillip with the arrival of official sponsorship, by London's Capital Radio among others. Then 'towards the end of the 1970s, with the rise of the second-generation, British-born, Afro-Caribbean population, a new period in the history of Carnival can be identified, associated with youth and, increasingly, with Jamaica. Many second-generation 'West Indians' had grown up to feel alienated and disillusioned with British society. They experienced London as 'Babylon' and their disaffection came to be expressed through the symbols and styles of Rastafarianism and reggae. The potential for conflict increased until, in 1976, violence finally erupted' (Jackson, 1988, pages 216–217).

On August Bank Holiday Monday 1976, more than 250 people were injured as scuffles in some sections of the Carnival became a full-scale riot, aimed particularly at the police.

Twenty-six officers were hospitalized, sixty-eight arrests were made, looting occurred, several police cars were burned – and an enduring image of the Notting Hill Carnival as dangerous, criminal-ridden, and possibly racially tense was coined by the popular media: 'Ulster in Portobello Road' (*Daily Mail* 31 August 1976). The next day the *Daily Express* (1 September 1976) contemplated 'race war in Britain', and complained 'are police, then, to keep a "low profile" in black areas of our own capital city?' Jackson (1988, page 220) does not let the burden of this last sentence elude him; he correctly takes note of 'The revealing contrast that the *Express* draws here in its juxtaposition of "our own" (implicitly white) metropolis and the (implicitly hostile) "black areas" within it'. Sir Robert Mark, the Metropolitan Police Commissioner, was reported by the *Daily Mirror* that same day (1 September 1976) as saying that, although not against the Carnival in principle, he would perhaps like to see it confined to a more manageable space such as a large stadium; that of nearby Chelsea football club had been suggested. The next day the *Daily Mirror* mused that 'the authorities do not like street life: it offends their sense of order and makes them nervous' (2 September 1976).

Unanticipated reactions

These three varied strands of evidence would seem to underline the importance of Notting Hill as a place of resonance, a particularly strong symbolic locale for black Londoners. But to my interviewees it was not so.

The possibility of relocating the Carnival is one that continually arises. Recall that London's police chief, after the 1976 disturbance, suggested nearby Chelsea stadium. Likewise, the use of Wembley stadium, in this sector of London, has also been mooted.

Notting Hill gentrifies

In the 1960s Notting Hill came in third on the list of stigmatized London places. Three respondents, however, pointed to Notting Hill in a yes-and-no manner: it *used* to be rough, 'but now it's coming up', or 'but now I hear it's gentrifying'. It was within ten years of the large Caribbean influx into Notting Hill that the area also began to take up a hip image of 'the underground, that anarchic counterculture that flowered in London in the 1960s ... Powis Terrace W11, *Private Eye*, Procol Harum, Psilocybin (see Drugs, LSD) ... (*The Economist*, 1988). The area still attracts some for such reasons: 'The trendiness of the downbeat, crumbling Ladbroke Grove, fissured by the Westway flyover and the Underground system ... the ideal strong cultural spot', wrote sociological researcher Bailey (1989, page 54). 'Young bohemians are the only people prepared to live in the All Saints', said Barbara Henderson, of the Notting Hill Housing Trust, to Bailey (page 54). And Niaill McMahon, of the Marsh and Parsons property firm, described today's Colville area to Bailey as 'a rich man's ghetto', 'bohemian', 'a real mixture of definite communities' such that 'it is hard to say which way it is going because there is such an assortment of types down there ... [Furthermore] the Afro-Caribbean presence deterred many people from buying perfectly reasonable property ... with identical architecture [but] with a price difference of over £100 000' (page 47).

Many studies have indicated that an area's perceivedly hip qualities seem to be associated with the early stages of gentrification – the subsequent career of San Francisco's Haight-Ashbury district of funky late 1960s fame is a noteworthy example, and most recently Anderson has detailed the dismay of aging counter-culturalists in inner west Philadelphia as the yuppies move in (1990, pages 38 and 139–145). I felt I was seeing something of the same in 1987–88 in Notting Hill. The neighbourhood mostly falls within the northern end of the Royal Borough of Kensington and Chelsea, the richest of all the Greater London boroughs. There has long been a fundamental dualism in the borough, the richer, leafier south ruling a poorer, scruffier, relatively neglected north. As living space has come to be at more and more of a premium in London, so have the monied sought residences ever further – and ever northward – from long-fashionable Kensington or Notting Hill Gate. To live on St Lukes Road in 1988 was to feel the edge of a gentrification wave lapping at one's feet from the south: just across Westbourne Park Road and down Ledbury Road toward the art and antiques galleries, or on St Lukes Mews right behind us, or on our own street. Always construction, always remodeling, scaffolding, skips (dumpsters) waiting to be filled with discarded bricks and beams and other materials, always the real-estate agents' signs. Through our letterbox came the free property circulars, celebrating the then upward-spiraling London prices. The April–May 1988 *Willmott's Property News* offered the unimproved house four doors down from us for £320 000 freehold:

> St Lukes Road is situated just off Westbourne Park Road in this rapidley [*sic*] improving area. The property is within walking distance to Portobello Road and

Ladbroke Grove Station. This imposing four storey Victorian home requires complete modernisation but could however be created into a fine family home or converted into flats.

Less than one hundred yards away was Rachman's clutch of holdings on Powis Terrace. No address more symbolic of the crowded rookeries of the early black London of the modern period could be found. However, the once preeminent north-of-the-river black focus of Notting Hill (as Brixton-Stockwell was the south-of-the-river focus) has become a mere sideshow: by the 1981 census sixteen of the thirty-two other London boroughs had greater numbers of households whose head was Caribbean-born than had Kensington and Chelsea (GLC, 1985). There has been a striking intercensal population decline in the borough, the greatest in all of London: 26.24% from 1971 to 1981. Many black people have moved out; the West Indian tincture of Notting Hill is fading.

In 1958 John Simmons lived in crowded squalor at number 10 Powis Terrace. Precisely thirty years later, here it is transformed as 'Hedgegate Court' (figure 6.2). Property appreciated over those thirty years in that area. Thus, in adjoining

Figure 6.2   Powis Terrace renovated, 1988.

Powis Square, in July 1988, one might for £73 000 acquire a 'delightful raised ground floor one-bedroomed flat offering superb accommodation for the first time buyer in the heart of fashionable Notting Hill Gate. Tel.: 01–221–3534.' The geography is a bit off (it is a good half mile from Notting Hill Gate), but the price is right.

Not surprisingly, some politically activist blacks perceive a pattern here. Lee Jasper of the National Black Caucus alleged:

> What has happened in Notting Hill is happening in black areas all over Britain. There seems to be a strategy of gentrification in the inner cities. Social engineering involving the police, local authority funding and agencies like the task forces are being used to move the black community out of inner city areas (*Independent* 1989, page 8).

Other blacks less activist than Jasper – which, emphatically, is a description that fits most of the Barbadian Londoners who were my informants – are evidently not greatly exercised by the apparently looming 'loss' of Notting Hill to the preponderantly white middle class. Why should 'the black community' want inner-city areas anyway? The people with whom I talked had forged, not without toil and for some not without racial stress, into well-set, mostly suburban neighbourhoods. They did not come from Barbados thirty or so years ago assertively bearing aloft the banner of black solidarity or whatever. They came as individuals, for themselves (and perhaps for their immediate families, yes), individuals who wanted to get on in the world. These particular ones have. They are prepared to be members of a small minority among white neighbors, because, as they often rationalize it, 'this England is a white country'. And they have done all right in it.

In the early days, as in the 1958 riots in Notting Hill, they did not do all right. In the early days, force of circumstance – an agreeably bland term for white racism – led them to congregate in the Notting Hills and the Brixtons. They are now nearly all convinced those days are past, so they no longer see any contemporary utility in such congregation. Is there any real point to staying in the inner city – to try to build up a black voting bloc, for example – if one is in such a small minority (5%) of the London population, and where quasi-exclusive black residential concentrations à la American ghettos just do not exist? Are these interviewees as a whole committed to some black solidarity anyway? Just as for the black middle class in the United States after Civil Rights, so for these interviewes the siren calls of the suburbs can be difficult to resist; Wilson (1978; 1987) observed the US pattern in general, Anderson (1990, page 65) among others has recently detailed it in the particular. In Britain, during the Thatcher decade, economic differentials widened within the population, blacks included. Afro-Caribbeans, for the past thirty or more years almost always associated with working-class jobs (or, latterly, with no jobs at all), are finding an as yet limited number of middle-class black households moving far away from the inner city. Such are the persons I interviewed. They seek what most others seem to seek in the materialist culture of advanced market economies: a pleasant detached house, amenable neighbors, an attractive garden, some disposable income, privacy, and solitude should they so wish. They are voting with their feet for the good, the quiet, and the apparently nonracial life.

There is, however, a most salient final point. To assume that all these achieving interviewees are treading an 'American Dream'-like immigrant-absorption path of embourgeoisement to the white suburbs – becoming 'Maltesers', as it were – would be utterly misleading, at least in certain cases. For the most unambiguous mode of voting with one's feet would be, of course, to quit Britain altogether. This is within their power. Owners of London real estate, they can sell up and return to Barbados. For them, the oft-alluded-to 'myth of return' may not necessarily prove a myth at all (Cottle, 1978; Foner, 1978; Fenton, 1988).

Notes

1 In the British context, 'Afro-Caribbean' refers to persons whose ancestry is visibly African. All Afro-Caribbeans are therefore 'black'. Other persons considered to be 'black' are ('nonwhite') Africans, and British persons of immediate ('nonwhite') African or Afro-Caribbean ancestry. Persons whose ancestry lies in South Asia, and Indo-Caribbeans, are not termed 'black' in this paper (see Modood, 1988).

2 Government-operated employment bureaus, today termed 'Job Centres'.

3 As Fryer's *Staying Power* (1984) has clearly demonstrated, there has been a black presence in Britain for hundreds of years. One risks espousing a belittling, 'foreshortened historical vision' if one states, as I now do, that in the broadest terms it is only since the arrival on 22 June 1948 of the immigrant ship *Empire Windrush* that Britain really became significantly multiracial (see Lawrence, 1982, page 113).

4 The parallel with Charlestown in Boston, as portrayed in Lukas's *Common Ground* (1985) is striking.

5 Teddy Boys were so called because their affected distinctive 'uniform' – long velvet-lined frock coats, bootlace ties, drainpipe trousers, and four-inch shoe soles – was borrowed from the Edwardian era. They appeared in the mid-1950s, were of working-class provenance, behaved in an 'antisocial' and violent manner, and gave rise to [in Hall's – (1978) celebrated phrase] a 'moral panic'.

6 See especially MacInnes's descriptions of the Notting Hill riots in *Absolute Beginners* (1959, pages 134–138 and 167–201). He could be seen as an artistic precursor of the seedily hip image that parts of Notting Hill were soon to acquire.

7 Informants in *Days in the Life* freely allege that Frank Crichlow's previous business, the El Rio café, saw drug dealing in the 1960s (Green, 1988, pages 31–32 and 50–51).

8 I coined this pseudonym for a particular reason. I have thus also changed the officer's name in the previous quotation from Keith.

References

Anderson, E. 1990: *Streetwise: Race, Class, and Change in an Urban Community* (University of Chicago Press, Chicago, IL).

Bacchra, K. 1988: 'Carnival – a story of opposition and survival' *Notting Hill Carnival Magazine* pp. 42–51 (Caribbean Times, London).

Bailey, B. 1989: 'The changing urban frontier: an examination of the meanings and conflicts of adaptation', unpublished MA thesis. Department of Sociology, University of Edinburgh, Edinburgh.

Burney, E. 1967: *Housing on Trial* (Oxford University Press, London).

Cohen, A. 1980: 'Drama and politics in the development of a London carnival' *Man* **15**(1) 65–87.

Cohen, A. 1982: 'A polyethnic London carnival as a contested cultural performance' *Ethnic and Racial Studies* **5**, 23–41.

Cottle, T. J. 1978: *Black Testimony: The Voices of Britain's West Indians* (Wildwood House, Aldershot, Hants).

Eyles, J. 1988: 'Interpreting the geographical world: qualitative approaches in geographical research', in *Qualitative Methods in Human Geography* Eds J. Eyles, D. M. Smith (Polity Press, Cambridge) pp. 1–16.

Fenton, S. 1988: 'Health, work, and growing old: the Afro-Caribbean experience' *New Community* **14**, 426–443.

Foner, N. 1978: *Jamaica Farewell: Jamaican Migrants in London* (University of California Press, Berkeley, CA).

Fryer, P. 1984: *Staying Power: The History of Black People in Britain* (Pluto Press, London).

Glass, R. 1960: *Newcomers: The West Indians in London* (Allen and Unwin, London).

GLC, 1985: *London's Ethnic Population* statistical series No. 44, Greater London Council, London.

Green, J. 1988: *Days in the Life* (William Heinemann, London).

Hall, S. 1978: 'Race and "moral panics" in postwar Britain', public lecture, British Sociological Association, Birmingham, 5 February; reprinted as 'Racism and reaction', in *Five Views of Multi-racial Britain* (Commission for Racial Equality, London) pp. 23–25.

Independent 1989: 'Resignation calls over drugs trial rejected by police', 23 June, page 8.

Jackson, P. 1988: 'Street life: the politics of Carnival' *Environment and Planning D: Society and Space* **6**, 213–227.

Kasinitz, P. 1992: *Caribbean New York: Black Immigrants and the Politics of Race* (Cornell University Press, Ithaca, NY).

Keith, M. 1986: *The 1981 Riots in London* unpublished DPhil thesis. Department of Geography, Oxford University, Oxford.

Keith, M. 1988: 'Racial conflict and the "no-go areas" of London', in *Qualitative Methods in Human Geography* Eds J. Eyles, D. M. Smith (Polity Press, Cambridge) pp. 39–48.

Lawrence, E. 1982: 'In the abundance of water the fool is thirsty: sociology and black pathology', in *The Empire Strikes Back: Race and Racism in 70s Britain* Centre for Contemporary Cultural Studies (Hutchinson Education, London) pp. 95–142.

Lee, T. R. 1977: *Race and Residence: The Concentration and Dispersal of Immigrants in London* (Clarendon Press, Oxford).

Lukas, J. A. 1985: *Common Ground: A Turbulent Decade in the Lives of Three American Families* (Alfred A. Knopf, New York).

MacInnes, C. 1959: *Absolute Beginners* (MacGibbon and Kee, London).

McIntosh, N., Smith, D. J. 1974: 'The extent of racial discrimination', Political and Economic Planning Report 547, Policy Studies Institute, London.

Miller, Holland, 1965: *Report of the Committee on Housing in Greater London* cmnd 2605 (HMSO, London).

Modood, T. 1988: '"Black", racial equality and Asian identity' *New Community* **14**, 397–404.

Peach, C. 1984: 'The force of West Indian island identity in Britain', in *Geography and Ethnic Pluralism* Eds C. Clarke, D. Ley, C. Peach (Allen and Unwin, London) pp. 214–230.

Pearson, R. 1987: 'Carnival time' *Notting Hill Carnival Magazine* pp. 12–14 (Caribbean Times, London).

Pilkington, E. 1988: *Beyond the Mother Country: West Indians and the Notting Hill White Riots* (I. B. Tauris, London).

Police Foundation, 1987: *Neighbourhood Policing Evaluation: The Background to Notting Hill* (The Police Foundation, 1 Glyn Street, London SE11 5HT).

Ramdin, R. 1987: *The Making of the Black Working Class in Britain* (Wildwood House, London).

The Economist 1988: 'Sex and after. A review of *Days in the Life* by Jonathan Green', 22 October, page 99.

Wilson, W. J. 1978: *The Declining Significance of Race* (University of Chicago Press, Chicago, IL).

Wilson, W. J. 1987: *The Truly Disadvantaged: The Inner City, the Underclass, and Public Policy* (University of Chicago Press, Chicago, IL).

7   Beatrix Campbell
Space and Power

Excerpts from: *Goliath: Britain's dangerous places*, pp. 166–87. London: Methuen (1993)

The recruitment of active citizens for law and order, by the engagement of communities in crime prevention, became the Government's primary law and order innovation in the Eighties. Neighbourhoods were not to take the law into their own hands, but they were to be the eyes and ears whose vigilance would be rewarded by a rapid response from the police, and whose cooperation would help catch criminals. That was the theory behind Neighbourhood Watch. The scheme was to borrow several politically neutral or mobile themes for the dominant discourse of law and order. It enclosed concepts of community, neighbourhood, self-help and fear within its own circuit. Crime was outside, outcast, other.

Neighbourhood Watch was launched in 1982. In terms of growth, it was phenomenally successful and reached eighty thousand schemes, covering four million homes, in 1990, although by then chief constables were resisting their expansion – apparently at a rate of six hundred a week – because of their claim on police resources.[1]

The watch schemes were one of the few contexts in which the Government conceded the concept of community. Like community care, a way of emptying hospitals, its ideological force was its anti-Statism. Its economic merit was that it cut costs, or rather that it redistributed costs from Government-funded institutions to individuals citizens who were increasing driven to fortify their own homes. The potency of Neighbourhood Watch was that it reverberated with nostalgic echoes of the night watch, it was an ideological wedding between community and protection, between security and surveillance. In practice it was neither about community nor about crime because it could not cope with the consequences for citizens and communities alike of crime in their midst.

Neighbourhood Watch could not navigate its way around the phantoms and realities of local crime. It was predicated on the assumption that crime against property was incident and opportunist and came from outside. It was, therefore, a pessimistic project which could not promise any greater police presence in besieged neighbourhoods. The paranoia behind Neighbourhood Watch and community policing was fixated on 'stranger danger'. It echoed the postwar panic that *public* places were the dangerous places, full of strangers and surprises. 'Stranger danger' minimised the sense of safety that could be promised by public space, by visibility, tolerance, cooperation and collectivity.

Stranger danger undoubtedly attracted real fears, but assigned them to *places*. The *perpetrators* could, of course, have been anywhere, everywhere. But popular dreads were assigned to open and shared space rather than the mythic sanctuary of the home. It was assumed that danger lived in public places, not in the private domain. The weirdo waving his willy in the park, who was the classic object of parental fear and loathing, was more likely to be an ordinary dad abusing his wife or children at home. Home may be a frightening prison for a battered woman. The neighbourhood may feel like a prison to residents, a local landscape is as likely to be endangered by members of its own community as by any intruders. The lads who take over the streets and refuse to share them are sponsoring a lacuna which is filled with such a sense of threat that it is evacuated by their neighbours. Typically, Ely, Blackbird Leys, Meadowell, Elswick and Scotswood had already been the site of a struggle over young men's criminality and control over their shared streets. Since Neighbourhood Watch was predicated on the protection of communities from intruders, not insiders, as a mode of crime prevention it had shown no stamina for the conflicts of interest which may shatter neighbourhoods' abilities to look after themselves. Nor could Neighbourhood Watch and community policing cope with the difficulties of places overwhelmed by economic desperation, and therefore by petty crime.

In the year before the 1991 riots, despite the impressive four million households covered by schemes, there was the sharpest increase in recorded crime since records began in 1857. The notion heralded by the Home Office minister, John Patten, that the schemes were reuniting communities and cutting crime at the same time was shattered – Neighbourhood Watch was, it seemed, neither here nor there. The postwar annual rate of increase in crime in general of about five per cent had grown, by the end of the Eighties, into a sixteen-per-cent increase in theft, an eighteen-per-cent increase in burglary and a twelve-per-cent increase in criminal damage.

Crime might be personal or it might be business. It could be casual, opportunist or organised. In the year before the riots in Tyneside there were forty ram-raids, a North Eastern speciality: warehouses, sports shops, electrical goods shops – located at industrial estates or in shopping centres – had their shutters and windows rammed. Ram-raids were an audacious modernisation of the old principle of smash-and-grab. One of the most spectacular ram-raids happened only months before the Tyneside riots, on 21 May 1991, when thieves drove their car through a domed mall at the Gateshead Metro Centre, having burst through the entrance and into the mall, and rammed the vehicle into the shutters of an electrical goods store. A group of men poured out, dashed into the store, aimed swiftly for the electrical goods they wanted, noticed the security camera on their way out and smashed that, too.

Neither Neighbourhood Watch nor community policing was designed to cope with domestic violence or strategic theft. It was known that ram-raiders were 'known' in their areas, but in the absence of a police service that was, in the words of the Scarman report, 'firm and flexible', these communities had enjoyed little or no service from the police. They felt engulfed by a criminalised coterie. Diligent detection depended on the cooperation of communities who had felt abandoned by the police, and who reciprocated by giving little or no information about the criminals of whom they were afraid. This is not to describe those entire communities as criminalised – police and state agencies had done precisely that and then walked away from them, leaving the neighbourhoods to survive as best they might against forces that the police themselves would not or could not confront. It is to say, however, that neither the constabulary nor community policing, nor crime prevention initiatives like Neighbourhood Watch, lent their support to citizens who were trying to survive crime and its oppressive cultures of domination. The riots were only a matter of scale: they were how scores were settled.

Meadowell did not need Neighbourhood Watch. People knew what was going on, they were familiar with the power of the criminal fraternity. The lads ensured that the graffiti announced who was innocent and who was guilty. Some man up for murder in the week of the riot was, of course, deemed to be innocent; some man up on sex offences was a pervert; some girl was a slag; some person was a grass.

The lads in neighbourhoods in every hard-pressed estate in Britain adhered to a cult of honour and loyalty which exempted them from everything that demanded responsibility. At the same time it conscripted a communal complicity – everyone kept *their* secrets for them. It was the ancient solidarity of silence. The injunction against being a grass was sustained by a long history of class solidarity that was contingent on an economic ethic: the working class could accommodate respectable villainy, it could forgive fugitives from the class enemy, those whose crimes came from poverty and whose pillage afflicted only the privileged.

The cult of honour also positioned the villain as a victim. The poor boys had a point, of course: they were victims. But the helpless heroes were also villains whose freedom of movement was never to be impaired, whose tyranny in the streets brooked no challenge. Their solidarity was exclusive. A measure of their power in these neighbourhoods was the extent to which almost everyone felt silenced. Active citizens, mainly women, who had witnessed the riots refused to give evidence against the rioters because they *knew* they could not name names and remain in their community. For these citizens their community was synonymous with their social being, they could not think of leaving it. The police would only offer them safe passage out, apparently they could not offer protection within. Their safety depended on the solidarity of the other women and the strong-arm of a man in the family. If they had neither, they were doomed.

Neighbourhood Watch and a community bobby combined could not take care of them.

Before and after the riots, residents who might bear witness became *suspects* in their own space; their property was attacked, their persons were abused in the streets. Potential witnesses in the riot trials needed no reminding of the rioters' power to damage and brutalise, because they had seen it all before. The lads' power, together with their paranoia, had well-known effects. Paranoia might be excited by nothing more than the facts of city life – having neighbours, being seen, being known. Everything that was supposed to be sublime about an urban community became unsafe. Being seen might mean being watched. Having a telephone might mean owning the means to grass. On one of the estates a young woman, among the few in her street to have a telephone, was fingered as a target after the riots, apparently because she had the means and because she was vulnerable. Her windows were broken and her children were threatened. On another estate dead animals – a cat, a bird – were dumped in one family's porch and a fire was lit at another family's front door. A woman well known in another neighbourhood was walking to the shop when a car aimed straight at her and knocked her down. 'Fucking grasser!' shouted the voices in the car. These women were targeted because they had at some time challenged the lads.

The police could not offer protection to the people *they* needed. Nor could they offer support to people whose status as victims often slipped into the status of symbolic culprits on the poor estates – single mothers. They became the point at which Conservative moralism met monetarism: they were the *undeserving* poor of their time, a part of the community whose stamina was unseen, whose vulnerability was rebuffed and whose needs were used as an accusation against them. The single mother, in her demonisation as 'problem family', united the theorists of the 'underclass' as the rough rabble outside society, on both the Right and the Left.

Single parents or women fleeing from their husbands are unloaded as emergency cases on the hard-pressed estates where they command no respect, especially from the lads who are their contemporaries. Joe Caffrey, a former shipyard worker turned community worker in Scotswood, concluded that there is 'a real problem about roving boys – nobody knows what to do with them. The youth and community service hardly recognise the problem in most places, so they don't target that problem group. In the Seventies the way the lads targeted their aggression was different; it would be directed somewhere else. Now they're screwing their own people. These lads gravitate towards the vulnerable young lasses.'

Often the lads become parochial itinerants, going home to change, or to eat, or collect a Giro, while nesting in other women's houses, often those of young mothers whose parental responsibilities exile them from the culture of their own generation, and who are glad of some company. A young woman in the West End of Newcastle regarded a group of lads as her mates. They used to gather in her back garden where they would bring out bags of glue and get high. They more or less moved in. They would come for a coffee, watch television or stay the night on the sofa. She knew that they were into stealing cars and this and that.

Often, women bringing up children alone had to put up with encampments of roving boys outside their homes shouting at them because, unlike this young mother, they would not welcome them inside. One young woman in the West End kept a bucket to store the stones thrown through her windows. If you could not beat them, it seemed, you could not join them either – because you were a girl – but you could let them in. A young woman who moved into Scotswood with her baby had her house squatted, in effect, by the lads. Six or ten of them would 'visit' with a bag of cans, drink, and throw the cans out of the window. 'She really tried to look after herself,' said one of her neighbours; but she could not get rid of the lads. To get clear of them she had to get clear of the neighbourhood.

No one was free from the culture of intimidation and hassle. Some Meadowell children talked about the things they put up with and the things which would make a difference to their lives:

'It would be better if people didn't fight, if there was no biting or kicking or pushing or strangling.'

'People go away when somebody tries to burgle them. Somebody burgled our house and they took the bread; smashed the window and the front door; they took every single thing out of the house – my computer, the beds, the curtains, all the ornaments.'

'The bad man was in the bushes and went shuffle shuffle. The bad man took our baby away. The bad man said, "Do you want a sweet?" and he said, "Come with me," and where they went he said "You're not getting a sweet, you're getting killed."'

'I saw a man taking a child away because he'd been shooting birds.'

'It would be better if the big lads would not kick or be naughty, and if people talked to each other and you could say, "Please don't nip me" and they would say, "I'm sorry."'

The criminal fraternities were well known on the troubled estates. They belonged to small networks, often only a handful of extended families, fortified by their access not only to arsenal – guns, crossbows, catapults – but also to a battalion of cousins and uncles, and, orbiting around them, their courtiers, admirers and apprentices. 'Newcastle has always had a reputation for hard crime,' says Elswick councillor Nigel Todd, 'and there have always been West End families connected with crime. But it's changing, it's recruiting more people. It's a business. Young people of sixteen and seventeen can't get social security benefit. They are supposed to be taken on by training schemes, but there aren't enough places, so there is a body of people who can be exploited – they're the criminals' footsoldiers. The way the syllabus works is from joyriding to ram-raiding.' Their alleged enemies were their neighbours, the school dinner ladies, the chip shop ladies and catalogue ladies, the men in the bookies, anyone who might challenge them. 'The lads can control the philosophy of an estate just by intimidation,' said a Meadowell youth worker.

Neighbourhood Watch and community policing rose with the economic decline of neighbourhoods all over Britain, but neither system responded to the impact of that crisis on estates whose social space was increasingly regulated by organised crime and masculine tyrannies.

This has created a crisis of spatial democracy in neighbourhoods. 'The lads thought they *owned* the shopping centre, it was their territory,' said one of the Scotswood women. 'After six o'clock very few people went out on foot, if only just to the chippie. Even the police used to *drive* through.'

There is always a spatial dimension to a power struggle, and here it was in the streets. 'Men and women are only as free as they are mobile,' says a geographer from London University, Bill Hillier, who has definitively mapped the daily movement of men and women across their communities. 'Men are always trying to immobilise women. Women create networks around a landscape: they tend to grow their networks outwards, by movement, by contacts. Men make theirs through formal associations, with rules of entry: you've got to be one of the boys. Men make formal associations, women make open associations.' Having hijacked public space, these local imperialists create lacunae, they sponsor vacant blots which everyone else evacuates. What was once shared space becomes a colony.

Community policing had not risen to this challenge of informal intimidation and spatial tyranny. A police tendency in some of the peripheral white estates has been to maximise the sense of danger to themselves and minimise the danger to the community itself. What they therefore did not seek was a more penetrating but subtle alliance with the community. The aura around these lads misted both their danger to their own kin and their dependence on their community in general and their mothers in particular. 'For every lad who is a nasty little shit there is a mother at home who loves him. So there will be no love lost for the police,' suggested a Tyneside lawyer. 'The reality is that the police find it very hard to make any meaningful links. The police don't think in terms of meaningful communication with the community in this kind of context. They blame the mothers.' The fact that communities and families do not ostracise or evict their criminalised children, or rather their sons, is forgotten in the lament, interminably invoked, that Britain's poor places are impoverished because of the failure of the family. But the failure of both politics and policing to support the mothers leaves them with *effects* of a mode of masculinity promoted by the powerful men in the lads' lives – their fathers, the police, the politicians, the prison officers and the judiciary.

'These men live in a twilight world,' said one of the men's lawyers. 'They're lying around on the sofa in their boxer shorts, watching videos; they have their tea when it's put in front of them; then they go out TWOCing and burgling.' When the men get into trouble, or when their wives want them out, it is their wives and mothers who make the arrangements. 'The men won't go to their solicitors, they won't liaise with the housing department, they won't liaise with their kids' schools. It's the women who make the appointments, it's the women who call to cancel the men's appointments, it's the women who make the apologies. We have women who ring up saying the men want to know what's happening to their case, or when he's due in court. What is absolutely astonishing about these tough men is that they have to have their slippers under some woman's bed. The men cannot make out on their own. The reality is that children in this community do not grow up seeing men do any of the coping, caring or standing on their own two feet.'

Neither the police nor most politicians challenged this mode of masculinity or the 'philosophy of intimidation' as a message by men to their own community as well as to outsiders. The locals knew just how dangerous life was and that the police, to be effective, would have to offer a protection that matched the power that endangered them.

Evacuation from Meadowell was the only solution for thirty households facing serious threats – including guns and firebombing – during the year before and after the riots. North Tyneside council's neighbourhood housing office used emergency powers to rehouse them away from the area.

One Scotswood resident complained that when the police came to her home they said, 'If you live in a swamp, what do you expect?' Towards the end of the decade people had already begun moving out of the area and within a couple of years there had been a mass evacuation. Residents were becoming refugees elsewhere in the city, doing anything to get away. In the summer of 1988 there were 2,350 council dwellings in Scotswood and the adjacent Fergusons Lane estate. Of these, 65 were empty. A year later that figure had almost trebled and in summer 1990 the number of empty dwellings had reached 278. By the summer of the riots it was 388. A year later, another hundred dwellings were empty, bringing the total to 488. An even higher proportion had been abandoned in the private sector – a hundred private properties were lying vacant, out of six hundred. Homes valued at between £30,000 and £40,000 were selling at £5,000 or even £3,000 – a £5,000 flat in Elswick was not uncommon. The exodus was caused by harassment, crime and the absence of police protection.

There was a sense of catastrophe in Scotswood. In May 1989, when the exodus from the neighbourhood was already well underway, citizens involved in the web of organisations anchored in Scotswood Community Project launched a campaign called Stop Crime Against Residents (SCARE). They wanted the authorities to deal with a minority within the community who were 'making life unbearable' by joyriding, burglaries and threats of reprisals. People felt that 'the police just don't seem interested in us' and thus 'the overwhelming feeling is that there is now a lack of trust in the police and despair that residents have been abandoned by them.'

Northumbria police responded negatively with the riposte, 'A lot of crime in Scotswood is committed by residents on residents, which is more difficult to detect. We rely a lot on public support which sometimes we don't get.' In May 1989 Chief Superintendent John Hillyer went to a meeting with Scotswood residents and heard about harassment, children being beaten up, bomb threats, bricks through windows, empty houses being left vulnerable to entry, joyriding, police taking two hours to respond to calls and telling victims, 'What do you expect? You're in Scotswood.'

Superintendent Hillyer listened to all this and then told the forty residents that he wasn't surprised at two-hour response times – resources were 'slim'. Most of the estate's problems came from the estate itself, he told them. The police problem was evidence. The residents, not unnaturally, felt that it was the police force's job to find evidence. He urged the residents to give Neighbourhood Watch another try.

SCARE was not welcomed because it was demanding action by the police rather than simply offering support to the police. Its genesis lay in local experience, its template was defined by fiercely-felt *people's* needs rather than *police* needs, and it was autonomous. That made it unacceptable. 'They weren't prepared to do something. Well, they did it in their own way, though not by supporting the police.'

SCARE was naturally no more acceptable to some of the lads in the neighbourhood. Indeed, they made their hostility plain. Anyone walking into the Scotswood Community Project building, home to SCARE as well as to a cluster of other campaigns and services – Newcastle's first Credit Union, a childcare scheme, a tenants' City Challenge team – was accused of being a grass, too. When well-known local activists stood at the bus stop, popped into the chip shop, dropped into the local post office, went for a bus, or leaned over their garden fence, they ran the hazard of a personalised chant of 'Grass, grass, grass'. If the lads were being imaginative the chant might vary: 'Get back in your cage', or simply, 'Cunt'.

A month after their first encounter, a hundred residents turned up for another meeting with the police, who told them that nothing more could be done. A year later, a baby, Richard Hartill, not yet a year old, was killed in his buggy when a car driven by lads known locally not as 'joyriders' but 'death riders' ran out of control. Before the tragedy the community had identified one of the drivers to the police. More than two hundred residents blocked the road where the baby had been killed. It took another year and somebody else's riot before Scotswood got a traffic calming scheme they had been asking for.

Note

1 Les Johnston, *The Rebirth of Private Policing*, Routledge, London, 1992.

8   Patrick Wright
The Ghosting of the Inner City

Excerpts from: *On living in an old country: the national past in contemporary Britain*, pp. 215–49. London: Verso (1985)

The refurbishment of ancestral memory

> Anyone seeking refuge in a genuine, but purchased, period-style house, embalms himself alive.
>
> Theodor Adorno (*Minima Moralia*)

There is a place some miles to the 'north and east' of a still existing Saint Pancras Station called Stoke Newington. In this area the terraced houses have a yard or two of space between their doors and the street. In Thatcher's years these incurably Labour voting inner-city areas have been deliberately deprived of what to start with were less than adequate public funds. The pavements seem more cracked and broken than in the recent past and the area's elderly people are falling down more often than they should be. Similarly, an increasing number of houses – even in the face of an intense housing shortage – are boarded up empty awaiting repairs which are currently beyond the means of the financially strapped local authority. As for the roads, they also lack repair; tarmac is being worn away and in many places the underlying cobbles have recently risen up into view again.

The past reappears in other ways too. Stoke Newington is fairly typical of many inner-city areas in which a white working-class coexists with a diversity of minority groups and an incoming middle class, and it is seeing changes which are familiar to other such areas as well. Thus an increasingly *preservational* emphasis has established itself in this area over recent years – an emphasis which is more contemporary and ambitious than the one rather tired blue plaque which was put up in 1932 to 'indicate' the spot where Daniel Defoe once lived and wrote *Robinson Crusoe*. This newer emphasis is closely connected with what is often called 'gentrification' – a process which has certainly taken place here between the nineteen forties (when the area's Victorian terraced houses were scarcely marketable at all) and a present day rich in mortgages (for those who can afford them), tax relief (for those who have mortgages) and (until very recently) improvement grants.

The houses which the planners of the nineteen sixties so loudly decried as slums are being refitted in more senses than one. As any observer who walked the streets of this area would be likely to notice, the whirring of industrial sewing machines (on which Asian and Cypriot homeworkers labour for the small sums which must be won in direct competition with third world manufacturers) has recently been augmented to produce a modified soundscape. There may indeed be reggae and funk in the air, but these days there is also more of the quiet purring associated with consumer durables – along with the resounding bangs and crashes of middle class self-sufficiency and house renovation. Robinson Crusoe may now be drawing a salary from a job in lecturing, teaching, local government or the probation service, but he's still doing it himself in his own time. His island in the late twentieth century may be no larger than a terraced house, but the same sky stretches out overhead and his garden – with a little extra planting, some additional fencing and the odd trellis – is still idyllic enough. What Raymond Williams has called 'mobile privatisation' – that moveable if not entirely atomised life which brings with it the interior styles of the new *biedermeir* – has certainly found its way into this area of late.[1]

If possessed of a remotely wry inclination, our observer might be impressed by the extent to which such paradoxical forms of coexistence have become the mode in places like this. The clapped out but still powerful old Daimler-Jaguar, even if it is just broken down and rusting at the side of the road (testimony, perhaps, to the long term financial impossibility of a dream-laden purchase), finds itself alongside increasing numbers of expensively renovated Morris Minors. (A 'Morris Traveller Centre' turns a brisk trade restoring these old eco-cars – one imagines green fields wherever they go – on the nearby Lea Bridge Road.) The traditional pushbike rider is liable to be overtaken – especially taking off wheezily from the traffic lights – by more and more correctly clad (sensible shoes included) and non-smoking owners of shining, light-weight Claud Butler or Peugeot racing bicycles. As for the area's many working-class dogs, they are more aggressively regarded by people for whom brown heaps on the pavement or in the park constitute a pressing social problem. The sweated immigrant textile industry has also seen some changes recently, now finding itself in co-existence with a white assertion of ethnicity as style – not just skin-heads but also a somewhat wealthier and more 'cultured' appropriation which finds expression in Palestinian scarves, central American or Indian rugs and fabrics, the hand-knitted sweaters and even the legwarmers of late seventies post-eroticism. So comes the time to diversify – into restaurants perhaps.

The middle classes, in short, have been moving in since the late sixties, and the signs are everywhere to be seen. This new population brings with it a market for wine bars and a new range of culturally defined shops (some of which appeal to tradition directly while others display their more discontinuous modernism against a surrounding background of tradition.) It enters the area with an attention of its own – with particular ways of appropriating the place in which it finds and must sustain and understand itself. Alongside the ethnic restaurants (the latest of which is 'Californian' and eking out a living one flight up from Seval and Son's sweatshop), therefore, this is also where a sense of the past comes in. Suddenly this hardpressed inner city area is a settlement again – a new town in the wood as the name Stoke Newington would suggest to those interested in the historical meanings of place names. Because this upwardly mobile slum is in an old country, we could easily enough push the reference back to the Domesday Book (which registers this place as 'Newtowne'), but a more informative starting point is provided by Sir Walter Besant who took an interested stroll through the area and reported back to his public in a book called *London North of the Thames* which was published in 1911.[2] Besant concluded that 'almost all there is of history in the Parish' was concentrated around Church Street. Seventy years later, (and if one excludes a cluster of Queen Anne houses on the north end of the High Street) many of the newcomers – owner-occupying residents of what Besant dismissed as the 'little villa houses which the modern builder strings up by the row' along the streets off Church Street – would agree. The new focus is certainly centred on Church Street, even though the commercial and retailing centres are firmly established elsewhere (on the more recession-struck High Street for example), and even though the old Borough of Stoke Newington has been integrated into the bigger administrative quagmire known as Hackney. Of course Church Street, being westerly, is just that little bit closer to respectable Highbury and further from the rather less gentrified wastes of Clapton and other places east. For this reason among others, perhaps it is not so surprising that a sense of local history should be making its way down to meet the newcomers along this particular road. This, after all, is the road in which the area's great names tend to congregate. Daniel Defoe lived on Church Street and Harriet Beecher Stowe stayed in a house at what used to be its junction with Carysfort Road; Isaac Watts wrote his hymns in a no longer existing mansion off to the north and as a young boy Edgar Allen Poe went to school at the connection with Edwards Lane. There is also the old Victorian Free Library in which this sort of information can be looked up, and eighteenth century 'Tall Houses,' as they were known locally, survive on the south side of the road. For those who want to round the whole atmosphere off, a tradition of dissent can with some limited (in this case that is to say seventeenth century) historical accuracy be imagined to hang over the whole village scene.

I'm not suggesting that these new settlers have exactly researched the history of the area, only that a certain appreciation of the remaining past has facilitated their settlement in the area. This appreciation has much more to do with attitudes towards surviving physical presences than with any formulated historiography. So it happens that old rotting bricks take on an aesthetic aura, testifying now to valued age rather than to bad manufacture, cheap building, dampness or recent urban dereliction and decay. So it happens that there is a market for those very nineteenth century fittings (cast iron fireplaces, sash windows, cheap pine – which can always be stripped) which the renovators were ripping out only a couple of years ago when *modernisation* was still the essence of conversion. So it happens that small voluntary associations with an interest in the area's architectural heritage spring up, organising tours of the area and producing booklets like The Hackney Society's *The Victorian Villas of Hackney*. So it happens that the few enamelled signs – 'Win her affection with A1 Confections' – remaining above shop fronts or the barely visible traces of wartime camouflage paint on the townhall are brought to a new kind of focus by some passers-by. So it happens that the junkshops in which a few ordinary but irreplaceable 'treasures' might be found among the general accumulation of undeniably local detritus are being augmented by a new kind of 'not-quite-antique' shop in which the selection (which includes nineteenth century engravings of noteworthy Stoke Newington buildings) has already been made by well-travelled proprietors who value old things. So it happens that one of the several building societies opening branches in the area (often on the back of booming estate agents) fills its windows with eighteenth and nineteenth century images of the place. So it also happens that this contemporary preservational emphasis comes to be linked with official policy. Stoke Newington Church Street has recently been declared a preservation area, and the newly rediscovered grave of the nineteenth century Chartist Bronterre O'Brien is said to be undergoing some sort of restoration in the local cemetery.

In this new perspective Stoke Newington is not so much a literal place as a cultural oscillation between the prosaic reality of the contemporary inner city and an imaginative reconstruction of the area's past as a dissenting settlement (which it was in Defoe's time) that even the plague, according to some accounts, never really entered. For those who want it, this imagined past will keep looming into view. In the midst of the greyness, the filth and the many evidences of grinding poverty, the incoming imagination can dwell on those redeeming traces which still indicate a momentary 'absence of modernity,' to use Besant's phrase. Thus in Oldfield Road (the northern end, which was formerly known as Cutthroat Alley and which is now bleak in a more municipal sense) there is an old, slightly buttressed brick wall which once marked the end of Daniel Defoe's four acre garden. The space now holds rows of badly built and subsiding Victorian terraces, cheap post-war council flats (some of which were built after the clearance of bomb damage) and a large depot for the varied machinery of municipal refuse collection. This wall is the last trace of Defoe's abode. Incongruous it may be, but it is there and still capable of supporting little moments of epiphany not just for those in the know – a tiny minority of those who pass by – but for people whose appreciation of pastness is not dependent on precise archival definition. The same old brick that testifies to the eighteenth century at one moment can speak of the early nineteen forties at the next.

The stroll down Church Street can be full of such moments for those who are inclined to tune into them. What looks like nothing more than a Spar supermarket redeclares itself on the second glance as an eighteenth century hostelry. And underneath the greengrocer's next door there are apparently cellars where the horses used to be stabled. Meanwhile those run-down and densely tenanted houses further to the east are actually four storey eighteenth century townhouses – the 'Tall Houses' of Besant's account. Small wonder that one has been bought, emptied of its tenants and done up by an art-historian – true 'restoration' indeed (and doubtless the new owner's pride in the place is only enhanced by the derelict state of an identical council owned house next door but one). Traces remain, as one might say, except that this preservationist attention is not automatically granted to residues which have survived neglect, decay and the barbarism of post-war planning. It has a more subjective side as well, involving as it does a contemporary *orientation* towards the past rather than just the survival of old things. As so few guide-books ever recognise, this is not merely a matter of noticing old objects situated in a self-evident reality: the present meaning of historical traces such as these is only to be grasped if one takes account of the doubletake or second glance in which they are recognised. The ordinary and habitual perspectives are jarred as the old declares itself in the midst of all this dross. There is active distantiation and even what some philosophers have called 'astonishment' to be found in their recognition.[3] Like the sculpture in Rilke's poem, 'The Archaic Torso of Apollo,' this past doesn't just endure: it displays itself against the tawdry present which it also actively indicts.

So while there are indeed some fine eighteenth century town houses, the odd bow window, and other such remains along Church Street, there is also on the subjective side an increased inclination to value the past, to notice and cherish it, to move into it and maintain it as a presence in our lives. To a considerable extent, I suggest, middle class incomers have brought with them Orwell's fond perspective on the prole quarter – a perspective which is not based on cherishing things and places which have been near and lived with for years, but which finds its basis in a more abstract and artificial aestheticisation of the ordinary and the old. These flashes of redemptive disclosure in which the past is glimpsed as both other and miraculously still present may seem, and in some cases may also be, innocent enough. But before anyone gets too excited about the modernist possibilities of an alienated past which keeps breaking into view in the present it should also be recognised that such moments of sudden disclosure can be deeply problematic in their significance. Some indication of this is to be found in the example of a largely derelict triangle of ground which lies behind the Red Lion – one of the few pubs on Church Street which hasn't been 'improved' lately, or at least fitted out with the usual selection of 'real ale' to go alongside the ever flowing lager. The familiar inner city wreckage is to be found on this patch of land, but as those who participate in the secret wisdom of the guidebook, the public reference library or even the winebar reverie may well know, this is also where the old village stocks used to stand – along with the cage, the watchhouse and the whipping post. There need not be any positive craving for the cagings and whippings of yore for it to become evident that there may be difficulties involved in the counterposing – however momentary – which displays this idealised imagination of an old village against the dingy urbanity of an admittedly less brutal welfare state. Redemption, critical distantiation and astonishment – these are all aspects of the Orwellian presentation of the past and if they come through directly enough to the inner city of the nineteen eighties they also bring pressing questions with them. For if the nineteen seventies brought the Habitat and Laura Ashley styles of interior decor into this area, they also saw the rise of the National Front. Likewise, if the early eighties have seen the intensified restoration of Victorian Villas in old inner city areas like Stoke Newington, they have also – and still almost unbelievably to many – heard a prime minister advocating what she was pleased to define as a return to Victorian values.

What exactly is it that keeps breaking through? The reappearance is not simply of the past as it 'really' was: indeed, sometimes the authentic trace of history is precisely what just has to go. In Thatcher's years old ideas of 'charity' and 'philanthropy' may indeed have been in the air again, but if the 'deserving poor' are still sometimes to be found living in 'model dwellings' of the sort Miss Savidge mused over so ambivalently, the terms of residence are different and a certain remodelling of the original nineteenth century edifice may also have become necessary. Towards the end of 1984, for example, a small improvement was made to Gibson Gardens, a large model dwelling tenement (dating from 1880) in which flats – among the very cheapest in the area – seem now to be permanently for sale. A chimney had to come down, although not exactly for structural reasons. In these times when the deserving poor are somewhat more likely to become modest owner-occupiers (happily relinquishing their status as tenants in council owned post-war tower-blocks?) the problem had far more to do with the old words which were still cut in bold letters on the public face of that chimney: 'The Metropolitan Association for the Improvement of the Dwellings of the Industrious Classes . . .'

Past against past

Although it can be imagined as an English settlement with roots in the Domesday Book ('There is land for two ploughs and a half . . . There are four villanes and thirty seven cottagers with ten acres'), we should remember that Stoke Newington has recently been administratively integrated into Hackney, a borough which in 1983 was declared to be the poorest in Britain. This suggests other perceptions of the place, and these certainly exist. For a start there are clear indications of the ongoing and customary practices of a white working-class – indications which may well be resignified and mythologised in the incoming middle class perspective, but which are not in themselves mere hallucinations.

Within this white working-class there is also a sense of the area and its past – one that is significantly different, going to the considerably less aesthetic High Street for its focus far more automatically than to Church Street. There are many older people who remember a more prosperous High Street as it was in the fifties – there was at least one large department store, and the pavements could be so crowded on Saturdays that people had to walk in the road. This remembered past exists in stark contrast with the present, for if the High Street is still the place to shop it has clearly also seen better days. A recently introduced one-way system drove some trade out of what is also an arterial road leading in the direction of Cambridge and other august locations. The big department store was demolished – after years of dereliction – by the beginning of 1984, and many of the more recent shops (like Marks and Spencer) have relocated in the current economic decline. As for the large supermarkets of recent years (Sainsbury's, Safeway and so on), these have started as they mean to go on – elsewhere. Little has improved from this point of view. Some pubs have been extensively smartened rather than gentrified – the old Rochester Castle, for example, used to be a cavernous dive with semi-derelict if not exactly lumpen drinkers at the front and a large, somewhat more purposeful Socialist Workers Party clientele further in. When its doors reopened after renovation this year, they were under a new sign as well as new ownership and management. The Tanners Hall, as it is now called (a name which appeals to the supposed traditions of the area while carefully cancelling out any association with the more grubby and immediate past of the Rochester Castle), has had a hundred or so thousand pounds shoved through it. Where there was Orwellian dinginess there are now stuffed pelicans, copious ferns, a Conservatory like bar, a beer garden and – a pleasantly florid touch – old (although distinctly not antiquarian) books everywhere: these signs of culture and the pre-television age are piled up in heaps as elements of an intensified and slightly manic interiority which must indeed look pretty 'vulgar' from many a Church Street perspective (people who persist in thinking that books should be *read*). And it certainly *is* a mixed crowd which drinks at the Tanners Hall, not least because the more derelict and SWP constituencies don't seem completely to have got the idea and moved on.

While some of the cultural relations of the white working class have survived the upheavals of the post-war period they have done so most strongly off either of the main streets, in amongst those rows of 'little village houses' which Besant found so meaningless. Here the pub may still be the 'local,' but it is also clear that cultural survival has often occurred alongside a growing sense of anxiety. It only took one person (an incomer) to walk into The Prince of Wales (a backstreet pub) not so long ago with a jug, and the intention of taking beer home, for conversation to pause before a torrent of memories started to flow: 'When I was a girl I used to fetch for my Grandad' and 'Of course that's a four pint jug – what do you think it was made for?' Such discussion forms an act of commemoration in a full sense of that word – a sharing of memories, certainly, but also tribute and testimony to a valued time which is increasingly alienated from the present and only to be recalled occasionally. The streets were cleaner and the children less disorderly: faces were whiter in those days, and doubtless the grass was greener too. There were indeed some people in that pub for whom this sort of replay was uneasy – the younger people, by and large, who distinctly didn't join in this ceremony of remembrance and who may have gone further saying that they wanted nothing to do with all that. A few weeks later during the 1983 election campaign, these same sceptics would sit at the bar scoffing at Michael Foot as the television showed him evading questions about defence and going on about the Forties and the traditions of the Labour Party. 'There he goes, Old Worzel . . .' And these weren't even the people who remembered that before 1945 Stoke Newington was a *Tory* Borough.

A similar sense of upheaval was evident during the encounters at-the-standpipe which were laid on for us during the water strike at the beginning of 1983. 'Just like the old days,' as one woman remarked in the cold but almost convivial misery of a momentarily reactivated street corner, an only partly ironic remark which also resounded with ambiguity, contradiction of feeling and (once uttered) embarrassment – perhaps at the way deep feeling and cliché can run so cruelly together. As for the embitterment and reaction which grows in such a context, it will take me a few years to forget the somewhat dishevelled elderly man who happened to make his way down the pavement as we were moving in. 'People moving in,' he said with affected surprise and a certain amount of contempt in his voice: 'people (by which he meant white people) still moving in.' He went on to comment that had he been able to he would have moved *out* years ago, along with the many he had known leave for the suburbs and new towns in the sixties and seventies. Too many 'rusty spoons' around here as it's sometimes put in rhyming slang. Encounters such as these indicate the extent to which this traditional white working class has been affected by a dislocation of culture and memory like that which Orwell describes in *Nineteen Eighty-Four*. In this situation the recourse to racism can provide easy compensation.

And of course things have changed, although not simply in the ways which compensatory understanding suggests. Early in 1983 there were police helicopters hovering overhead, and shoppers emerged onto the High Street to find it filled with men and women in blue engaged in community policing. There have been demonstrations and arrests, and graffiti has been appearing on the walls: 'Police scum killed Colin Roach – We are getting angry' and 'Police pigs murdered Colin Roach.' Graffiti has also been disappearing, the large and loud letters sprayed along the length of Church Street quickly silenced by someone who followed the same route with brush and yellow paint – not a 'memory hole' exactly but another act, no doubt, of community policing.[4]

There is a long established Jewish population in the area with its Hasidic community concentrated slightly to the north of Stoke Newington in Stamford Hill. There are Irish, African, Italian, Asian, Cypriot (both Greek and Turkish) and West Indian people in the area – people who have their own routes through the place, although not necessarily ones that move in any easy accordance with the imaginative reconstructions and memories which hold the measure of the place for many white inhabitants. Given the prevailing white criteria, which measure belonging and cultural authenticity in terms of continuity of place and an imaginary valuation of the remaining trace, it is entirely consistent that the belonging for these people should seem (and I speak here from a dominant point of view) more makeshift and improvised. For people excluded from conventional identification with the area's historical geography, the traditional structure of the place is still there to be dealt with. Sometimes, courtesy not least of imperial history, old centres can be adopted – like the sixteenth century church on the edge of Clissold Park which, while it looks exactly like the fragment of old Elizabethan village that in one limited respect it is, seems now to be used most actively by a West Indian congregation – but at other times the connections seem more strained. Thus on the High Street there is a Turkish mosque and community centre. This is housed in a building which until recently was being used as a cinema and which was originally built as an entertainment palace with exotic domes thrown in for orientalist effect. While transcultural forms of identification should certainly not be judged inauthentic, there is no conventionally sanctioned dream of ancestral continuity or home-coming about that connection.

The different populations in the Stoke Newington area have different senses of the place and these are certainly not always congruent with one another. Thus, for example, I recall walking through Stoke Newington with Annette, a twenty one year old white woman who has lived her entire life in Hoxton (a more thoroughly working class area which lies a few miles south of Stoke Newington). Annette comes from what years of ethnographic sociology have treated as the classic East End background and she finds it hard to believe that anyone would ever *choose* to live in this borough: she herself has eyes set on Kent or, failing that, Essex. As we walk through streets of Besant's 'little villa houses,' she comments approvingly on a house which from any culturally sanctioned perspective is a complete eyesore. Its bricks have recently been covered with a fake stone cladding, the sash windows have been replaced with cheap louvres, and the whole place is painted up in gloss so that it shines like a birthday cake. These 'improvements' certainly stand out, and it seems very likely that the quieter neighbours (subtler greys and whites, carefully restored wooden shutters . . .) might be among those who see them as acts of vandalism. Annette, meanwhile, sees it differently. Somebody *owns* this place and their renovation of it speaks of pride, self-determination and freedom to this woman who has lived her whole life in council flats. For her it is exactly the point that this house stands out from the rest. The uniformity of the street is merely the grey background against which this improved house glows – a well packaged interior and 'home.' As for all that stripped pine to be seen in other houses around here, what about painting it? And what are all these enormous plants doing in some front windows – the ones without net curtains (now *there's* a sign of a likely haul – insured videos, stereos etc. Talk about inviting trouble . . .)?

These tensions between different appropriations of the place are articulated around many different phenomena or issues. Thus, for example, middle class incomers value Abney Park cemetery precisely because it is overgrown and four-fifths wild – a good place for a Gothic stroll. A very different view is taken by some working class people (far more likely to have relatives buried in the place), who find the unknown and neglected appearance of this nineteenth century cemetery a mark of decay, and argue that it should definitely be tied up.[5] The same contest of views occurs over the bay windows in the area's terraced houses. Many house owners – working class, black, Cypriot (although very rarely if ever incoming middle class) rip these out, throw the whole front of the house forward with an extension which opens up considerably increased space inside. None of this looks good from the point of view of the Hackney Society. In his preservationist pamphlet on *The Victorian Villas of Hackney* Michael Hunter (himself a relatively early settler in the area) celebrates every feature of these little villa houses – from plaster cornices, ceiling roses, moulded skirting boards and door frames through to the floral capitals above the bay windows and those touches of Italianate influence which come to these slum houses 'from the Renaissance palaces of Venice, Florence and Rome' no less. The plea is straightforward and direct.[6]

> Above all, if you live in a Victorian house and are considering altering it in any way, please respect the aesthetic merits of this type of building which have been outlined here. Each Victorian house is, in its way, a period piece and worth respecting as such; many form part of terraces which have an aesthetic unity that is easily destroyed by piecemeal, unsympathetic alterations to individual houses.

The point should be clear enough. People live in different worlds even though they share the same locality: *there is no single community or quarter.* We are not dealing with any free plurality either. The distance between different people or groups is not just a matter of neutral space – of relations which are merely absent or yet to be made – for one can't consider them without also raising the larger question of domination and subordination: without, in other words, recognising that relations *do* exist even though they are not always directly or deliberately experienced. For the worlds inhabited by some groups work against the needs and interests of others, defining them according to an imported logic, romanticising and mythologising them, confining them to the margins of public life and defining them as 'makeshift'. The sense of history plays its part in all this. For in the midst of all this romantic attachment to old brick and earth, the large and mixed ethnic minority and black populations in the Hackney area are still struggling against formidable odds for the basic constitutional and cultural rights of a citizenship which is itself far from secure. History from this point of view remains to be made, as do the cultural means of developing and expressing a different past or, in wider terms, a different experience of the same imperialist history. From this perspective (a perspective which, far from being a 'minority' matter, is central to the democratic development of the borough) the other appropriations of the area's past may well constitute part of the problem, valuing as they do a time before much recent immigration took place.

Notes

1 Raymond Williams, *Towards 2000*, London 1983.
2 Sir Walter Besant, *London North of the Thames*, London 1911, pp. 573–87.
3 See Agnes Heller, *A Radical Philosophy*, Oxford 1984, pp. 15–17.
4 Twenty one year old Colin Roach died of shotgun wounds in the foyer of Stoke Newington police station late on January 12th 1983. Considerable protest followed and demands were made for a full public enquiry into the incident. Roach's death raised, once again, serious questions about police conduct towards the black population in the area. Was this another case like that of the White family who in 1982 were awarded £50,000 exemplary damages against the Stoke Newington police who illegally entered their home in 1976, beat them up and prosecuted them on rigged charges? While the inquest into Roach's death (held after considerable controversy in June '83) delivered a verdict of suicide, the jurors wrote to the Home Secretary expressing their 'deep distress' at the police's handling of the case. True to form, Leon Brittan decided against a public enquiry, referring the letter instead to the police so that they could investigate themselves.
5 Paul Joyce's *A Guide to Abney Park Cemetery* was published in 1984 by the London Borough of Hackney and a voluntary association of fairly recent origin called Save Abney Park Cemetery. Joyce's booklet treats the cemetery as part historical testimony and part wilderness – preservation and conservation together. The Hackney Society has also found occasion to comment on Abney Park Cemetery in its survey of the *Parks and Open Spaces in Hackney* (London 1980). There is little satisfaction here for those who make the error of mistaking this 'open space' for a cemetery rather than a park. As it is put, 'any attempt to "tidy up" this park should be treated with suspicion. Yet soon after Hackney took the cemetery over, flowerbeds were inserted into its main avenues as if no public park could be complete without them, thus displaying a simplistic and uniform attitude to park design which it is hoped will in future be avoided.' (p. 7) The real point, however, is not to do with 'park design' so much as with the use of this 'park' as a cemetery at all. The Hackney Society laments the fact that burials have taken place 'on the grass verges and other unsuitable positions' and hopes with winning sympathy that 'these could perhaps be deprived of their headstones in due course.' (p. 18) Fortunately, however, no burials are now taking place 'except for insertions into existing family graves,' so the era of disgraceful and predominantly working-class indiscretion is coming to an end. In future, strolling incomers will be less and less disturbed in their musings by the vulgar glare of new stone, crudely exposed soil, plastic flowers or any well-trimmed and excessively decorous verge. As somebody who enjoys the occasional Gothic stroll himself, I find the Hackney Society's aesthetic remarkably sympathetic. What remains problematic is the way this aesthetic is forced over other forms of life with such stunning disregard for what it would disconnect. With blithe confidence the Hackney Society elevates its own particular choice to the level of aesthetic and universal law. 'In the name of memory history has been abolished.'
6 Michael Hunter, *The Victorian Villas of Hackney*, The Hackney Society, 1981.

9 Benjamin Forest

West Hollywood as Symbol: The Significance of Place in the Construction of a Gay Identity

Excerpts from: *Environment and Planning D: Society and Space* 13, 133–57 (1995)

November 6, 1984: In California, West Hollywood becomes the first 'gay city' in the United States, after voters there decide to turn the previously unincorporated area into a self-governing municipality and elect a largely gay city council to run it.

The Gay Decades Rutledge (1992, page 231)

In this study I trace the definition of a gay identity in West Hollywood, California as expressed in a series of articles appearing in the gay press during the campaign for municipal incorporation between mid-1983 and late 1984. These articles tied the physical and social characteristics of the new city to the physical, mental, and moral character of gays living in West Hollywood. I identify seven elements of this new gay male identity – creativity, aesthetic sensibility, an affinity with entertainment and consumption, progressiveness, responsibility, maturity, and centrality – but do not claim that this is an exhaustive or exclusive list. Incorporation of the city was portrayed as a way to consolidate and legitimize this identity, so that the 'cityhood' movement acted to reduce the 'marginal' status of gays, and to draw them closer to the 'center' of US society (Shils, 1975, pages 3–16). As presented in the gay press, the incorporation of West Hollywood was less a radical project than an attempt to achieve recognition of gays as members of civil society. Indeed, several authors argue that community debates over citizenship within 'identity groups' are one of the necessary conditions for the creation of a civil society (Shklar, 1991; Shotter, 1993). As such, the effort for incorporation resembles an ethnic group strategy (Epstein, 1987). It differs sharply from more radical strategies for oppressed groups, such as those suggested by hooks (1990), that use marginal status as a strategy of empowerment.

The conflation of place attributes with the personal qualities of gay men contributes to what Knopp (1992, page 652) describes as 'sexual identity formation', and in turn 'symbolic and representation struggles over the sexual meanings associated with particular places'. Using place as a symbol tends to mask the socially constructed quality of gay identity, so that it takes on a 'natural' existence.[1] The narrative construction of a 'gay city', and thus the attempt to create an identity based on more than sexual acts, suggests that the gay press sought to portray gayness as akin to ethnicity, in contrast to homophobic characterizations of gayness as a perversion, sickness, or moral failure.

The 'geographies' of gays and lesbians have received considerable attention in the last four years, although studies by geographers date back to Weighman

(1980). For the most part these recent studies have been of the political economy of gay neighborhoods, or have been ethnographic studies of gays and lesbians. Generally, those working within the framework of political economy note that symbolic struggles over the meaning of places often coincide with economic and political struggles, but concentrate on these latter issues (Knopp, 1990a; 1990b; 1992). Ethnographic studies document how different places, such as the home, workplace, etc are experienced by gays and lesbians, with particular attention to the constraints imposed by '(hetero)sexed' space (Valentine 1993a). Both of these approaches share a view that places take on importance primarily as sites of routine activities, so that the important issues are how the daily lives of gays and lesbians are constrained or empowered in particular localities. More generally, studies of the relationship between place and identity have focused on political, rather than symbolic-cultural, issues (Agnew, 1992). Humanistic geographers have discussed the importance of place as a center of meaning, as well as simply a site of routine activities (Tuan, 1977). The symbolic element is especially important to the normative importance of place because morally valued ways of life are often created, shaped, and reinforced through the construction of real and imagined places. Hence, in this study I do not focus on the social history of West Hollywood, nor do I make any claims about the experience or perceptions of gays and lesbians. Rather, I examine the symbolic representation of West Hollywood in the gay press during the 'cityhood' campaign. This emphasis on the creation of place through material transformation follows from both Tuan (1991) and Barnes and Duncan (1992). I argue that two characteristics of place – its capacity to 'concretize' an idea or culture, and its holistic quality – make it a particularly effective means to create social identities.

I discuss the holistic quality of place to argue that the capacity to experience place as a whole helps to resolve the internal contradiction of identity. Massey's (1991, page 276) description of place identity as 'frequently riven with internal tensions and conflicts' suits the actual city of West Hollywood, and even (to some degree) the symbolic city described by the gay press. I argue, however, that the use of a holistic symbol was an attempt to resolve these contradictions. Thus the use of place in this fashion was a political decision, one which exploited the unique capacity of place to be experienced holistically. This is emphatically not the only possible or actual experience of West Hollywood.

Except where noted, 'gay' refers specificlly to gay men. The 'geographies' of gay men in West Hollywood certainly differ from the 'geographies' of West Hollywood lesbians. The highly visible public expression of gay men in West Hollywood in contrast to lesbians is a case in point. Authors of some studies argue that lesbians are generally less visible in urban areas because they have a relatively private orientation (Castells, 1983). More compelling are studies that have focused on the fact that lesbians, like women in general, have less access to capital, earn lower incomes, and have a higher risk of violence in public places (Adler and Brenner, 1992; Valentine, 1993a; 1993b). The symbols and meanings used in the gay press are also almost certainly class specific, an issue which has been addressed by relatively few geographers (Geltmaker, 1992, page 633; Knopp, 1987; Lynch, 1987). The gay papers used for the study targeted relatively affluent gays, which no doubt largely determined the generally nonradical nature of West Hollywood gay identity.

Place and identity

Although place in general has long been a concern of humanistic geographers (Relph, 1976, for example), Agnew (1989) argues that place as a concept has been devalued because contemporary social science has conflated it with 'community', and concern with community has been supplanted by concern for the social. Agnew believes that place can be 'rehabilitated' though the study of social relations in place, a project which has been taken up by locality studies (Massey, 1991; 1993). One attempt to integrate locality, social and political relations, and moral evaluations focuses on the idea of citizenship. Smith (1989) argues that locality is the basis on which to advance the normative dimension of citizenship. It is not inconsequential then, that (1) the attempt to form a new identity was centred around a political incorporation, and that (2) this new identity included characteristics of 'good citizens'. It is not clear from Smith (1989), however, why locality is an effective vehicle for realizing the normative values of citizenship. The answer, I believe, lies in the role that place can play in moral narratives.

Morally valued ways of life are evaluated in the narratives of individuals and groups, and form, in part, the basis of self-identity. These narratives, however, have increasingly been seen as constructed and reflective, so much so that Giddens (1991, page 31) describes the modernistic conception of the self as a 'reflexive project'. Place is particularly significant when it is used to construct the normative conventions of a group. Thus 'the cohesiveness of social groups is related to the constitution of individual and communal identity, which cannot be removed from the question of valued ways of life' (Entrikin, 1991b, page 82). This quality of place is recognized by Tuan (1977, page 178), who writes that place can embody a culture, and achieves an identity 'by dramatizing the aspirations, needs, and functional rhythms of personal and group life'. It is the symbolic value of place that makes it an effective organizer of identity.

Authors of recent works (Duggan, 1992; Stein, 1992, for example) draw on the literary critic Sedgwick (1990) to question the value and legitimacy of fixed categories of identity, particularly those related to gender and sexuality. Sedgwick (1990, page 25) argues that conceptions of sexual identity can be disrupted, even among 'people of identical gender, race, nationality, class, and "sexual orientation"'. While addressing the question of gender identity, Butler (1990, page 16) makes a related point when she suggests that 'identity' is more 'a normative ideal rather than a descriptive feature of experience'. Butler also argues that the ways in which gender and sexual identities have been constructed are not politically neutral, but are intimately connected to heterosexual domination: 'The univocity of sex, the internal coherence of gender, and the binary framework for both sex and gender are ... regulatory fictions that consolidate and naturalize the convergent power regimes of masculine and heterosexist oppression' (page 33).

The questions raised by Sedgwick and Butler present significant problems to claims of homogeneous, hegemonic identities, and would seem to undermine attempts to demonstrate the ontological status of group identities. Such a demonstration would require an ethnographic study to document empirically the 'success' of the gay identity advanced in the gay press, that is, the degree to which gay men in West Hollywood internalized this identity. As noted above, my ambitions for this study are more modest. I seek only to evaluate a particular 'normative ideal' of gay identity, and not to make a claim about the influence of that ideal.

Gays and public space

Geographers have tied the construction of gay identity to material and symbolic transformations of public spaces. Knopp (1987; 1990a; 1990b; 1992) and Valentine (1993a; 1993b; 1993c) have been among the most active contributors to this literature, but not the only ones (Bell, 1991). Knopp's ultimate ambition is 'to identify specific ways in which sexuality is implicated in the spatial constitution of society, and, simultaneously, specific ways in which space and place are implicated in the constitution of sexual practices and sexual identity' (Knopp, 1992, page 652). Valentine (1993b; 1993c) argues that the focus on the transformation of public spaces stems largely from the concentration by geographers on gay men in urban neighborhoods. Her studies concern the social networks of lesbians, the strategies used by lesbians to negotiate and develop 'multiple sexual identities', and the availability of sites for gays and lesbians to meet. Hence most work to date has concerned the effect of gays on particular places, or has treated place largely as a location for social activities. The focus in this study differs in that my primary concern is not on the forces that resulted in the historical concentration of gays in West Hollywood, nor on the dynamics that led to an economic and political power base, nor on the social networks of gays and lesbians in West Hollywood. Rather, I analyze how conceptions of place served to organize a model identity in the gay press.

In addition to providing a gathering place, public spaces created by gays provide for relative safety, for the perpetuation of gay subcultures, and, most important to this study, provide symbols around which gay identity is centered (Castells, 1983; Godfrey, 1988; Jackson, 1989; Knopp, 1992). In the case of West Hollywood, the designation of a gay area as an independent city was important symbolically because it lent legal legitimation to gay identity, and, for a time at least, raised gays' level of political activity in West Hollywood (Moos, 1989, page 357). The coverage of the cityhood movement in national gay newsmagazines, such as *The Advocate*, suggests that the symbolic influence of West Hollywood extended to the national level.

A consequence of the struggle for public expression is the close tie between gay identity and themes of repression and confrontation. The degree of contrast with heterosexual norms, however, may vary between gay groups. Groups such as ACTUP (AIDS Coalition to Unleash Power) and Queer Nation are part of the 'liberationist' sector of the gay liberation movement which seeks to challenge the heterosexual assumption in public and semipublic spaces in a direct and confrontational fashion (Davis, 1991; Geltmaker, 1992). Because the incorporation movement worked within, and indeed glorified, existing political, economic, and social systems, the model gay identity associated with the effort is much closer to what Davis (1991) characterizes as the 'assimilationist' tradition (which seeks acceptance by heterosexual society). Additionally, public confrontation is not the only strategy used by gays and lesbians. Rather they may 'negotiate' multiple identities which depend in part on place and location, because most gays and lesbians spend most of their time in heterosexual environments (Valentine, 1993c, page 246). Even in Valentine's study of lesbians in the United Kingdom, however, occasional (if routine) informal gatherings of lesbians in 'marginal' spaces helped foster a sense of collective identity. For gay men in the United States, however, gay social space has been particularly critical to 'coming out', so that it is not surprising that there is a close tie between gay identity and the redefinition of public and social space.

West Hollywood: the historical context[2]

Covering less than two square miles, West Hollywood is an irregularly shaped entity bordering the city of Los Angeles on the north, east, and south, and Beverly Hills on the west (see figure 9.1). In 1924 the area voted against annexation to the city of Los Angeles, apparently because Los Angeles County had fewer restrictions on night-clubs. By the 1960s West Hollywood had become known as an area tolerant of 'alternative' life-styles – particularly for those associated with the music industry. As an unincorporated area, West Hollywood was policed by the county sheriff office, which (at the time) was thought to take a less oppressive stance toward gays. Geltmaker's (1992, pages 640–642) description of the current relationship between gays and the county sheriffs in Los Angeles suggests that this is no longer true. In either case, police raids on gay bars led to the formation of a militant gay group, PRIDE, which in turn led to the founding of *The Advocate* (now one of the largest gay newsmagazines in the United States), and a failed incorporation effort in 1969. Shortly after this period, several social service agencies supporting gays and lesbians were established, and relatively affluent gays began to concentrate in the area. An annual gay and lesbian pride festival began in 1970, and is now one of the largest of its kind in the world. By 1984 gays and lesbians were estimated to constitute 30–40% of West Hollywood's population.

The only published works on the incorporation of West Hollywood (Christensen and Gerston, 1987; Moos, 1989; Waldman, 1988) do not address the symbolic connection between gay spaces and identity. Moos (1989, page 366) writes that gays 'desired to capture the local state in order to establish a territory where their sexual orientation could not be used as a weapon against them'. He suggests, however, that gays became interested in the incorporation campaign only after the county had approved the election, when 'the question for the gay community became one of local control – control of issues that affected and concerned gays – and having a direct voice in how those issues would be handled' (page 357). Their initial lack of support for the cityhood campaign reflected a political calculation meant to retain the patronage of a county supervisor. Moos's position ignores the very early support for incorporation evident in the gay press beginning in mid-1983, and a long article in the *LA*

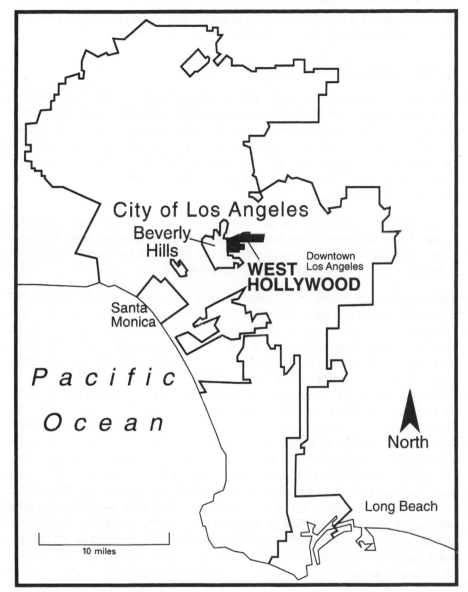

Figure 9.1 West Hollywood in the Southern California region.

Weekly (6/24/83) focusing on the role of gays in West Hollywood, especially as related to incorporation.

The incorporation campaign was a logical step in the creation of gay spaces, because it added the element of legal recognition and power. Since there is a strong tie between gay identity and gay territories, an opportunity to redefine the legal and symbolic character of an area, through incorporation, would also provide the opportunity to redefine an identity for gays. Thus the incorporation movement provided a context in which place became a particularly powerful way to organize the meaning of West Hollywood and gay identity.

The West Hollywood gay identity

One might expect that an active local paper should encourage a strong sense of community identity, but this has been difficult to demonstrate in practice. Stamm (1985) makes a relatively comprehensive, but largely unsuccessful attempt. This difficulty in part explains the limited progress in conceptualizing the ties between the local press and community since the work of Park (1922; 1940) and Janowitz (1952). With regard to the link between local newspapers and a sense of community, Stamm writes:

> The difficulty ... stems not only from the intransigence of the phenomenon, but also from the incompleteness of the model. The most we can hope to produce with such a model is evidence that newspaper reading is associated with community ties some of the time for some kinds of persons (1985, pages 40–41).

The problems associated with developing a *general* model of newspaper–community ties is that there is little reason to think that the press has the same role in different sorts of communities (geographic, ethnic, racial, gender, sexual, etc). Face-to-face interactions and mundane political concerns, rather than the local paper, may have far more influence on a geographically defined community than on a widely dispersed ethnic 'community'.

Cultural studies of mass communication have raised significant questions about how the audience 'reads' newspapers and other texts, arguing that texts always have multiple meanings, that is, they are polyphonic. See Streeter (1989), Bird (1992), and Crang (1992) for a general treatment of this question, and Cohen (1991), for an attempt to analyze the role of 'gay discourse' in the textual interpretations of gay men. Alternatively, one can follow Burgess (1985) and hold that places created by the media are mythical. She argues that like a myth, the meaning of the places created in newspapers is already complete, and does not require interpretation. The problems raised by the polysemy of texts are similar to the ones raised earlier regarding identity. My argument is not that the gay press is representative of a coherent block of gay opinion (it may be), but that as an institution in the gay community, it developed an ideal identity tied to place. It did not seek to portray itself as an objective recorder of events but actively developed a model of a 'gay city'.

As a relatively accessible public information source, as opposed to private friendship-networks, the gay press can provide an important means to socialize individuals into a gay identity. The empirical extent of this socialization is a somewhat different matter, and it would require an extensive ethnographic study to document. At least one gay publication, however, sees an analogy between its role and the role of public space in the creation of gay culture. The editor and cofounder of *Christopher Street* writes:

> Magazines have been a peculiarly modern device for bringing a public space into existence. Like a town meeting, a magazine enables people to be in each other's

company by sharing talk about matters that concern them. It is through talking with each other that most of us start to make some sense out of the world, and begin to discover who we are and what we think ...

We always thought our task was to open a space, a forum, where the developing gay culture could manifest and experience itself. For people who have been excluded from the social world ... this access to public space is basic and urgent (Denneny et al, 1983, page 13).

Although this does not address the issue of how gays might 'read' these articles differently, it supports the notion that the gay press – by providing a more stable, reliable, and accessible means for socializing gays – was more likely to encourage geographically centered communities than were informal networks. For individuals unwilling or unable to 'come out', a freely distributed paper also provides anonymity.

In addition, the gay press reported the incorporation campaign quite differently from the nongay ('mainstream' or mass) papers. Both the local paper, *The West Hollywood Post*, and the *Los Angeles Times* portrayed the incorporation campaign primarily as an issue of rent control, whereas the gay press, *The Advocate*, *Edge*, *Frontiers*, *Update*, and others, emphasized symbolic rather than economic issues (Forest, 1991). The articles also indicate that these latter papers were highly aware of their role, and were adept at both rejecting and incorporating stereotypes about gay men. The gay press was sensitive to incidents or statements which portrayed gays in a negative light during the incorporation campaign. Gay writers did not simply reject stereotypes of gays but rather sought to construct an alternative, positive identity, one that accepted and co-opted some existing stereotypes. The narrative of incorporation in the gay press is reminiscent of Foucault's – (1978, page 101) 'reverse discourse', that is, a discourse in which 'homosexuality began to speak in its own behalf, to demand that its legitimacy or "naturality" be acknowledged'.

The issue of the multiple meaning of texts does raise some difficult questions about my own reading of the gay press, because, if one takes the polysemy of texts seriously, the seven characteristics I have identified may be thought of as merely idiosyncratic, or generated by (unstated) ideological, political, or personal concerns. This is an important issue, but not one which undermines the study.

With a 'centrist' reading of the press, I emphasized characteristics which are important in the discourse of US civil society. Alexander and Smith (1993) argue that the 'discursive structure' of civil society can be broken down into democratic and counterdemocratic codes. Their 'democratic' codes – rationality, equality, reasonableness, lawfulness, etc – correspond roughly to what I have called the center of US society. Furthermore, I would argue that the same 'codes' that informed my study also provided basic vocabulary for the West Hollywood gay press. Additionally it is important to note that four of the characteristics I have identified – creativity, aesthetic sensibility, an orientation toward entertainment or consumption, and progressiveness – are elements of gay stereotypes, reflected in sources as different as opinion surveys (Taylor, 1983) and literary criticism (Sedgwick, 1985, page 173). Hence the selection of my seven characteristics was not arbitrary, but drew on existing stereotypes. The gay press did not simply reject all stereotypes, but rather it adopted those aspects that it deemed positive. By using the city as a symbol for gays, the gay press could incorporate many diverse characteristics into this identity, some which were closely related, and others which seem contradictory.

Creativity

Creativity was a prevalent theme in the gay press throughout the incorporation campaign. Articles sometimes directly stated that gays are more creative than non-gays: the 'motion picture industry . . . has attracted thousands of artistically motivated people . . . [who] include a rather high percentage of folks of the gay persuasion, one of those stereotypes solidly based on fact' (*Frontiers* 5/11/83).

It is not so much the 'fact' of having or attracting an artistically talented population, but the symbolic connection between West Hollywood and creativity that is important. Articles used 'creativity' to describe the city in the past, present, and future, suggesting that the relationship between the two is not one to be measured by documentation or empirical fact. The symbolic connection implies that gays in West Hollywood are more creative, innovative, and talented than the nongay population. Creativity is spoken of in a general way, creating the impression that the type of creativity needed to run a city is the same type of creativity used in artistic projects. Such references to creativity comprise part of a 'constitutive narrative' of West Hollywood, a narrative which sought to embody gay identity (Bellah et al, 1985).

Aesthetic sensibility

Gays' aesthetic sensibility was an attribute closely related to creativity. An article describing public space in the future city of West Hollywood was particularly rich in its use of this characteristic, and it is worth quoting at length:

> Colorful open-air mini-buses take residents from their offices to their homes. The new city's system of 'pocket parks' comes alive with . . . lovers strolling past the fountains and through the gardens. . . . The lights go on along the center strip of West Hollywood Concourse (formerly called Santa Monica Boulevard) which is now lined with tall palm trees, birds of paradise flowers and sculptures. . . . The City of West Hollywood has commissioned local architects and designers to create special bus benches, sidewalks and refuse containers for the area. Flower planters now line the entire Melrose district. (*Frontiers* 3/7/84)

The emphasis on public, as opposed to private, spaces should be noted. This is a city which is highly conscious of its visual appearance, where even the garbage cans are 'created' by designers. The city would not just 'landscape' the barren gravel-covered median strip along Santa Monica Boulevard (which had been a point of contention with the county for some time), but transform it into a 'concourse', lighted for display.

The gay press sometimes represented beauty as the rightful domain of gays. An article entitled 'Beautiful W. Hollywood' (*Update* 5/30/84) described the preincorporation landscaping of the Santa Monica median strip for the 1984

Olympic games in this way: 'The problem, say longtime Gay community activists in the area, is that before [Los Angeles County Supervisor] Edelman's tree could be planted, trees put in place four years ago by a group of Gay women and men were bulldozed into oblivion'. The suggestion was that gays were the only group truly concerned with the beauty of the area, and that nongay politicians should simply let them take care of the city. In other articles, however, the connection between gays and beauty was less direct, merely suggesting that the coalition of 'seniors, gays, business people, homeowners and renters . . . can build a more beautiful, safer city' (*Frontiers* 9/19/84b).

Entertainment and consumption

The affinity for entertainment and consumption, particularly centered around performance and bars often occurred in conjunction with 'creativity', and is evident in several passages noted in that discussion. Gay bars frequently played a critical role as social centers, especially before the gay liberation movement allowed for more open meeting places (Weightman, 1980). Although gays now have numerous meeting centers, including those not directly involved with entertainment and consumption activities, the gay press continued to use this image.

Progressiveness

The gay press portrayed gays, and gays in West Hollywood in particular, as progressive trendsetters, both culturally and politically. The names of gay publications reflect the importance of this: *Frontiers*, *The Edge*, *Update*. These all suggest that gays explore the limits of what is acceptable in society. The importance of gays in design, and thus in determining new clothing, architectural, and theatrical fashions is discussed above. This progressiveness seems to be at the very core of gay identity, because the gay press feared its loss.

Like creativity, 'progressiveness' was used to describe both cultural affairs – fashion, style, etc – and a political orientation. Hence the gay press created a strong symbolic connection between cultural production and politics, but never explicitly documented the material ties between these two kinds of activities. In terms of a community narrative, however, the existence of an actual affiliation between the two was far less important than the symbolic link.

Responsibility

The concern in the gay press for having openly gay elected officials illustrates the characteristic of responsibility. Unlike some of the preceding characteristics, responsibility seemed to have a fairly narrow meaning: the wise use of political power. *The Advocate* suggested that this was the primary concern of the Harvey Milk Democratic Club: ' "We're looking first of all for gay elected officials" ' (*The Advocate* 5/29/84). The same article highlighted a quote by Stone, 'It [cityhood] would also provide an excellent forum for openly gay candidates to seek and hold public office'. Yet these groups do not seek just exposure of gays, but gays who could be trusted to hold office. 'According to Stone . . . cityhood would definitely be beneficial to gay people. "It means that the leaders of the city would be directly accountable to its residents"' (*The Advocate* 5/29/84).

The concern with this type of conventionally defined responsibility illustrates the essentially assimilationist position of the incorporation campaign.

Maturity

Many articles closely related maturity and responsibility. The clearest statement appeared in an article by Stone, 'Cityhood simply gives us the right to manage our own affairs, just as other cities do. . . . West Hollywood will enter the world as an adult and will be expected to act like one' (*Frontiers* 3/14/84). He used this to suggest that the new city will not embark on radical changes, and that West Hollywood already had the characteristics of a 'real' city.

The gay press was particularly critical of candidates who suggested that gays were not dealing with the campaign in a mature way. In a report on the destruction of a gay candidate's campaign posters, which *Frontiers* characterized as 'attempts to squash cityhood', the paper quoted an anti-cityhood candidate's campaign manager as saying that the poster incidents 'are equivalent to little boys and girls playing in the sand box and saying "it's time to take our marbles and go home"' (*Frontiers* 10/24/84). The newsmagazine sought to show that the incident in question was indeed part of a larger, important campaign issue, and to deal with the incident in a 'mature' fashion.

Centrality

The concern with centrality reflected the 'ethnic strategy' of the gay press. According to Epstein (1987, page 38), like other ethnic groups, 'gay ethnicity functions typically through appeals to the professed beliefs of the dominant culture, emphasizing traditional American values such as equality, fairness, and freedom from persecution'. (The limits to the analogy of gays as an ethnic group are discussed below.) The gay press promoted the centrality of gays to US society by countering the idea that gays are frivolous, and are only concerned with activities on the fringe of American culture.

It is important to note that the gay press endorsed some nongay candidates, although this was relatively rare. These endorsements, however, always referred to nongay candidates' relationship to gays: 'Craig Lawson – his outstanding sensitivity to gay and lesbian issues is particularly remarkable for one not a member of that community' (*Frontiers* 10/31/84). The gay press, and his own campaign literature, often used Candidate Bud Siegel's relationship with his gay children to illustrate his positive attitude toward gays and his ability to work with the gay community (*Frontiers* 9/5/84). This shows that, although the gay press sought to portray gays as integrated into mainstream society, it did not want to ignore sexual identity. Candidates could retain their gayness, yet be close to traditional centers of power and legitimacy.

'Constitutive narratives' and gay communities

It is tempting to draw parallels between gay and ethnic communities, particularly in light of postmodern treatments of ethnicity that emphasize the unstable and reflective nature of identity. Smith (1992, page 512), for example, defines ethnicity as a 'provisional, historically conditioned social construct ... A dynamic mode of self-consciousness, a form of selfhood reinterpreted if not reinvented generationally in response to changing historical circumstances'. One notes, however, that the transformation of ethnic identity is not a particularly new phenomenon, and that even early urban sociologists observed that immigrant identities are remade from local to ethnic groupings (Park, 1955, pages 157–158). Epstein (1987, page 38) suggests that, in contrast to traditional forms, contemporary 'new ethnicities' are all 'future oriented' and are concerned with 'an instrumental goal of influencing state policy and securing social rewards on behalf of the group', a statement which also aptly describes the goal of many gay advocates.

There are limits, however, to the resemblance between ethnic and sexual identity, and one must not blindly apply concepts developed in the context of ethnic studies to gays (Valentine, 1993c, page 247, note 3). This caution applies to traditional ethnic studies in particular: The use of 'ghetto idea' to study gay neighborhoods for example, can mask the role gay territories play in the development of gay identity (for example, Levine, 1979). Godfrey (1988, page 215) argues that, although nonconformist communities (gays and lesbians) are more fluid than traditional ethnic communities, the 'morphological evidence' indicates that they can be regarded as minorites that cluster in 'special-identity neighborhoods'. These identifiable neighborhoods 'serve as both the symbols and substance of subcultural expression ... These places create, maintain, and reinforce group identities' (pages 45–46). For Godfrey the resemblance between gay and ethnic communities is largely a matter of 'urban form and spatial structure', but this is hardly a sufficient comparison. A focus on urban morphology may serve to marginalize lesbians further, given Adler and Brenner's (1992) contention that gay women exercise less control over urban space than gay men. Epstein (1987) treats the question of gays as an ethnic group more comprehensively. He argues that, as a strategy, this characterization has a political utility in the context of the US civil-rights movement: Ethnicity provides social groups with legitimacy.

The issue of citizenship underlies much of West Hollywood's incorporation vis-à-vis gay identity. In the United States after the civil rights movement, ethnicity provides one of the foundations for citizenship claims. Shotter (1993, pages 194–195) argues that the 'topic' of citizenship supplies the basis for a community discussion of identity without imposing an unduly restrictive 'narrative order'. A tradition of 'argumentation' over citizenship creates conditions in which personal issues of belonging and identity can be freely, and reflexively, addressed. This claim is somewhat questionable because the debate over citizenship itself has not been free of constraint and, as Kearns (1992) shows, appeals to 'active citizenship' can serve a variety of political interests. In the United States the struggle for citizenship has focused on efforts to break down barriers to recognition and inclusion in the polity, rather than being 'an aspiration to civic participation as a deeply involving activity' (Shklar, 1991, page 3). Smith (1989) believes that the idea of 'citizenship as critique' can measure the degree to which civil, political, and social rights have been achieved, and that this critique is best manifest in localities. Hence it is possible to see the incorporation campaign as an attempt by gays to achieve the entitlements of citizenship by (symbolically) creating themselves as a social group.

The question of whether gays are an ethnicity is unlikely to be settled, because, as Epstein (1987) points out, the answer depends largely on how one defines an ethnic group. I argue that a useful approach is to draw a comparison between ethnic groups and gays using what Bellah (1985) calls the 'constitutive narrative' of communities. This focus on the narratives of identity groups is particularly useful for geographers because it reveals how places come to be morally valued. 'Taking language seriously shows, moreover, that the "quality" of place is more than just aesthetic or affectional, that it also has a moral dimension, ... for language – ordinary language – is never morally neutral' (Tuan, 1991, page 694). Gay areas like West Hollywood can be described as 'communities of memory' as opposed to 'life-style enclaves' (Bellah et al, 1985). Members of a community of memory are tied to both the past and the future of the group; this makes the community 'genuine' or 'real'. 'They carry a context of meaning that can allow us to connect our aspirations for ourselves and those closest to us with the aspirations of a larger whole' (Bellah, 1985, page 153). Ethnic and racial communities exemplify this sort of community. There is a limit, of course, to the access one has to narratives of any community. As feminist and postcolonial writers have made clear, the relationships between ethnography, the construction of the 'Other', and power are quite complex (Butler, 1990; Katz, 1992; Keith, 1992; Said, 1979; 1989).

Conclusion

There is an intimate connection between the social process that forms personal and group identities, and the symbolic aspect of place. This connection is now attracting the attention of disciplines such as anthropology (Gupta and Ferguson, 1992). Places, since they are experienced as wholes, organize meaning in such a way that contradictory ideas can be held simultaneously. This quality is particularly important for the component of identity formed by place. Characteristics (of an identity) with conflicting normative values, centrality and progressiveness for example, can be more easily combined if they are embodied in a place rather than in a person. The holistic quality of place also allowed the gay press to create symbolic, 'natural' connections between cultural and political activities, when these activities may not be related in practice, for example, artistic creativity and administrative innovation, or political and cultural progressiveness. The connection between West Hollywood and gays tends to disguise the constructed nature of gay identity. Thus the use of place encourages the 'common-sense' perception that gays are a social group as natural, and therefore as legitimate, as ethnicities.

The question of identity has received tremendous attention in the past several years, but little consensus has been reached beyond the relatively widespread agreement about its constructed, rather than natural or essential, nature. Studies in geography, as well as other fields, have tended to focus exclusively on the ways in which the 'normative ideal' of identity fails to correspond to the particular experience of individuals within that group (Butler, 1990; Sedgwick, 1990). Humanistic geography can complement this perspective by casting the question of identity in terms of place and morally valued ways of life. Place, I have argued, plays a fundamental role in the creation of a particular 'normative ideal' of gay identity. Approaches which take note of the perspectival quality of place, and which emphasize the role of language in the creation of place can address the cultural dimensions of these broader relationships of place and identity (Entrikin, 1991; Tuan, 1991). While the ontological status of identity groups such as gays may remain indeterminate, studies of links between place and identity provide an important means to show how ideal identities are created and perpetuated.

Valentine (1993a), Knopp (1992), and Bell (1991) offer agendas for geographical studies of sexuality generally and of gays and lesbians in particular. This study will be seen, I hope, as complementary to these programs, but as was noted in the introductory section, certain limitations need to be addressed through further research. In particular, humanistic work concerning the role of place in the symbolic construction of lesbian identity needs to be further pursued. If lesbians – or gay men of lower economic status – indeed do not exercise the kind of control over public space that upper-class gay men enjoy, it would be especially important for researchers to heed Bell's (1991, page 328) call to conduct investigation' "close" to the study group,' for example, Valentine (1993b; 1993c). Ethnographic studies and/or surveys of gay men in West Hollywood would also be useful, particularly work that investigates how closely the identity realized by the gay press in 1984 conforms to the self-identity of gay men currently associated with the city.

Notes

1 It is also important to note that heterosexual identities are as much a social construction as homosexuality (Katz, 1990; Peake, 1993). For example, one can easily envision a study of the role of place in the construction of heterosexual male identity.

2 Except where noted, the historical information on West Hollywood is drawn from Kepner and Williams (1985), Envicom (1986, pages 1.1–1.3), and an interview with Robert Vulcan, president of the West Hollywood Historical Society. For an account of incorporation and annexations in Los Angeles up to 1950, see Bigger and Kitchen (1952); for information after 1950, see Miller (1981).

Newspaper references

The Advocate
5/29/84, 'Move for cityhood gains steam in West Hollywood', Christine S Shade

Frontiers
5/11/83, 'West Hollywood's gay roots'
3/7/84, 'What would a "City" of West Hollywood be like?'
3/14/84, 'Cityhood should make plenty of dreams come true', Ron Stone
9/5/84, 'Bud Siegel: West Hollywood city council candidate'
9/19/84, 'Valerie Terrigno runs as qualified candidate for West Hollywood council'
10/24/84, 'Tactics muddy W. H. Campaign issues'
10/31/84, 'Gay and lesbian alliance rates highly qualified candidates'

LA Weekly
6/24/83, 'Gay pride: will success spoil west Hollywood', Ron Stone

Update
5/30/84, 'Beautiful W. Hollywood'

References

Adler, S., Brenner, J. 1992: 'Gender and space: lesbians and gay men in the city' *International Journal of Urban and Regional Research* **16**, 24–34.

Agnew, J. 1989: 'The devaluation of place in social science', in *The Power of Place* Eds. J. A. Agnew, J. S. Duncan (Unwin Hyman, Winchester, MA) pp. 9–29.

Agnew, J. 1992: 'Place and politics in post-war Italy: a cultural geography of local identity in the provinces of Lucca and Pistoia', in *Inventing Places: Studies in Cultural Geography* Eds. K. Anderson, F. Gale (Longman Cheshire, Melbourne) pp. 52–71.

Alexander, J., Smith, P. 1993: 'The discourse of American civil society: a new proposal for cultural studies' *Theory and Society* **22**, 151–207.

Barnes, T. J., Duncan, J. S. 1992: *Writing Worlds: Discourse, Text and Metaphor in the Representation of Landscape* (Routledge, London).

Bell, D. J. 1991: 'Insignificant others: lesbian and gay geographies' *Area* **23**, 323–329.

Bellah, R. N., Madsen, R., Sullivan, W. M., Swidler, A., Tipton, S. M. 1985: *Habits of the Heart: Individualism and Commitment in American Life* (Harper and Row, New York).

Bird, S. E. 1992: 'Travels in nowhere land: ethnography and the "impossible audience"' *Critical Studies in Mass Communication* **9**, 250–260.

Burgess, J. 1985: 'News from nowhere: the press, the riots and the myth of the inner city', in *Geography, the Media, and Popular Culture* Eds. J. Burgess, J. R. Gold (Croom Helm, London) pp. 192–228.

Butler, J. 1990: *Gender Trouble: Feminism and the Subversion of Identity* (Routledge, New York).

Castells, M. 1983: 'Cultural identity, sexual liberation and urban structure: the gay community in San Francisco', in *The City and the Grassroots* (Edward Arnold, London) pp. 138–172.

Christensen, T., Gerston, L. 1987: 'West Hollywood: a city is born' *Cities* **4**, 299–303.

Cohen, J. R. 1991: 'The "relevance" of cultural identity in audiences' interpretation of mass media' *Critical Studies in Mass Communication* **8**, 442–454.

Crang, P. 1992: 'The politics of polyphony: reconfigurations in geographical authority' *Environment and Planning D: Society and Space* **10**, 527–549.

Davis, T. H. 1991: '"Success" and the gay community: reconceptualizations of space and urban social movement', paper presented at the First National Student Conference on

Lesbian and Gay Studies, Milwaukee, WI, April; copy available from the author 26 Hemenway Street, #29, Boston, MA 02115.

Denneny, M., Ortleb, C., Steele, T. 1983: *The Christopher Street Reader* (Coward-McCann, New York).

Duggan, L. 1992: 'Making it perfectly queer' *Socialist Review* **22**(1), 11–32.

Entrikin, J. N. 1991: *The Betweenness of Place: Toward a Geography of Modernity* (Johns Hopkins University Press, Baltimore, MD).

Epstein, S., 1987: 'Gay politics, ethnic identity: the limits of social construction' *Socialist Review* **17**(3–4), 9–54.

Forest, B. 1991: 'Political territory and symbol: West Hollywood, gays and rent control', unpublished masters thesis, Department of Geography, University of California, Los Angeles, CA.

Foucault, M. 1978: *The History of Sexuality: Volume 1* translated by R. Hurley (Pantheon, New York).

Geltmaker, T. 1992: 'The Queer Nation Acts Up: health care, politics, and sexual diversity in the County of Angels' *Environment and Planning D: Society and Space* **10**, 609–650.

Giddens, A. 1991: *Modernity and Self-identity: Self and Society in the Late Modern Age* (Stanford University Press, Stanford, CA).

Godfrey, B. J. 1988: *Neighborhoods in Transition: The Making of San Francisco's Ethnic and Nonconformist Communities* (University of California Press, Berkeley, CA).

Gupta, A., Ferguson, J. 1992: 'Beyond "culture": space, identity, and the politics of difference' *Cultural Anthropology* **7**, 6–23.

hooks, b. 1990: 'Choosing the margin as a space of radical openness', in *Yearnings: Race, Gender, Cultural Politics* (South End Press, Boston, MA) pp. 145–154.

Jackson, P. 1989: *Maps of Meaning* (Unwin Hyman, London).

Janowitz, M. 1952: *The Community Press in an Urban Setting* (Free Press, Glencoe, IL).

Katz, C. 1992: 'All the world is staged: intellectuals and the projects of ethnography' *Environment and Planning D: Society and Space* **10**, 495–510.

Katz, J. N. 1990: 'The invention of heterosexuality' *Socialist Review* **20**(1), 7–34.

Kearns, A. J. 1992: 'Active citizenship and urban governance' *Transactions of the Institute of British Geographers, New Series* **17**, 20–34.

Keith, M. 1992: 'Angry writing: (re)presenting the unethical world of the ethnographer' *Environment and Planning D: Society and Space* **10**, 551–568.

Knopp, L. 1987: 'Social theory, social movements and public policy: recent accomplishments of the gay and lesbian movements in Minneapolis, Minnesota' *International Journal of Urban and Regional Research* **11**, 243–261.

Knopp, L. 1990a: 'Some theoretical implications of gay involvement in an urban land market' *Political Geography Quarterly* **9**, 337–352.

Knopp, L. 1990b: 'Exploiting the rent-gap: the theoretical significance of using illegal appraisal schemes to encourage gentrification in New Orleans' *Urban Geography* **11**, 48–64.

Knopp, L. 1992: 'Sexuality and the spatial dynamics of capitalism' *Environment and Planning D: Society and Space* **10**, 651–669.

Levine, M. P. 1979: 'Gay ghetto', in *Gay Men: The Sociology of Male Homosexuality* Ed. M. P. Levine (Harper and Row, New York) pp. 182–204.

Lynch, F. 1987: 'Non-ghetto gays: a sociological study of suburban homosexuals' *Journal of Homosexuality* **13**(4), 13–42.

Massey, D. 1991: 'The political place of locality studies' *Environment and Planning A* **23**, 267–281.

Moos, A. 1989: 'The grassroots in action: gays and seniors capture the local state in West Hollywood, California', in *The Power of Geography* Eds J. Wolch, M. Dear (Unwin Hyman, Winchester, MA) pp. 351–369.

Park, R. E. 1922: *The Immigrant Press and its Control* (Harper and Brothers, New York).

Park, R. E. 1940: 'News as a form of knowledge: a chapter in the sociology of knowledge' *Americal Journal of Sociology* **46**, 669–686.

Park, R. E. 1955: 'Immigrant community and immigrant press', in *The Collected Papers of Robert Park, Volume 3. Society: Collective Behaviour, News and Opinion, Sociology and Modern Society* Eds E. Charrington Hughes, L. Wirth, R. Redfield (The Free Press, Glencoe, IL) pp. 152–164.

Peake, L. 1993: 'Race' and sexuality: challenging the patriarchal structuring of urban social space' *Environment and Planning D: Society and Space* **11**, 415–432.

Relph, E. 1976: *Place and Placelessness* (Pion, London).

Rutledge, L. W. 1992: *The Gay Decades: From Stonewall to the Present: The People and Events that Shaped Gay Lives* (Penguin Books, New York).

Said, E. 1979: *Orientalism* (Vintage Books, New York).

Said, E. 1989: 'Representing the colonized: anthropology's interlocutors' *Critical Inquiry* **15**, 205–225.

Sedgwick, E. K. 1985: *Between Men: English Literature and Male Homosocial Desire* (Columbia University Press, New York).

Sedgwick, E. K. 1990: *Epistemology of the Closet* (University of California Press, Berkeley, CA).

Shils, E. 1975: *Center and Periphery: Essays in Macrosociology* (University of Chicago Press, Chicago, IL).

Shklar, J. N. 1991: *American Citizenship: The Quest for Inclusion* (Harvard University Press, Cambridge, MA).

Shotter, J. 1993: *Cultural Politics of Everyday Life* (University of Toronto Press, Toronto).

Smith, M. P. 1992: 'Postmodernism, urban ethnography, and the new social space of ethnic identity' *Theory and Society* **21**, 493–593.

Smith, S. J. 1989: 'Society, space and citizenship: a human geography for the "new times"?' *Transactions of the Institute of British Geographers, New Series* **14** 144–156.

Stamm, K. R. 1985: *Newspaper Use and Community Ties: Toward a Dynamic Theory* (Ablex Publishing Corporation, Norwood, NJ).

Stein, A. 1992: 'Sisters and queers: the decentering of lesbian feminism' *Socialist Review* **22**(1), 33–55.

Streeter, T. 1989: 'Polysemy, plurality, and media studies' *Journal of Communication Inquiry* **13**, 88–106.

Taylor, A. 1983: 'Conceptions of masculinity and femininity as a basis for stereotypes of male and female homosexuals', in *Homosexuality and Social Sex Roles* Ed. M. Ross (Haworth Press, New York) pp. 37–53.

Tuan, Y-F, 1974: 'Space and place: a humanistic perspective', in *Progress in Geography: International Reviews of Current research* Eds C. Board, R. J. Chorley, P. Haggett, D. Stoddart (Edward Arnold, London) pp. 211–252.

Tuan, Y-F, 1977: *Space and Place: The Perspective of Experience* (University of Minnesota Press, Minneapolis, MN).

Tuan, Y-F, 1980: 'Rootedness versus sense of place' *Landscape* **24**(1), 3–8.

Tuan, Y-F, 1982 *Segmented Worlds and Self: Group Life and Individual Consciousness* (University of Minnesota Press, Minneapolis, MN).

Tuan, Y-F, 1984: 'In place, out of place' *Geoscience and Man* **24**, 3–10.

Tuan, Y-F, 1989: *Morality and Imagination: Paradoxes of Progress* (University of Wisconsin Press, Madison, WI).

Tuan, Y-F, 1991: 'Language and the making of place: a narrative-descriptive approach' *Annals of the Association of American Geographers* **81**, 684–696.

Valentine, G. 1993a: '(Hetero)sexing space: lesbian perceptions and experiences of everyday spaces' *Environment and Planning D: Society and Space* **11**, 395–413.

Valentine, G. 1993b, 'Desperately seeking Susan: a geography of lesbian friendships' *Area* **25**, 109–116.

Valentine, G. 1993c, 'Negotiating and maintaining multiple sexual identities' *Transactions of the Institute of British Geographers, New series* **18**, 237–243.

Weightman, B. 1980, 'Gay bars as private places' *Landscape* **24**(1), 9–16.

SECTION THREE

OUT OF PLACE: ESCAPE ATTEMPTS

Editorial introduction

As I have suggested in the introductions to the previous sections, from the 1950s onwards the development of mass consumption had a significant impact on the living standards of US and British households and on their relationships to place. While the main emphasis so far has been on the assumed decline of a communal solidarism and a retreat to the privacy of the home, the Fordist era with its relatively secure employment for the working class, as well as growing opportunities to move into middle-class occupations as school leaving ages rose and higher education expanded, led to a new market for the products of expanding industries: this was the youth market. As with the earlier puritanical reactions to the goods that made a more comfortable family life possible for growing numbers of households, here too the development of mass markets for clothes, and for 'office furniture, scooters, expresso machines, coffee bars, interior design, generated a hostile reaction in various quarters in Britain' (Chambers, 1986, p. 51). As with the arguments about the decline of class politics, a similar argument was voiced in Britain about national identity and the distinctiveness of British culture. As Hebdige (1988) explains:

> The singularity of British culture is felt to be increasingly threatened in the post-war period by the conditions under which consumption values and popular culture are disseminated. For critics pledged to defend 'authentic' British values, mass-produced commodities aimed at specific target groups begin to function as symbols of decadence. They are seen to pose a threat to native traditions of rugged self-reliance, self discipline and the muscular puritanism of the stereotyped (male) workforce, thereby leading to a softening up and 'feminisation' of the national stock. (p. 9)

The main influences leading to the supposed decadence (in Britain at least) were from two directions: a generalised European influence, reflected in those commodities singled out in the quote above by Chambers, and from across the Atlantic. It was from the USA, where the development of a mass consumer market was a decade or so in advance of Great Britain, beginning in the 1930s but interrupted by the Second World War, that the greatest threat was perceived.

Beginning in the war, when American men purportedly seduced British womanhood with nylon stockings, chewing gum and chocolate (the problem with Americans was that they were 'overpaid, oversexed and over here') and then deepening in influence in the postwar period through clothes, styles and popular culture from jazz to the cinema, these new forms of consumption apparently produced an homogeneous youth who were placeless. This observation by J.B. Priestley, from his novel *Angel Pavement* (1968) and also quoted in Chambers (1986), nicely sums up distaste for US and European influences, as well as their spatial implications:

> They were the children of the Woolworth stores and the moving pictures. Their world was at once larger and shallower than that of their parents. They were less English, more cosmopolitan. Mr Smeeth could not understand George and Edna, but a host of youths in New York, Paris and Berlin would have understood them at a glance. Edna's appearance, her grimaces and gestures, were temporarily those of an Americanised Polish Jewess, who, from her mint in Hollywood, had stamped them on these young girls from all over the world. George's knowing eye for a machine, his cigarettes and drooping eyelid, his sleek hair, his tie, and shoes and suits, the smallest detail of his motor-cycling and dancing, could be matched almost exactly around every corner in any American city or European capital. (Priestley, 1968, p. 77)

While the specifics of Priestley's comments now seem dated (a tie and suit!) the sentiments have echoed through the decades as generations of parents lament the shallowness and consumerism of youth culture and its decadence. The fears of a homogenised culture and the decline of local particularity are also a key thread in anti-postmodernist sentiments, as I suggested in the introduction.

What I want to explore in this section is the ways in which youth cultures are based on a desire to escape the bounds of a localised community, to deny the specificity both of locality and national identity. I also want to explore the ways in which style is an important part of youth culture as particular items become transformed into symbolic statements about identity. A sub-theme is an exploration of Hebdige's comment that youth culture and street style often challenge accepted notions of masculinity. We have already broached these ideas in Part I where Ehrenreich argued that the 1950s saw a challenge to hegemonic notions of adult masculinity being based on the 'good husband, good provider' role with a hedonistic and selfish masculinity providing an alternative. Here, however, the emphasis is on the extent to which a new form of 'feminised masculinity' based around consumption and style developed in the post-war period. Uniting this dual emphasis are the same questions that lie behind the reader as a whole: what are the consequences of post-war social changes for the nature of place-based identities? And how far are specific places important sites of symbolic meaning in youth cultures?

To the extent that youth culture is seen as a reaction to the settled stable and domestic worlds of adulthood, it has been the road and the street, public rather than private spaces, that have been the focus of research. Here the key figure is that of the outlaw, the gang member, or the rebel (who as I noted earlier also attracted the attention of the

Chicago school of urban sociologists). The apocryphal imaginary figure of youthful rebellion is probably Marlon Brando, who with Lee Marvin made *The Wild Ones* in 1953, based on an incident between outlaw motorcycle gangs in Hollister, California in 1947. James Dean runs Brando a close second as the icon of 1950s and early 1960s youth. (When asked what he was against in *Rebel without a Cause*, he replied 'What have you got'?) And both stars, of course, wore that iconic item of youthful rebellion, the black leather jacket, in these films.

The actual motorcycle gangs, the outlaws, the bikers and Hell's Angels who roamed the open roads in the USA and their somewhat paler reflections in the rockers in the UK (Cohen, 1972), were a seamier side of these imaginary rebels. For the outlaws being 'on the road' was a permanent way of life but as Columbus Hopper and Johnny 'Big John' Moore (1983) have shown in their work about these gangs, they are also geographically specific, often with only fifteen or twenty members, although some gangs have chapters in different parts of the USA. Outlaw members are usually male, aged between 21 and 45, and the gangs are usually racially unmixed: there are black gangs, white gangs, and Mexican and other Spanish-speaking gangs. These gangs had a more cultured counterpart in the United States in the 'beats', most notably Jack Kerouac and his friends who cast off the shackles of 1950s domesticity in the US and periodically went 'on the road'. Like the gangs they drank and had sex, but they also wrote poetry and novels. Indeed, the 'novelisation' of their life, *On the Road* (Kerouac, 1957) is still a cult book among young readers and has recently attracted the attention of geographers (Cresswell, 1993, 1996; McDowell, 1996; Rycroft, 1996).

The Hell's Angels and bikers, like other more recent gang cultures in the US, as well as being divided by ethnicity, were also a particularly masculinist or patriarchal group. Although they rejected settled domesticity, they maintained a strict gender division of labour in gang activities and an extremely possessive attitude towards their 'old ladies'. In Chapter 10, a more recent paper by **Columbus Hopper and Johnny Moore**, the ties that held these men together in peripatetic outlaw gangs and the reasons why women join such misogynist groups are explored. Women who join these gangs are in a sense double outsiders as not only have they rejected settled lives but they have also challenged conventional associations between femininity and space. At least for men the association between masculinity and freedom was accepted. And as Hopper and Moore also point out, many of the women members of the gangs also suffered the 'double jeopardy' of male violence in childhood and as gang members from the very men whom they looked to for protection.

Although the possession and display of symbols that identify group membership – leather jackets, chains, tattoos and long hair – are a key way of constructing a symbolic and placeless community, the bikers, like the beats before them, seem to counter assertions that the construction of a group identity through the possession of particular consumption goods might lead to the feminisation of Britain and North America's youth. These men were 'real men' or so they seemed from the outside. The beats did, after all, write and quote poetry, and have sex with men as well as women, but they were in the long tradition of the North American pioneer, who had to move on. That the cramped, urban islands of the UK rather restricted the scope to go on the road was overlooked by the chapters of Hell's Angels and rockers who formed in the post-war decades. An accusation of femininity, however, had more meaning in the case of that other group – the 'mods': a short-lived British phenomenon that led to group rivalry and outbreaks of ritual violence between mods and rockers in the early 1960s. **Rob Shields** in Chapter 11 describes the ritual confrontations that occurred in Brighton on bank holiday Mondays in the early 1960s. These set battles are now a part of folk history and it is surprising to remember that in fact they happened on only two occasions in 1964.

It was later in the 1960s, however, that the establishment's fear of feminised men seemed to become reality with the antiwar movements, student protest, the hippy rejection of 'mindless consumerism', the rise of green politics and the women's movement at the end of the 1960s. This is where the origins of identity politics and postmodernism are sometimes located, in the 'teeming and boiling society of the 1960s' (Berman, 1992, p. 44) as much as in the deconstructivist moves in French social theory. For these groups, place was a significant element in the cultures of resistance that grew up. The names of the sites of the free festivals in the 1960s – Woodstock in the USA, the Isle of Wight and Hyde Park in the UK – still resonate in the 1990s (at least to those who are old enough to have been there), as well as Haight Ashbury in San Francisco, Grosvenor Square in London, the site of a big anti-Vietnam war demonstration, or more recently peace camps at Greenham Common in England and in the USA. These places are what Shields terms liminal spaces, places on the edge: often literally on the edge of countries or continents and metaphorically on the edge of the central values systems of the nation. Many of these sites are places of carnival and excess (Stallybrass and White, 1986), of dubious pleasures and easy virtues, dedicated not to the protestant ethic of work but to alternative pleasures. Other spaces mark the memory of political and social resistances.

In the 1990s, the significance of place in a local politics of resistance has taken a different form. In Britain, there are an increasing number of groups living, either permanently or temporarily, outside the conventional ways of living. In *Senseless Acts of Beauty*, a survey of the different cultures of resistance that have been significant in Britain since the 1960s, George McKay (1996) examines the extent to which contemporary movements are united as a politics of disenfranchisement by the youth who were left out of the Thatcher 'revolution'. He quotes an Observer article summarising a conventional view of those involved the new 'DIY' lifestyles:

> [They] are typically categorised as creatures of the social wildlife: travellers, ravers, squatters, crusties, hunt saboteurs, animal liberationists; or more overtly pejoratively, rioters, trespassers, or just plain good-for-nothing layabouts. (Hill, 1995 quoted in McKay, 1996, p. 1)

The single issue politics of the 1990s echo many of the currents of the 1960s and raise some similar questions too. How far, for example, is it possible to argue that movements based on style and pleasure are actually resistance. Is rave culture a radical culture of resistance; was hippy and punk culture before it? Like McKay, here in this reader I want to challenge conventional notions of periodisation. The 1950s were not solely a decade of settled domesticity as nostalgic reconstructions now suggest. Similarly the 1960s was not the last radical generation, the end of politics, but rather as McKay suggests 'a beginning not an end ... youth revolution or counterculture is not by itself a new thing, indeed it extends far back beyond 1945 and the end of the Second World War' (p. 3).

McKay argues that the current cultures of resistance are not what Jon Savage (1991) called 'an explosion of negatives' (p. 195) or what those on the Left, yearning the decline of class politics, see as a descent into postmodern relativism where 'anything goes', but part of the long tradition of radical idealism or utopianism, with historic roots in Britain and the USA, back through bohemians, romantics, Quakers, Shakers, to witches and heretics. I shall return in the final section (Section Six) to questions about the possibility of a new form of politics based on the tolerance of difference and transgressions rather than a single view of a progressive future. I have included **George McKay's** chapter about New Age travellers as Chapter 12. At first, this may seem an example of a culture of resistance that challenges the significance of place. As their name indicates, these are people who once again are 'on the road', rejecting, although perhaps of necessity rather than choice, a settled life, 'economic refugees' as a traveller quoted in the extract claims. But despite their travelling culture, certain sites and locations have a key significance for this group. Partly based on sympathetic landowners, but also on the symbolic significance of places like Glastonbury, a 'folk myth' about place has developed. McKay also explores aspects of everyday life in camps and convoys, including dirt and domesticity. While he does not explore gender relations in detail, one of my own students, in his undergraduate dissertation work in Pollock Free State, commented on the rigidity of gender divisions in New Age settlements (Featherstone, 1996). He suggested that there were a number of poles of identification which combined to reinforce these divisions: the 'macho' attitudes of the dispossessed male working class, the masculine tradition of 'nomadology' exemplified by Kerouac and the beats and the identification of (primarily middle-class) women with essentialist ideas about femininity being close to Nature, valorising notions of female self-sacrifice and service (Featherstone, 1996, p. 27–8) Perhaps ironically, the rigidity of gender divisions is a link between outlaw biker gangs and travellers, two groups who probably would deny that they had anything in common.

It might be argued that not only are many of the typical examples of youth cultures predominantly masculinist but also that the research has mirrored this emphasis. This partly reflects the almost unchallenged associations between public and visible activities and acts of resistance. The women Hopper and Moore studied were active in public spaces. But in general young women have more spatially restricted lives than their male counterparts and so, as McRobbie argues in her book *Feminism and Youth Culture* (1991), a rich vein of 'girls' culture' that takes place 'indoors' in private or quasi-public places has been ignored. McRobbie's own work has provided an exceptional corrective in Britain. She has investigated girls' magazines and dance cultures, for example. But it is still the lure of the 'deviant' and the traveller that has dazzled the academic commentators on youth culture.

As well as gender, ethnicity is often a key divide in youth culture. Although there has been an enormously significant transfer of cultural ideas between groups from Jazz onwards to current African-Asian music, there are also ethnically specific youth cultures. Bikers and mods in Britain, for example, were white youth, partly, of course, as West Indian Britons were more concerned with economic survival than escape attempts in the 1950s and 1960s. But for young British Blacks, creating and exerting a separate cultural identity was an important political gesture in the 1980s. I shall turn in more detail to questions about Black identity in the next section. Here, however, I want to look at the links between public acts of resistance and the behaviour of young Black men. For these young men the 'freedom' of the streets and their association with rebellion and resistance through the ritualised adoption of 'street style' and particular forms of behaviour raises more problematic issues. Young Black men are seen as threatening and have been demonised as muggers (Hall, 1978). They may also be excluded from that public ritual celebration of working class masculinity – the football match – by racist language and attitudes. In her paper (Chapter 13), **Sallie Westwood** explores the ways in which young Black and Asian men negotiate their masculinity on the streets and construct for themselves an oppositional identity that challenges British stereotypes of Black as dangerous or the Orientalism of discourses of Asian identities. As Westwood notes, echoing the article by bell hooks in Section One, the home, as well as the streets, is also a site of resistance against racism but one that is perhaps more significant for women, given the sets of associations between gender and location.

In recent cultural criticism, in anthropology and in geographical writing, travel as a metaphor has emerged as a significant way of theorising the disruption of settled notions of place that is part of the shift in socio-spatial relations at the end of the century. Clifford (1992), for example, has argued for the construction of 'travelling theory' in his discipline of anthropology. As he notes, the huge global migrations of the century that have re-sited peoples and altered the relationships between identity and place mean that the journey is as significant as either places of origin and arrival. Among geographers, travel and travel writing has become a key new area of emphasis (Blunt and Rose, 1995). Although in the more recent work, the significance of gender differences is no longer ignored, **Janet Wolff**'s paper (reprinted here as Chapter 14) is still an important critique of the problems inherent in the use of metaphors of travel. Her paper forms a bridge between Sections Three and Four as in Section Four I shall turn to a specific examination of the impact of migration on post-colonial subjects. It is worth keeping this change of emphasis in mind while reading Wolff's paper, as well as relating her comments about masculinity and metaphors of travel, nomadic subjects and pioneers to the earlier extracts that are included in this section of the reader.

References and further reading

Berman, M. 1992: Why modernism still matters. In Lash, S. and Friedman, J. (eds) *Modernity and Identity*. Oxford: Blackwell, 33–58.

Blunt, A. and Rose, G. (eds) 1995: *Women, writing and travel*. London: Routledge.

Cassady, C. 1990: *Off the road*. London: Black Spring Press.

Chambers, I. 1986: *Popular culture: the metropolitan experience*. London: Routledge.

Clifford, J. 1992: Travelling cultures. In Grossberg, L. et al. (ed.) *Cultural studies*. London: Routledge.

Cohen, S. 1972: *Folk devils and moral panics: the creation of the mods and rockers*. London: McGibbon and Kee.

Cresswell, T. 1993: Mobility as resistance: a geographical reading of Kerouac's 'On the road'. *Transactions of the Institute of British Geographers* **18**, 249–62.

Cresswell, T. 1996: Writing, reading and the problem of resistance. *Transactions of the Institute of British Geographers* **21**, 420–4.

Featherstone, D. 1996: *'Pollock Free State', Glasgow: lived space as the universe of dreams, ideas and creativity*. Unpublished dissertation submitted as part of the final examinations, Department of Geography, Cambridge (available from the library).

Hall, S. 1978: *Policing the crisis: mugging, the state and law and order*. London: Macmillan.

Hall, S. and Jefferson, T. (eds) 1976: *Resistance through rituals: youth subcultures in postwar Britain*. London: Hutchinson.

Hebdige, D. 1988: *Hiding in the light: on images and things*. London: Comedia.

Hill, D. 1995: The new righteous. The *Observer* Life Magazine. 12 February, p. 20.

Hopper, C. and Moore, J. 1983: Hell on wheels: the outlaw motorcycle gangs. *Journal of American Culture* **6**, 58–64.

Kerouac, J. 1957: *On the Road*. Harmondsworth: Penguin.

McDowell, L. 1996: Off the road: alternative views of rebellion, resistance and 'the beats'. *Transactions of the Insitute of British Geographers* **21**, 412–19.

McKay, G. 1996: *Senseless acts of beauty: cultures of resistance since the sixties*. London: Verso.

McRobbie, A. (ed.) 1989: *Zoot suits and second hand dresses*. London: Routledge.

McRobbie, A. 1991: *Feminism and youth culture*. London: Macmillan.

Mort, F. 1988: Boy's own? Masculinity, style and popular culture. In Chapman, R. and Rutherford, J. (eds), *Male order*. London: Routledge.

Polhemus, T. 1994: *Streetstyle*. London: Thames and Hudson.

Ross, K. 1995: *Fast cars, clean bodies: decolonisation and the reordering of French culture*. Cambridge, Mass.: MIT Press.

Rycroft, S. 1996: Changing lanes: textuality on and off the road. *Transactions of the Institute of British Geographers* **21**, 425–8.

Stallybrass, P. and White, A. 1986: *The politics and poetics of transgression.* Ithaca, New York: Cornell University Press.

Savage, J. 1991: *England's dreaming: Sex Pistols and punk rock.* London: Faber.

Thompson, H. 1967: *Hell's Angels.* Harmondsworth: Penguin.

10 Columbus Hopper and Johnny Moore
Women in Outlaw Motorcycle Gangs

Excerpts from: *Journal of Contemporary Ethnography* **18**, 363–87 (1990)

This article is about the place of women in gangs in general and in outlaw motorcycle gangs in particular. Street gangs have been observed in New York dating back as early as 1825 (Asbury, 1928). The earliest gangs originated in the Five Points district of lower Manhattan and were composed mostly of Irishmen. Even then, there is evidence that girls or young women participated in the organizations as arms and ammunition bearers during gang fights.

The first gangs were of two types: those motivated primarily as fighters and those seeking financial gain. Women were represented in both types and they shared a remarkably similar reputation with street gang women more than 100 years later (Hanson, 1964). They were considered 'sex objects' and they were blamed for instigating gang wars through manipulating gang boys. The girls in the first gangs were also seen as undependable, not as loyal to the gang, and they played inferior roles compared to the boys.

The first thorough investigation of youth gangs in the United States was carried out by Thrasher (1927) in Chicago. Thrasher devoted very little attention to gang girls but he stated that there were about half a dozen female gangs out of 1,313 groups he surveyed. He also said that participation by young women in male gangs was limited to auxiliary units for social and sexual activities.

Short (1968) rarely mentioned female gang members in his studies, which were also carried out in Chicago, but he suggested that young women became gang associates because they were less attractive and less socially adequate compared to girls who did not affiliate with gangs.

According to Rice (1963), girls were limited to lower status in New York street gangs because there was no avenue for them to achieve power or prestige in the groups. If they fought, the boys thought them unfeminine; if they opted for a passive role, they were used only for sexual purposes.

Ackley and Fliegel (1960) studied gangs in Boston in which girls played both tough roles and feminine roles. They concluded that preadolescent girls were more likely to engage in fighting and other typically masculine gang actions while older girls in the gangs played more traditionally feminine roles.

Miller (1973, 1975) found that half of the male gangs in New York had female auxiliaries but he concluded that the participation of young women in the gangs did not differ from that which existed in the past. Miller also pointed out that girls who formed gangs or who were associates of male gangs were lower-class girls who had never been exposed to the women's movement. After studying black gangs in Los Angeles, Klein (1971) believed that, rather than being instigators of gang violence, gang girls were more likely to inhibit fighting.

The most intensive studies of female gang members thus far were done by Campbell (1984, 1986, 1987) on Hispanic gangs in New York City. Although one of the three gangs she studied considered itself a motorcycle gang, it had only one working motorcycle in the total group. Therefore, all of the gangs she discussed should be thought of as belonging to the street gang tradition.

Campbell's description of the gang girls was poignant. The girls were very poor but not anomic; rather, they were true believers in American capitalism, aspiring to success as recent immigrants always have. They were torn between maintaining and rejecting Puerto Rican values while trying to develop a 'cool' streetwise image.

As Campbell reported, girl gang members shared typical teenage concerns about proper makeup and wearing the right brands of designer jeans and other clothing. Contrary to popular opinion, they were also concerned about being thought of as whores or bad mothers, and they tried to reject the Latin ideal that women should be totally subordinate to men. The basic picture that came out of Campbell's work was that gang girls had identity problems arising from conflicting values. They wanted to be aggressive and tough, and yet they wished to be thought of as virtuous, respectable mothers.

Horowitz (1983, 1987) found girls in Chicago gangs to be similar in basic respects to those that Campbell described. The gang members, both male and female, tried to reconcile Latin cultural values of honor and violence with patterns of behavior acceptable to their families and to the communities in which they existed.

The foregoing and other studies showed that girls have participated in street gangs as auxiliaries, as independent groups, and as members in mixed-gender organizations. While gangs have varied in age and ethnicity, girls have had little success in gaining status in the gang world. As reported by Bowker (1978, Bowker and Klein, 1983), however, female street gang activities were increasing in most respects; he thought that independent gangs and mixed groups were increasing more than were female auxiliary units.

Unlike street gangs that go back for many years, motorcycle gangs are relatively new. They first came to public attention in 1947 when the Booze Fighters, Galloping Gooses, and other groups raided Hollister, California (Morgan, 1978). This incident, often mistakenly attributed to the Hell's Angels, made headlines across the country and established the motorcycle gangs' image. It also inspired *The Wild Ones*, the first of the biker movies released in 1953, starring Marlon Brando and Lee Marvin.

Everything written on outlaw motorcycle gangs has focused on the men in the groups. Many of the major accounts (Eisen, 1970; Harris, 1985; Montegomery, 1976; Reynolds, 1967; Saxon, 1972; Thompson, 1967; Watson, 1980; Wilde, 1977; Willis, 1978; Wolfe, 1968) included a few tantalizing tidbits of information about women in biker culture but in none were there more than a few paragraphs, which underscored the masculine style of motorcycle gangs and their chauvinistic attitudes toward women.

Although the published works on outlaw cyclists revealed the fact that gang members enjoyed active sex lives and had wild parties with women, the women have been faceless; they have not been given specific attention as functional participants in outlaw culture. Indeed, the studies have been so one-sided that it has been difficult to think of biker organizations in anything other than a masculine light. We have learned that the men were accompanied by women but we have not been told anything about the women's backgrounds, their motivations for getting into the groups or their interpretations of their experiences as biker women.

From the standpoint of the extant literature, biker women have simply existed; they have not had personalities or voices. They have been described only in the contemptuous terms of male bikers as 'cunts', 'sluts', 'whores', and 'bitches'. Readers have been given the impression that women were necessary nuisances for outlaw motorcyclists. A biker Watson (1980: 118) quoted, for example, summed up his attitude toward women as follows: 'Hell,' he said, 'if I could find a man with a pussy, I wouldn't fuck with women. I don't like 'em. They're nothing but trouble.'

In this article, we do four things. First, we provide more details on the place of women in arcane biker subculture, we describe the rituals they engage in, and we illustrate their roles as money-makers. Second, we give examples of the motivations and backgrounds of women affiliated with outlaws. Third, we compare the gang participation of motorcycle women to that of street gang girls. Fourth, we show how the place of biker women has changed over the years of our study and we suggest a reason for the change. We conclude by noting the impact of sex role socialization on biker women.

Methods

The data we present were gathered through participant observation and interviews with outlaw bikers and their female associates over the course of 17 years. Although most of the research was done in Mississippi, Tennessee, Louisiana, and Arkansas, we have occasionally interviewed bikers throughout the nation, including Hawaii.[1] The trends and patterns we present, however, came from our study in the four states listed.

During the course of our research, we have attended biker parties, weddings, funerals, and other functions in which outlaw clubs were involved. In addition, we have visited in gang clubhouses, gone on 'runs' and enjoyed cookouts with several outlaw organizations.

It is difficult to enumerate the total amount of time or the number of respondents we have studied because of the necessity of informal research procedures. Bikers would not fill out questionnaires or allow ordinary research methods such as tape recorders or note taking. The total number of outlaw motorcyclists we studied over the years was certainly several hundred. In addition to motorcycle gangs in open society, we also interviewed and corresponded with male and female bikers in state and federal prisons.

The main reason we were able to make contacts with bikers was the background of Johnny Moore, who was once a biker himself. During the 1960s, 'Big John' was president of Satan's Dead, an outlaw club on the Mississippi Gulf Coast. He participated in the rituals we describe, and his own experience and observations provided the details of initiation ceremonies that we relate. As a former club president, Moore was able to get permission for us to visit biker clubhouses, a rare privilege for outsiders.[2]

Most of our research was done on weekends because of our work schedules and because the gangs were more active at this time. The bikers usually had a large party one weekend a month, or more often when the weather was nice, and we were invited to many of these.

At some parties, such as the 'Big Blowout' each spring in Gulfport, there were a variety of nonmembers present to observe the motorcycle shows and 'old lady' contests as well as to enjoy the party atmosphere. These occasions were especially helpful in our study because bikers were 'loose' and easier to approach while partying. We spent more time with three particular 'clubs,' as outlaw gangs refer to themselves, because of their proximity.

In addition to studying outlaw bikers themselves, we obtained police reports, copies of Congressional hearings that deal with motorcycle gangs, and indictments that were brought against prominent outlaw cyclists. Our attempt was to study biker women and men in as many ways as possible. We were honest in explaining the purpose of our research to our respondents. They were told that our goal was only to learn more about outlaw motorcycle clubs as social organizations.

Dilemmas of biker research

Studying bikers was a conflicted experience for us. It was almost impossible to keep from admiring their commitment, freedom, boldness, and fearlessness; at the same time, we saw things that caused us discomfort and consternation because bikers' actions were sometimes bizarre. We saw bikers do things completely foreign to our personal values. Although we did not condone these activities, we did not express our objections for two reasons. First, we would not have been able to continue our study. Second, it was too dangerous to take issue with outlaws on their own turf.

Studying bikers was a risky undertaking for us, even without criticizing them. At times when we were not expecting any problems, conditions became hazardous. In Jackson, Tennessee, for example, one morning in 1985 we walked into an area where bikers had camped out all night. Half asleep and hung over, several of them jumped up and pulled guns on us because they thought we might be members of a rival gang that had killed five of their 'brothers' several years earlier. If Grubby, a biker who recognized us from previous encounters, had not interceded in our behalf, we could have been killed or seriously injured.

For practical purposes, both male and female bikers worship the Harley Davidson motorcycle. One Mississippi group that we studied extensively had an old flathead 'Hog' mounted on a high tree stump at the entrance to their clubhouse. When going in or out, members bowed to the old Harley or saluted it as an icon of the highest order. They took it very seriously. Had we not shown respect for their obeisance, our relationship would have been terminated, probably in a violent manner.

It was hard to fathom the chasm between bikers and the rest of us. Outlaw cyclists have no constraints except those their club mandates. When a biker spoke of something being 'legal,' he was referring to the bylaws of his club rather than to the laws of a state or nation. A biker's 'legal' name was his club name that was usually inscribed on his jacket or 'color'. Club names were typically one word, and this was how other members and female associates referred to a biker. Such descriptive names as Trench Mouth, Grimy, Animal, Spooky, and Red sufficed for most bikers we studied. As we knew them, bikers lived virtually a tribal life-style with few restraints. The freedom they enjoyed was not simply being 'in the wind'; it was also emotional. Whereas conventional people fear going to prison, the bikers were confident that they had many brothers who would look out for them inside the walls. Consequently, the threat of confinement had little influence on a biker's behavior, as far as we could tell.

Perhaps because society gave them so little respect, the bikers we studied insisted on being treated with deference. They gave few invitations to non-members or 'citizens' and they were affronted when something they offered was refused. Our respondents loved to party and they did not understand anyone who did not. Once we were invited to a club party by a man named Cottonmouth. The party was to begin at 9:00 p.m. on a Sunday night. When we told Cottonmouth that we had to leave at seven in the evening to get back home, we lost his good will and respect entirely. He could not comprehend how we could let anything take precedence over a 'righteous' club party.

Bikers were suspicious of all conversations with us and with other citizens; they were not given to much discussion even among themselves. They followed a slogan we saw posted in several clubhouses: 'One good fist is worth a thousand words.' Studying outlaw cyclists became more difficult rather than easier over the course of our study. They grew increasingly concerned about being investigated by undercover agents. In 1989, '1%'[3] of bikers *never* trusted anyone except their own kind. At times, over the last years of our study, respondents whom we had known for months would suddenly accuse us of being undercover 'pigs' when we seemed overly curious about their activities.

Our study required much commitment to research goals. We believed it was important to study biker women and we did so in the only way open to us – on the terms of the bikers themselves. We were field observers rather than critics or reformers, even when witnessing things that caused us anguish.

Problems in studying biker women

Although it was difficult to do research on outlaw motorcycle gangs generally, it was even harder to study the women in them. In many gangs, the women were reluctant to speak to outsiders when the men were present. We did not hear male bikers tell the women to refrain from talking to us. Rather, we often had a man point to a woman and say, 'Ask her,' when we posed a question that concerned female associates. Usually, the woman's answer was, 'I don't know.' Conse-quently, it took longer to establish rapport with female bikers than it did with the men.

Surprisingly, male bikers did not object to our being alone with the women. Occasionally, we talked to a female biker by ourselves and this is when we were able to get most of our information and quotations from them. In one interview with a biker and his woman in their home, the woman would not express an opinion about anything. When her man left to help a fellow biker whose motorcycle had broken down on the road, the woman turned into an articulate and intelligent individual. Upon the return of the man, however, she resumed the role of a person without opinions.

The place of women in outlaw motorcycle gangs

Although national[4] outlaw motorcycle clubs of the 1980s had restricted their membership to adult males (Quinn, 1983), women were important in the outlaw life-style we observed. We rarely saw a gang without female associates sporting colors similar to those the men wore.

To the casual observer, all motorcycle gang women might have appeared the same. There were, however, two important categories of women in the biker world: 'mamas' and 'old ladies.' A mama belonged to the entire gang. She had to be available for sex with any member and she was subject to the authority of any brother. Mamas wore jackets that showed they were the 'property' of the club as a whole.

An old lady belonged to an individual man; the jacket she wore indicated whose woman she was. Her colors said, for example, 'Property of Frog'. Such a woman was commonly referred to as a 'patched old lady.' In general terms, old ladies were regarded as wives. Some were in fact married to the members whose patches they wore. In most instances, a male biker and his old lady were married only in the eyes of the club. Consequently, a man could terminate his relationship with an old lady at any time he chose, and some men had more than one old lady.

A man could require his old lady to prostitute herself for him. He could also order her to have sex with anyone he designated. Under no circumstances, however, could an old lady have sex with anyone else unless she had her old man's permission.

If he wished to, a biker could sell his old lady to the highest bidder, and we saw this happen. When a woman was auctioned off, it was usually because a biker needed money in a hurry, such as when he wanted a part for his motorcycle, or because his old lady had disappointed him. The buyer in such transactions was usually another outlaw.

Rituals involving women

Outlaw motorcycle gangs, as we perceived them, formed a subculture that involved rituals and symbols. Although each group varied in its specific cere-monies, all of the clubs we studied had several. There were rites among bikers that had nothing to do with women and sex but a surprising number involved both.

The first ritual many outlaws were exposed to, and one they understandably never forgot, was the initiation into a club. Along with other requirements, in some gangs, the initiate had to bring a 'sheep' when he was presented for membership. A sheep was a woman who had sex with each member of the gang during an initiation. In effect, the sheep was the new man's gift to the old members.

Group sex, known as 'pulling a train,' also occurred at other times. Although some mamas or other biker groupies (sometimes called 'sweetbutts') occasionally volunteered to pull a train, most instances of train pulling were punitive in nature. Typically, women were being penalized for some breach of biker conduct when they pulled a train.

An old lady could be forced to pull a train if she did not do something her old man told her to do, or if she embarrassed him by talking back to him in front of another member. We never observed anyone pulling a train but we were shown clubhouse rooms that were designated 'train rooms'. And two women told us they had been punished in this manner.

One of the old ladies who admitted having pulled a train said her offense was failing to keep her man's motorcycle clean. The other had not noticed that her biker was holding an empty bottle at a party. (A good old lady watched her man as he drank beer and got him another one when he needed it without having to be told to do so.) We learned that trains were pulled in vaginal, oral, or anal sex. The last was considered to be the harshest punishment.

Another biker ritual involving women was the earning of 'wings,' a patch similar to the emblem a pilot wears. There were different types of wings that showed that the wearer had performed oral sex on a woman in front of his club. Although the practice did not exist widely, several members of some groups we studied wore wings.

A biker's wings demonstrated unlimited commitment to his club. One man told us he earned his wings by having oral sex with a woman immediately after she had pulled a train; he indicated that the brothers were impressed with his abandon and indifference to hygiene. Bikers honored a member who laughed at danger by doing shocking things.[5]

The sex rituals were important in many biker groups because they served at least one function other than status striving among members. The acts ensured that it was difficult for law enforcement officials, male or female, to infiltrate a gang.

Biker women as money-makers

Among most of the groups we studied, biker women were expected to be engaged in economic pursuits for their individual men and sometimes for the entire club. Many of the old ladies and mamas were employed in nightclubs as topless and nude dancers. Although we were not able to get exact figures on the proportion of 'table dancers' who were biker women, in two or three cities almost all of them were working for outlaw clubs.

A lot of the dancers were proud of their bodies and their dancing abilities. We saw them perform their routines in bars and at parties. At the 'Big Blowout' in Gulfport, which is held in an open field outside of the city, in 1987 and 1988 there was a stage with a sound system set up for the dancers. The great majority of the 2,000 people in attendance were bikers from around the country so the performances were free.

Motorcycle women who danced in the nightclubs we observed remained under the close scrutiny of the biker men. The men watched over them for two reasons.

First, they wanted to make sure that the women were not keeping money on the side; second, the cyclists did not want their women to be exploited by the bar owners. Some bikers in one gang we knew beat up a nightclub owner because they thought he was 'ripping off' the dancers. The man was beaten so severely with axe handles that he had to be hospitalized for several months.

While some of the biker women limited their nightclub activities to dancing, a number of them also let the customers whose tables they danced on know they were available for 'personal' sessions in a private place. As long as they were making good money regularly, the bikers let the old ladies choose their own level of nightclub participation. Thus some women danced nude only on stage; others performed on stage and did table dances as well. A smaller number did both types of dances and also served as prostitutes.

Not all of the money-making biker women we encountered were employed in such 'sleazy' occupations. A few had 'square' jobs as secretaries, factory workers, and sales persons. One biker woman had a job in a bank. A friend and fellow biker lady described her as follows: 'Karen is a chameleon. When she goes to work, she is a fashion plate; when she is at home, she looks like a whore. She is every man's dream!' Like the others employed in less prestigious labor, however, Karen turned her salary over to her old man on payday.

A few individuals toiled only intermittently when their bikers wanted a new motorcycle or something else that required more money than they usually needed. The majority of motorcycle women we studied, however, were regularly engaged in work of some sort.

Motivations and backgrounds of biker women

In view of the ill treatment the women received from outlaws, it was surprising that so many women wanted to be with them. Bikers told us there was never a shortage of women who wanted to join them and we observed this to be true. Although it was unwise for men to draw conclusions about the reasons mamas and old ladies chose their life-styles, we surmised three interrelated factors from conversations with them.

First, some women, like the male bikers, truly loved and were excited by motorcycles. Cathy was an old lady who exhibited this trait. 'Motorcycles have always turned me on,' she said. 'There's nothing like feeling the wind on your titties. Nothing's as exciting as riding a motorcycle. You feel as free as the wind.'

Cathy did not love motorcycles indiscriminately, however. She was imbued with the outlaw's love for the Harley Davidson. 'If you don't ride a Hog,' she stated, 'you don't ride nothing. I wouldn't be seen dead on a rice burner' (Japanese model). Actually, she loved only a customized bike or 'chopper.' Anything else she called a 'garbage wagon.'

When we asked her why she wanted to be part of a gang if she simply loved motorcycles, Cathy answered:

> There's always someone there. You don't agree with society so you find someone you
> like who agrees with you. The true meaning for me is to express my individuality as
> part of a group.

Cathy started 'putting' (riding a motorcycle) when she was 15 years old and she dropped out of school shortly thereafter. Even with a limited education, she gave the impression that she was a person who thought seriously. She had a butterfly tattoo that she said was an emblem of the freedom she felt on a bike. When we talked to her, she was 26 and had a daughter. She had ridden with several gangs but she was proud that she had always been an old lady rather than a mama.

The love for motorcycles had not dimmed for Cathy over the years. She still found excitement in riding and even in polishing a chopper. 'I don't feel like I'm being used. I'm having fun,' she insisted. She told us that she would like to change some things if she had her life to live over, but not biking. 'I feel sorry for other people; I'm doing exactly what I want to do,' she concluded.

A mama named Pamela said motorcycles thrilled her more than anything else she had encountered in life. Although she had been involved with four biker clubs in different sections of the country, she was originally from Mississippi and she was with a Mississippi gang when we talked to her. Pamela said she graduated from high school only because the teachers wanted to get rid of her. 'I tried not to give any trouble, but my mind just wasn't on school.'

She was 24 when we saw her. Her family background was a lot like most of the women we knew. 'I got beat a lot,' she remarked. 'My daddy and my mom both drank and ran around on each other. They split up for good my last year in school. I ain't seen either of them for a long time.'

Cathy described her feelings about motorcycles as follows:

I can't remember when I first saw one. It seems like I dreamed about them even when I was a kid. It's hard to describe why I like bikes. But I know this for sure. The sound a motorcycle makes is really exciting – it turns me on, no joke. I mean really! I feel great when I'm on one. There's no past, no future, no trouble. I wish I could ride one and never get off.

The second thing we thought drew women to motorcycle gangs was a preference for macho men. 'All real men ride Harleys,' a mama explained to us. Generally, biker women had contempt for men who wore suits and ties. We believed it was the disarming boldness of bikers that attracted many women.

Barbara, who was a biker woman for several years, was employed as a secretary in a university when we talked to her in 1988. Although Barbara gradually withdrew from biker life because she had a daughter she wanted reared in a more conventional way, she thought the university men she associated with were wimps. She said:

Compared to bikers, the guys around here (her university) have no balls at all. They hem and haw, they whine and complain. They try to impress you with their intelligence and sensitivity. They are game players. Bikers come at you head on. If they want to fuck you, they just say so. They don't care what you think of them. I'm attracted to strong men who know what they want. Bikers are authentic. With them, what you see is what you get.

Barbara was an unusual biker lady who came from an affluent family. She was the daughter of a highly successful man who owned a manufacturing and distributing company. Barbara was 39 when we interviewed her. She had gotten into a motorcycle gang at the age of 23. She described her early years to us:

> I was rebellious as long as I can remember. It's not that I hated my folks. Maybe it was the times (1960s) or something. But I just never could be the way I was expected to be. I dated 'greasers', I made bad grades; I never applied myself. I've always liked my men rough. I don't mean I like to be beat up, but a real man. Bikers are like cowboys; I classify them together. Freedom and strength I guess are what it takes for me.

Barbara did not have anything bad to say about bikers. She still kept in touch with a few of her friends in her old club. 'It was like a family to me,' she said. 'You could always depend on somebody if anything happened. I still trust bikers more than any other people I know.' She also had become somewhat reconciled with her parents, largely because of her daughter. 'I don't want anything my parents have personally, but my daughter is another person. I don't want to make her be just like me if she doesn't want to,' she concluded.

A third factor that we thought made women associate with biker gangs was low self-esteem. Many we studied believed they deserved to be treated as people of little worth. Their family backgrounds had prepared them for subservience.

Jeanette, an Arkansas biker woman, related her experience as follows:

> My mother spanked me frequently. My father beat me. There was no sexual abuse but a lot of violence. My parents were both alcoholics. They really hated me. I never got a kind word from either of them. They told me a thousand times I was nothing but a pain in the ass.

Jeanette began hanging out with bikers when she left home at the age of 15. She was 25 when we talked to her in 1985. Although he was dominating and abusive, her old man represented security and stability for Jeanette. She said he had broken her jaw with a punch. 'He straightened me out that time,' she said. 'I started to talk back to him but I didn't get three words out of my mouth.' Her old man's name was tattooed over her heart.

In Jeanette's opinion, she had a duty to obey and honor her man. They had been married by another biker who was a Universal Life minister. 'The Bible tells me to be obedient to my husband,' she seriously remarked to us. Jeanette also told us she hated lesbians. 'I go in lesbian bars and kick ass,' she said. She admitted she had performed lesbian acts but she said she did so only when her old man made her do them. The time her man broke her jaw was when she objected to being ordered to sleep with a woman who was dirty. Jeanette believed her biker had really grown to love her. 'I can express my opinion once and then he decides what I am going to do,' she concluded.

In the opinions of the women we talked to, a strong man kept a woman in line. Most old ladies had the lowly task of cleaning and polishing a motorcycle every day. They did so without thanks and they did not expect or want any praise. To them, consideration for others was a sign of weakness in man. They wanted a man to let them know who was boss.

Motorcycle women versus street gang girls

The motorcycle women in our study were similar to the street gang girls described by Campbell and Horowitz because their lives were built around deviant social organizations that were controlled by members of the opposite sex. There were, however, important differences that resulted from the varying natures of the two subcultures.

As our terminology suggests, female associates of motorcycle gangs were women as opposed to the teenage girls typically found in street gangs. The biker women who would tell us their age averaged 26 years, and the great majority appeared to be in their mid-20s. While some biker women told us they began associating with outlaws when they were teenagers, we did not observe any young girls in the clubs other than the children of members.

Male bikers were older than the members of street gangs and it followed that their female companions were older as well. In one of the outlaw clubs we surveyed, the men averaged 34 years old. Biker men also wanted women old enough to be legally able to work in bars and in other jobs.

All of the biker women we studied were white, whereas street gang girls in previous studies were predominantly from minority groups. We were aware of one black motorcycle gang in Memphis but we were unable to make contact with it.[6]

Biker women were not homogeneous in their backgrounds. While street gangs were composed of 'home boys' and 'home girls' who usually grew up and remained in the areas in which their gangs operated, the outlaw women had often traveled widely. Since bikers were mobile, it was rare for us to find a women who had not moved around a lot. Most of the biker women we saw were also high school graduates. Two had attended college although neither had earned a degree.

While Campbell found girls in street gangs to be interested in brand name clothes and fashions, we did not notice this among motorcycle women. In fact, it was our impression that biker ladies were hostile toward such interests. Perhaps because so many were dancers, they were proud of their bodies but they did not try to fit into popular feminine dress styles. As teenagers they may have been clothes-conscious, but as adults biker women did not want to follow the lead of society's trend setters.

Biker women were much like street gang girls when it came to patriotism. They were proud to be Americans. Like biker men, they had conservative political beliefs and would not even consider riding a motorcycle made in another country.

As another consequence of the age difference, biker women were not torn between their families and the gang. Almost all of the old ladies and mamas were happy to be rid of their past lives. They had made a clean break and they did not try to live in two worlds. The motorcycle gang was their focal point without rival. Whereas street gang girls often left their children with their mothers or grandparents, biker women did not, but they wanted to be good mothers just the same. The children of biker women were more integrated into the gang. Children went with their mothers on camping trips and on brief motorcycle excursions or 'runs.'

When it was necessary to leave the children at home, two or three old ladies alternately remained behind and looked after all of the children in the gang.

The biker men were also concerned about the children and handled them with tenderness. A biker club considered the offspring of members as belonging to the entire group, and each person felt a duty to protect them. Both male and female bikers also gave special treatment to pregnant women. A veteran biker woman related her experience to us as follows:

> Kids are sacred in a motorcycle club. When I was pregnant, I was treated great. Biker kids are tough but they are obedient and get lots of love. I've never seen a biker's kid who was abused.

As mentioned, the average biker woman was expected to be economically productive, a trait not emphasized for female street gang members or auxiliaries. It appeared to us that the women in motorcycle gangs were more thoroughly under the domination of their male associates than were girls described in street gang studies.

The changing role of biker women

During the 17 years of our study, we noticed a change in the position of women in motorcycle gangs. In the groups we observed in the 1960s, the female participants were more spontaneous in their sexual encounters and they interacted more completely in club activities of all kinds. To be sure, female associates of outlaw motorcycle gangs have never been on a par with the men. Biker women have worn 'property' jackets for a long time, but in the outlaw scene of 1989, the label had almost literally become fact.

Bikers have traditionally been notoriously active sexually with the women in the clubs. When we began hanging out with bikers, however, the men and the women were more nearly equal in their search for gratification. Sex was initiated as much by the women as it was by the men. By the end of our study, the men had taken total control of sexual behavior, as far as we could observe, at parties and outings. As the male bikers gained control of sex, it became more ceremonial.

While the biker men we studied in the late 1980s did not have much understanding of sex rituals, their erotic activities seemed to be a means to an end rather than an end in themselves, as they were in the early years of our study. That is to say, biker sex became more concerned with achieving status and brotherhood than with 'fun' and physical gratification. We used to hear biker women telling jokes about sex but even this had stopped.

The shift in the position of biker women was not only due to the increasing ritualism in sex; it was also a consequence of the changes in the organizational goals of motorcycle gangs as evidenced by their evolving activities. As we have noted, many motorcycle gangs developed an interest in money; in doing so, they became complex organizations with both legal and illegal sources of income (McGuire, 1986).

When bikers became more involved in illegal behavior, they followed the principles of sex segregation and sex typing in the underworld generally. The low place of women has been well documented in the studies of criminal organizations (Steffensmeier, 1983). The bikers did not have much choice in the matter. When they got involved in financial dealings with other groups in the rackets, motorcycle gangs had to adopt a code that had prevailed for many years; they had to keep women out of 'the business.'

Early motorcycle gangs were organized for excitement and adventure; money-making was not important. Their illegal experiences were limited to individual members rather than to the gang as a whole. In the original gangs, most male participants had regular jobs, and the gang was a part-time organization that met about once a week. At the weekly gatherings, the emphasis was on swilling beer, soaking each other in suds, and having sex with the willing female associates who were enthusiastic revelers themselves. The only money the old bikers wanted was just enough to keep the beer flowing. They did not regard biker women as sources of income; they thought of them simply as fellow hedonists.

Most of the gangs we studied in the 1980s required practically all of the members' time. They were led by intelligent presidents who had organizational ability. One gang president had been a military officer for several years. He worked out in a gym regularly and did not smoke or drink excessively. In his presence, we got the impression that he was in control, that he led a disciplined life. In contrast, when we began our study, the bikers, including the leaders, always seemed on the verge of personal disaster.

A few motorcycle gangs we encountered were prosperous. They owned land and businesses that had to be managed. In the biker transition from hedonistic to economic interests, women became defined as money-makers rather than companions. Whereas bikers used to like for their women to be tattooed, many we met in 1988 and 1989 did not want their old ladies to have tattoos because they reduced their market value as nude dancers and prostitutes. We also heard a lot of talk about biker women not being allowed to use drugs for the same reason. Even for the men, some said drug usage was not good because a person hooked on drugs would be loyal to the drug, not to the gang.

When we asked bikers if women had lost status in the clubs over the years, their answers were usually negative. 'How can you lose something you never had?' a Florida biker replied when we queried him. The fact is, however, that most bikers in 1989 did not know much about the gangs of 20 years earlier. Furthermore, the change was not so much in treatment as it was in power. It was a sociological change rather than a physical one. In some respects, women were treated better physically after the transition than they were in the old days. The new breed did not want to damage the 'merchandise.'

An old lady's status in a gang of the 1960s was an individual thing, depending on her relationship with her man. If her old man wanted to, he could share his position to a limited extent with his woman. Thus the place of women within a gang was variable. While all women were considered inferior to all men, individual females often gained access to some power, or at least they knew details of what was happening.

By 1989, the position of women had solidified. A woman's position was no longer influenced by idiosyncratic factors. Women had been formally defined as inferior. In many biker club weddings, for example, the following became part of the ceremony:

> You are an inferior woman being married to a superior man. Neither you nor any of your female children can ever hold membership in this club or own any of its property.

Although the bikers would not admit that their attitudes toward women had shifted over the years, we noticed the change. Biker women were completely dominated and controlled as our study moved into the late 1980s. When we were talking to a biker after a club funeral in North Carolina in 1988, he turned to his woman and said, 'Bitch, if you don't take my dick out, I'm going to piss in my pants.' Without hesitation, the woman unzipped his trousers and helped him relieve himself. To us, this symbolized the lowly place of women in the modern motorcycle gang.

Conclusion

Biker women seemed to represent another version of what Romenesko and Miller (1989) have referred to as a 'double jeopardy' among female street hustlers. Like the street prostitutes, most biker women came from backgrounds in which they had limited opportunities in the licit or conventional world, and they faced even more exploitation and subjugation in the illicit or deviant settings they had entered in search of freedom.

It is ironic that biker women considered themselves free while they were under the domination of biker men. They had the illusion of freedom because they lived with men who were bold and unrestrained. Unlike truly liberated women, however, the old ladies and mamas did not compete with men; instead, they emulated and glorified male bikers. Biker women thus illustrated the pervasive power of socialization and the difficulty of changing deeply ingrained views of the relations between the sexes inculcated in their family life. They believed that they should be submissive to men because they were taught that males were dominant. While they adamantly stated that they were living the life they chose, it was evident that their choices were guided by values that they had acquired in childhood. Although they had rebelled against the strictures of straight society, their orientation in gender roles made them align with outlaw bikers, the epitome of macho men.

Notes

1 We briefly observed the Alii ('Chiefs'), a native Hawaiian gang, located on the island of Hawaii, the 'Big Island', Motorcycle gangs on the island of Oahu may have developed before the California clubs. Lord (1978) described a club that volunteered its services during the attack on Pearl Harbor in 1941. The bikers, wearing their colors, carried messengers and officers on motorcycles from one place to another because automobiles could not move efficiently during the traffic jams caused by the battle. This display of patriotism was not the only instance. Sonny Barger, the president of the Oakland chapter, sent President Lyndon Johnson a letter offering to send the Hell's

Angels to Vietnam. Some outlaws were active in 'Toys for Tots' drives and blood drives. These activities were seldom noted, which led bikers to the slogan: 'When we do right, nobody remembers; when we do wrong, nobody forgets'.
2 Bikers presented 'courtesy cards' to people that they believed deserved the privilege of biker acceptance. 'Big John', as Moore was called by outlaws, had cards from outlaw motorcycle clubs throughout the country.
3 Some years ago, an officer of the American Motorcycle Association, in defense of motorcycling, stated that 99% of all motorcyclists were decent and honorable people and that only 1% gave the rest of a bad name. Since that proclamation, outlaws have proudly worn the 1% patch.
4 A national club had chapters in different regions of the country. For more details on gang vocabulary and argot and the distribution of specific gangs, see Hopper and Moore (1983, 1984).
5 Although the initiation ceremony was the culmination of a biker's efforts to become a member of a club, it usually required a period of a year or more before a man would be made a member in full standing. During this time, a person was a 'probate.' He rode with the gang, and to the public he appeared to be a regular member, but a probate was not trusted until he proved himself worthy of complete membership. He did this by showing his courage and disregard for danger.
6 Motorcycle gangs in the 1980s emphasized race. White gangs often were members of the Aryan Brotherhood. Earlier bikers were sometimes racially integrated. They developed their racial division after many bikers got into prisons where they adjusted to intraracial groups that banded together for protection. Being white, we were unable to develop relationships with a black group.

References

Ackley, E. and B. Fliegel (1960) 'A social work approach to street-corner girls'. *Social Problems* **5**, 29–31.

Asbury, H. (1928) *The Gangs of New York*. New York: Alfred A. Knopf.

Bowker, L. (1978) *Women, Crime, and the Criminal Justice System*. Lexington, MA: D. C. Heath.

Bowker, L. and M. Klein (1983): 'The etiology of female juvenile delinquency and gang membership: a test of psychological and social structural explanation'. *Adolescence* **8**, 731–751.

Campbell, A. (1984) *The Girls in the Gang*. New York: Basil Blackwell.

Campbell, A. (1986) 'Self report of fighting by females.' *British J. of Criminology* **26**, 28–46.

Campbell, A. (1987) 'Self-definition by rejection: the case of gang girls'. *Social Problems* **34**, 451–466.

Eisen, J. (1970) *Altamont*. New York: Avon Books.

Hanson, K. (1964) *Rebels in the Streets*. Englewood Cliffs, NJ: Prentice-Hall.

Harris, M. (1985) *Bikers*. London: Faber & Faber.

Hopper, C. and J. Moore (1983) 'Hell on wheels: the outlaw motorcycle gangs.' *J. of Amer. Culture* **6**, 58–64.

Hopper, C. and J. Moore (1984) 'Gang slang'. *Harpers* **261**, 34.

Horowitz, R. (1983) *Honor and the American Dream*. New Brunswick, NJ: Rutgers Univ. Press.

Horowitz, R. (1986) 'Remaining an outsider: membership as a threat to research rapport'. *Urban Life* **14**, 238–251.

Horowitz, R. (1987) 'Community tolerance of gang violence'. *Social Problems* **34**, 437–450.

Klein, M. (1971) *Street Gangs and Street Workers*. Englewood Cliffs, NJ: Prentice-Hall.

Lord, W. (1978) *Day of Infamy*. New York: Bantam.

McGuire, P. (1986) 'Outlaw motorcycle gangs: organized crime on wheels'. *National Sheriff* **38**, 68–75.

Miller, W. (1973) 'Race, sex and gangs'. *Society* **11**, 32–35.

Miller, W. (1975) *Violence by Youth Gangs and Youth Groups as a Crime Problem in Major American Cities*. Washington, DC: Government Printing Office.

Montegomery, R. (1976) 'The outlaw motorcycle subculture'. *Canadian J. of Criminology and Corrections* **18**, 332–342.

Morgan, R. (1978) *The Angels Do Not Forget*. San Diego: Law and Justice.

Quinn, J. (1983) *Outlaw Motorcycle Clubs: A Sociological Analysis*. M.A. thesis: University of Miami.

Reynolds, F. (1967) *Freewheeling Frank*. New York: Grove Press.

Rice, R. (1963) 'A reporter at large: the Persian queens'. *New Yorker* **39**, 153.

Romenesko, K. and E. Miller (1989) 'The second step in double jeopardy: appropriating the labor of female street hustlers'. *Crime and Delinquency* **35**, 109–135.

Saxon, K. (1972) *Wheels of Rage* (privately published).

Short, J. (1968) *Gang Delinquency and Delinquent Subcultures*. Chicago: Univ. of Chicago Press.

Steffensmeier, D. (1983) 'Organization properties and sex-segregation in the underworld: building a sociology theory of sex differences in crime'. *Social Forces* **61**, 1010–1032.

Thompson, H. (1967) *Hell's Angels*. New York: Random House.

Thrasher, F. (1927) *The Gang: A Study of 1,313 Gangs in Chicago*. Chicago: Univ. of Chicago Press.

Watson, J. (1980) 'Outlaw motorcyclists as an outgrowth of lower class values'. *Deviant Behavior* **4**, 31–48.

Wilde, S. (1977) *Barbarians on Wheels*. Secaucus, NJ: Chartwell Books.

Willis, P. (1978) *Profane Culture*. London: Routledge & Kegan Paul.

Wolfe, T. (1968) *The Electric Kool-Aid Acid Test*. New York: Farrar, Straus & Giroux.

Mods, Rockers and Turf Gangs: Carnivals of Violence

Excerpt from: *Places on the margin*, pp. 101–5. London:
Routledge (1991)

Brighton's reputation as a town which accommodated both the wealthy and the poor, both the upright industrial bourgeoisie and the prostitutes and hucksters living by their wits, contributed to a lasting aura of petit-criminality. Inland from the smart Crescents and Parades were narrow terraces where slum-conditions reigned. Some local historians argue that this contrast is a persistent theme in the historiography of Brighton.[1] During the First World War, the impossibility of visiting the Continent had stimulated the Brighton hotel industry. A considerable number of Londoners took refuge from Zeppelin-raids in Brighton and the large Regency houses which had begun to go out of fashion before the war were again easily sold or let (Gilbert 1954: 219). But up-turns in the local economy continued to be accompanied by a seedy underside and the striking contrast the poor provided to the very wealthy.

In the inter-war years, Brighton's race-tracks and gambling establishments became the haunt of rival 'turf gangs', who feuded at the race-track and on the front, slashing their victims and enemies with razors. Extortion, loan-sharking, and protection rackets persisted into the 1930s, giving Brighton an unpleasant reputation as a 'nice place to visit' on a day-trip but not a nice environment to live in. It became dangerous to walk on the sea-front with assaults taking place even in broad daylight (*Southern Weekly News* 26 May 1928). These gangs reached their apogee in June 1936 when a gang of thirty men, the 'Hoxton Mob', attacked a bookmaker and his clerk but were detained after a mêlée with the police, who had anticipated violence. Graham Greene's *Brighton Rock* (1936) immortalised this period of gang violence. His Pinkie's gang was an invention and *Brighton Rock* fiction, but it was closely based on the actual reports of track violence. Apart from the slashings, the bizarre 'Trunk Murders' added to Brighton's unsavoury reputation. In one famous case, a man murdered and dismembered his wife, and sent different parts by rail to different destinations as baggage (see Lustgarten 1951: 187–238).

But Brighton had long been known as a 'miniature Marseilles' (Lustgarten 1951: 188). The Trunk Murders of 1934 were not the first of their kind, the classic case having also taken place in Brighton in 1831 (see Hindley 1875 and Lina Wertmuller's *Seven Beauties* movie version, set in Italy). Lowerson and Howkins argue that Brighton attracted 'rough cultures', a vertical banding of culture in contrast to the horizontal banding of socio-economic classes (1981: 72). However, this poses the problem of why Brighton with its liminal and carnivalesque image attracted these groups. On the one hand, the loosening of restraints on violence is a constitutive part of the carnivalesque, exaggerating violent tendencies that might have emerged anywhere. On the other hand, the scene of relatively wealthy people, with a less cautious hand on their wallets, must have been irresistible to con artists and small-time hucksters. With the structuring elements of everyday life removed or destabilised and the primacy of enjoyment and adventurousness in Brighton, the bases on which judgements could be formed were eroded: people would spend more, more impulsively and take more risk.

The status of Brighton as a liminal zone made any and all rumours of transgression, decadence, crime, and degeneration the basis for sensational newspaper reporting, which formed a staple diet of 'Brighton stories' which have circulated in the British press for almost two centuries. These reports served to restate and confirm the spatialisation of Brighton, on the margins of the orderly sphere of 'good governance' which reigned over other parts of the nation. The 'Brighton Rock' criminal image coexisted with the place-images of decadent, risqué and glamorous Brighton, reinforcing Brighton's liminal place-myth.

The beach fights between groups of 'Mods' and 'Rockers' of the early 1960s are another example of the liminal breakdown of social order (see Figure 11.1). From the late eighteenth century onwards, Brighton had continued to be a destination for anyone searching for escape: a liminal zone and social periphery in a marginal geographic location 'separated-off by the South Downs' (Brighton Tourism Committee 1954: 8).[2] Mods (from the bebop term 'Modernists') originated in the suburbs of London. Youth from a largely working-class but white-collar background attempted to abstract themselves from traditional class identities. Upward mobility was indicated with a neat, hip image adopted from the dandyism of black Harlem and European jazz artists. Stylistically the culture of the Mods was largely masculine in its trademarks. These included, for the men, suits with narrow trousers and pointed shoes, anoraks for scootering; and for the women, short hair and a cultivated, dead-pan elegance (Brake 1985: 75). Rockers, who eschewed the fashionable snobbery of the Mods (Nuttall 1969: 333), presented their alter-ego opposite: class-bound and masculine, low-paid, unskilled manual workers (Barker and Little 1964). Motorcycles became Rockers' symbols of freedom from authority, of mastery and intimidation (Willis 1978), while Mods glamorised accessory-bedecked Italian scooters. Clubs allowed the Mods a glamorous dream world (being in a sense proto-discos) in which class background could be rejected.

The two groups came together on Bank Holidays in the established motoring destination of Brighton (from London, literally the end of the road) with its clubs, glamorous past and reputation for freedom from moral and class restraint. From London, there was no more appropriate Bank Holiday weekend destination, for none of the other seaside resorts in reach of London shared this combination of images. In fact, Mods and Rockers clashed at Brighton on only two occasions (the May and the August Bank Holidays, 1964). In the original news photographs we see scenes of, on the one hand, youths being kicked by assailants, deck chairs been thrown and, on the other hand, grins on the faces of some of the participants in the mêlée (see Figure 11.1). The 'riots' were boisterous and violent but the bitterness one associates with rioting is missing. After the media hysteria, those smiles leave the different impression of a boisterous carnival of violence rather than planned attacks by groups of bitter criminal enemies, or the frustration of

Figure 11.1 Beach fights between Mods and Rockers.
Source: *Brighton and Hove Argus*, East Sussex County Library.

disenfranchised ethnic groups or the poor. 'Media coverage of a small amount of damage and violence on British seaside beaches on a rather dismal national holiday led to a situation of deviancy amplification' (Brake 1985: 64). Only after the notoriety of being media 'folk devils', was there a conscious embracing of the two deviant roles by large numbers of youth. The 'indiscriminate prosecution, local overreaction and media stereotyping' created a 'moral panic' (Cohen 1972) and implied a type of conspiracy: 'the solidifying of amorphous groups of teenagers into some sort of conspiratorial collectivity, which had no concrete existence' (Brake 1985: 64). On Brighton beach, Mods and Rockers became visible and socially identifiable groups. The combination of alcohol, drugs, and the release from the restraints of everyday domestic surroundings combined to make the beach an appropriate and available stage for an explosion of the tensions between the two groups.

Notes

1 This argument has been advanced by my colleague at Sussex University, Kevin Meetham.
2 See also the movie *Mona Lisa* (1987) where the charact scape from London gangs to Brighton where they are caught up with and where seduction, betrayal, and show-down takes place.

References

Barker, P. and Little, A. 1964. 'The Margate offenders – a survey' in *New Society* 4: 96 (July) 189–192.
Brake, M. 1985. *Comparative Youth Culture* (London: Routledge & Kegan Paul).

Brighton Tourism Committee 1937–1968. *Official Guidebook to Brighton* Yearly Pamphlet (Brighton: Brighton Council).

Cohen, S. 1972. *Folk Devils and Moral Panics* (London: MacGibbon & Kee).

Gilbert, E. W. 1954. *Brighton: Old Ocean's Bauble* (London: Methuen).

Greene, G. 1936. *Brighton Rock* (London: Heinemann).

Hindley, C. 1875. *The Brighton Murder. An Authentic and Faithful History of the Atrocious Murder of Celia Holloway* (no publisher).

Lowerson, J. and Howkins, A. 1981. 'Leisure in the Thirties' in Tomlinson 1981. 72–85.

Lustgarten, E. 1951. *Defender's Triumph* (London: no publisher).

Southern Weekly News 26 May 1928. 'Local News' column (Brighton).

Willis, P. 1978. *Profane Culture* (London: Routledge & Kegan Paul).

12   George McKay
O Life Unlike to Ours! Go for It! New Age Travellers

Excerpts from: *Senseless acts of beauty*, pp. 45–71. London: Verso (1995)

They call themselves new age travellers . . . we call them new age vermin.

Paul Marland MP

By the end of the 1970s a regular summer circuit had been established. From May Hill at the beginning of May via Horseshoe Pass, Stonehenge, Ashton Court, Inglestone Common, Cantlin Stone, Deeply Vale, Meigan Fair, and various sites in East Anglia, to the Psilocybin Fair in mid-Wales in September, it was possible to find a free festival or a cheap community festival almost every weekend.

Young people from traditional travelling families began to come into the festival scene and people from the cities began to convert vehicles and live on the road . . . [T]he habit of travelling in convoy caught on. . . . So the New Traveller culture was born, emerging into public view at Inglestone Common in 1980 with the 'New Age Gypsy Fair'.

Festival welfare worker Don Aitken[1]

Clearly, one way of explaining New Age travellers is to identify them as a product of times of domestic crisis and increasing legislation in Britain: for example, during the mid-seventies the legal procedures for evicting squatters were eased, forcing more people onto the roads, and squatting itself was made even more difficult under the 1977 Criminal Law Act.[2] Long-standing squatting communities, as in areas of Hackney in London, or Argyle Street in Norwich until early 1985, often extended to include truck- and bus-lined streets where travellers park up for the winter – literally parallel communities of houses and vehicles. The boom in urban squatting from the late sixties was itself a response to an ongoing housing crisis and, like squatters, travellers produce their own solutions to homelessness when few others are offered by the government.[3] When Tory minister Norman Tebbit suggested in the eighties that the unemployed should 'get on their bikes' to look for work as his own father had done during an earlier period of mass unemployment, little did he realize the swelling of nomadic lifestyles that was actually taking place in Britain.[4] Ironically, in terms of the economy, Jay points out that he and other travellers are

> probably the least affected people in the whole country by recession. I mean we live in a permanent recession anyway so it doesn't make a lot of difference. . . . Everything we buy apart from food and petrol just about is second-hand or traded, the whole black market economy.[5]

The New Age travellers of the nineties commonly describe themselves as 'economic refugees' . . . 'refugees seeking shelter. That's how I see myself, as a refugee.' Sociologist Kevin Hetherington mentions that 'there are even some [New Age travellers] who have taken up living in caves in Spain'. But despite such perceptions of neo-barbarian difference, the actions of the New Age travellers are not historically unfamiliar, even in recent history – for instance, during the Second World War Londoners and Bristolians fleeing aerial bombardment damage lived in nearby caves.[6]

Despite being sprayed with silage by a farmer; despite putting up with 'fuck-off looks' as well as far more physical manifestations of hostility; despite being harassed for 'producers' by police, that is, to produce vehicle documents at a local station within seven days; despite suffering periodic attacks of what one calls 'straight sickness' – why aren't I in a house, with a flushing loo, running water, a telly, pubs and cinemas nearby? – despite all that it's not all negative. As well as rejection and exclusion, travellers emphasize the attraction of the New Age traveller lifestyle, as in the Romantic pleasure of rural living. While 'rural Britain is for the rich', says traveller Shannon, 'if I want to live with space around me and trees and hills and woods, the only possible way apart from sleeping out is to buy a vehicle and live in that'. Jay concurs: 'your outside space is just as important as your inside space in my view. . . . That feeling of space and privacy to me is worth a lot.' Space and privacy, precious commodities, simple desires. Jeremy of the Levellers, himself a traveller, says that 'the appeal is quite romantic. It's the English dream really, isn't it? – the fantasy most English people have: trees, fields, all those images from *Tess of the d'Urbervilles'* – including Stonehenge again.[7] There's an image here of New Age travellers as contemporary scholar-gypsies, retreating from 'this strange disease of modern life' as Matthew Arnold put it in the poem 'The Scholar-Gypsy', or as updates from the fiction of George Borrow.[8] It is ironic that while the Thatcher government championed Victorian values it was giving free rein to the violent harassment of groups of such atavists. The government's imperfect grasp on history helped Thatcher out here, though: the Peace Convoy was described by her home secretary in June 1986 – one year after the Battle of the Beanfield – as 'a band of *medieval* brigands'.

The fact that early East Anglian fairs had seen the revival of horse fairs at Bungay links the fairs as well with a traditional travelling community, gypsies. So, are New Age travellers related to traditional gypsies, and if so, how?[9] Under the Caravan Sites Act 1968 gypsies are defined as 'persons of nomadic habit of life, whatever their race or origin, other than a group of travelling showmen, travelling together as such'. As the National Council for Civil Liberties (NCCL) has observed, 'It is vital to note that this is a definition *by lifestyle*', not by race or other origin. In 1986 the government reported that a total of 10,592 gypsy caravans were recorded in the country, and there were around 4,000 places for those caravans on legal sites. Less than 40 per cent of traditional gypsies were supplied with legitimate spaces, though the Caravan Sites Act puts a legal duty on councils to provide 'adequate accommodation' for 'persons of nomadic habit of life'.[10] Add, by official estimates, 8,000 New Age travellers and their 2,000 vehicles in the 1980s and 1990s and the already inadequate provision comes under even greater strain. When approached by the NCCL to explain how it could distinguish between New Age travellers and gypsies for the purpose of providing sites, the Department of the Environment replied, 'it is not considered that there is any duty under the 1968 Act for local authorities to provide a site specifically for the Peace Convoy'. The main reason given was that 'the Act is concerned with the provision of *caravan sites* not tenting and camping sites' – the sheer variety of convoy vehicles and living spaces became the government's let-out clause.[11] Kevin Hetherington observes that:

> Just as has been the case with Jews and gypsies down the centuries, the 'New Age travellers' are hated not because they are always on the move but because they might stay and 'contaminate' through their ambivalence and bring all manner of horrors upon the 'locals'.[12]

Do we need to interrogate further the casual comparison between ancient and modern wanderers? Does, for instance, the fact that there may be a degree of self-marginalization on the part of the New Age travellers contribute to their difference from traditional gypsies? The variety of names for the New Age 'trailer trash' betrays the extent of majority culture's distaste for and distrust of them: crusties, drongos, mutants, hedge monkeys, brew crew, soap dodgers, giro gypsies, brigands. In a sense, as with the gypsies complaining about their flaunted dirt and dreadlocks, New Age travellers attract such abuse and invective. In sociological terms, this means that they '*deliberately* assume "risk identities", . . . celebrate chaotic and expressive lifestyles', as Kevin Hetherington puts it.[13] The gypsies I talked to are correct about the plural nature of New Age travellers. Simon of Bedlam rave sound system neatly distinguishes ravers from the older generation of hippy travellers: 'We're different from them. They sit there admiring their crystal and we'll sit there admiring our vinyl, know what I mean?' Crystal power of hippies versus material vinyl of ravers' records. Traveller Shannon elaborates on generational distinctions and motivations:

> Whereas in the early eighties it was mostly alternative-y type people who picked an alternative lifestyle, now it's people who are basically fucked off with the city. . . . It's not the rosy rainbow hippy scene it was in the seventies and eighties, people trying to live in love and peace. . . . [T]hat isn't going to exist in the nineties.[14]

Addressing the local majority culture in rural Glastonbury, travellers Grig and Tosh also highlight urban space as the source of contemporary tension, not just between travellers and the majority culture, but between groups of travellers themselves:

> A lot of the travellers are coming out of the cities with nowhere to go. . . . They get full of the 'brew' and go around making a nuisance. . . . We can't chuck out our 'drongos', any more than you can yours.[15]

It seems as though at one extreme are the idealists from the sixties, still living in horse-drawn wagons, or in tipis, living the organic life in touch with the land and with local communities, and at the other are the 'brew crew', the 'drongos', young urban homeless who merely fulfil the prophecy of the Sex Pistols' 'No future for you!', and their injunction 'Get pissed. Destroy!'

But what is this New Age that these people are travelling through, or taking us to? The complexities of the term themselves need unpacking. In case you didn't know it, we are currently in the New Age. A global ritual called the Harmonic Convergence took place on 17 August 1987 to signal – among other things – the end of the nine cycles of hell which followed the thirteen cycles of heaven in ancient Mayan cosmology.[16] Apparently postponed from the late sixties, the dawning of the Age of Aquarius was finally here. New Age travellers are the gloriously downmarket end of the commodification spectrum.

But political critic Stuart Hall asks:

> How new are these 'new times'? Are they the dawn of a New Age or only the whisper of an old one? What is 'new' about them? How do we assess their contradictory tendencies – are they progressive or regressive? . . . If we take the 'new times' idea apart, we find that it is an attempt to capture, within the confines of a single metaphor, a number of different facets of *social change* . . .[17]

Pat Kane identifies 'a much wider and deeper culture of the irrational: a culture which we often dignify with the term "New Age", but which should properly be called occult'.[18] I'm not sure how far *New Age* can be called a term of dignity (any more), but possibly we should indeed be making greater use of the term *occult*, in its original sense of *hidden (from sight)*, *concealed*. Though their etymologies are in fact entirely different, occult sounds like it ought to be connected to *culture*, too, even to *counterculture*.[19] Occulture as a term for New Age's culture of resistance? Kane continues:

> After the convulsions of the sixties and seventies, the new right gained power by turning the counterculture's individualism against itself: freedom redefined as the freedom to consume aggressively. New Age and occult cultures represent an alternative politics of the individual.

Like Pat Kane, Nigel Fountain locates the impulse originally in the sixties:

> In the 1960s the young dropped out, in the 1980s they are dropped out. Two decades after drugs were supposedly a tool to heighten reality, and offer visions of the future, in the 1980s they are an escape from the present, and a replacement for the future. Across the decades the lost army of travellers still makes its pilgrimages to Glastonbury.[20]

Why so negative? Why must they be described as 'lost'? Apart from the romantic construction of drug use here by Fountain – writings about the sixties by people from the sixties are full of such nostalgic acid flashbacks – there is the larger issue about the degree of active choice involved. For Fountain one difference between 1960s and 1980s youth is that in the sixties there was the clear possibility of conscious decision on the part of dropouts. Dropping out could be seen as a critical indication of power. In the eighties and beyond, for Fountain, dropping out is passive, a symptom of powerlessness. But the standpoint he refers to on the extreme fringes of society is also that of other marginal groups who've been pushed rather than dropped out, maybe – some of the squatters, travellers.

From Tipi Valley to the peace camps

In Dyfed in South Wales there is a valley of traveller-type people, living in a community that's one of the longest-surviving alternative traveller lifestyle sites in Britain. This is Tipi Valley, Teepee Valley, sometimes Talley Valley, north of the A40, the old Roman road. Like Stonehenge, Tipi Valley has strong associations for New Age travellers, many of whom have made their pilgrimage or simply ended up there at some time while on the road. Before terms like *New Age travellers* or *the Convoy* were in use, groups travelling from festival to festival were known simply as Tipi People, whether they lived in tipis or not. Tipi Valley became more established in 1976 when some derelict farmland was bought by a group of festival-goers aiming to extend the festival season by establishing a long-term eco-friendly community. The Tipi People originally thought (hoped?) they would not need planning permission since their dwellings were not permanent structures. The nowadays 'world-famous'[21] Tipi Valley is a precarious community, partly by virtue of its autonomy, but also because of the regular efforts on the part of the authorities to end it. For instance in the mid-eighties the local district council tried to used the planning laws to evict the inhabitants of the entire village, claiming an unauthorized shift from agricultural to residential land use.

It may be that the adoption of the tipi lifestyle as model is a more self-consciously politicized choice than that of gypsies: after all, the Native Americans' great struggle has been against the dominant white culture of progress, industrialization and capitalist grasping in the USA, and for equitable distribution of land and respect for traditional skills and folk rituals. A repoliticized living space contributes to the romantic construction of alienated people, of marginals. Possibly the sheer otherness of this chosen living space – in racial, geographical and historical terms – helps to signal the degree of rejection the inhabitants of Tipi Valley wish to display to mainstream society; or maybe it's just easier to feel you're on the social warpath if you actually live in a tipi.

The utopian possibility of the version of Tipi Valley I've just given, from the outside, is undercut rather by an insider's recent description of it: internally, it's a working site of tension, which is seen even in the layout.

> There's something pretty magnificent about the sight of a load of tipis stuck together. In Talley, people were split at the time. There's the centre village, with about twenty people in it, which has been there for a long time and is pretty much self-sufficient. But there was a lot of friction between the centre village and the people on the outside where there were vehicles and first-time tipi dwellers and a lot of people with romantic

ideas of what it was going to be like. . . . If you go there and walk into the middle, it's lovely, but if you walk towards the outside there's bags of rubbish, just left lying around, and more shit around.[22]

Tipi Valley thus duplicates some of the social problems and structures it has sought to remove itself from: deliberately placing itself at the margins, away from the centre of majority culture – almost from *any* culture – it has developed into a central space inhabited by *authentic* veterans and idealists, surrounded by its own marginal types, problem cases.

More insistently nomadic New Age travellers were those in the Peace Convoy – the most-publicized and largest group of travellers during the 1980s. In the media it was usually 'the so-called Peace Convoy': the amount of sheer venom packed in to that small hyphenated word 'so-called' when used in connection with the convoy is extraordinary. The Peace Convoy interrogated what had become the comfortable binary framework of social establishment (parliament, landowners, the military and police) and established alternatives (the Campaign for Nuclear Disarmament, the fairs). New spaces were opening, were parked up on for a couple of months as participants waited for the inevitable eviction to come through. A circuit of temporary autonomous zones. A movable feast of folk devilry, one ripe for demonization by the many, for romanticizing by the few.

The convoy that took its name from the peace camp movement meandered and broke down around England and Wales in the 1980s. It showed itself to be less a considered political event or zone than a regular irritant to majority culture and, as I've suggested, even to majority culture of resistance.

Rainbow Fields Village at Molesworth RAF base became a traveller-related community signalling solidarity with a pre-existing peace camp. RAF Molesworth in Cambridgeshire had been a little-used airbase since the Second World War until the decision to transform it into an American cruise missile base. The peace camp had been established two and a half years before a Green Gathering held at Molesworth in late August 1984. Green Gatherings themselves grew out of summer meetings organized by the original Ecology Party in 1980; until Molesworth they were all legitimate – often as part of the Glastonbury Festival, which still maintains the influence of the Green Gatherings through the Green Field. Unable to find an independent site that would give them permission for the 1984 Green Gathering, the organizers decided to step outside the legal framework, effectively radicalizing themselves by this decision. Rainbow Fields Village was formed in the wake of the Molesworth Green Gathering, on the opposite side of the proposed cruise missile base from the peace camp. In fact, the village was founded on an earlier site of the peace camp. Here again some New Age travellers were locating their activities and lifestyle firmly within a framework of resistance.

Bruce Garrard gave this report at the end of October 1984, when the village was still in the process of being developed:

The village is a string of tipis, tents, vans, buses and benders, mostly pitched along the hedge which forms a natural windbreak down one side of the village field. . . . Between all these things, the gaps; filled with possibilities. . . . The village is slowly growing and changing (and if you want to join us, what we need most is more settlers). There's

Figure 12.1a   Rainbow Fields Village, Molesworth air base, 1985.

about 80 of us here now – travellers, peace campers, greens, tipi-dwellers. . . . Some are here to oppose cruise missiles, some to build a new free community, and some just to live here for the winter or for longer.[23]

Brig Oubridge, veteran of Tipi Valley and long-time thorn in the side of authorities up and down the land, was living at Rainbow Fields Village when the (even by their standards) massive mobilization of police and troops – yes, troops – came in the night in early February 1985 to evict them. The villagers' first response to imminent eviction was to argue among themselves about what their

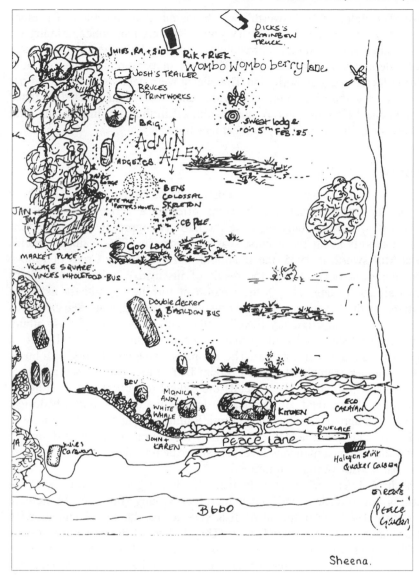

Figure 12.1b   Rainbow Village site plan.

reaction should be. Bruce Garrard sees the arguments as highlighting 'the same old division: the convoy wanted to survive, to get away from the aggro. The greens wanted to stay, to make a non-violent stand.'[24] One of the villagers called it 'the military occupation of Molesworth Common', and certainly the sheer scale of military action was the focus of media reports. Brigadier-General Edward Furdson, defence correspondent of the *Daily Telegraph*, described it as 'the largest single Royal Engineer operation since the Rhine crossing in 1944'. Another right-wing newspaper, the *Daily Express*, had a front-page article

headlined 'The Battle of Molesworth', which contained an inventory of personnel: 1,500 Royal Engineers, 100 military police, 600 civilian police, infantry on
reserve just in case, huge numbers of vehicles including bulldozers to clear trees,
hedges, benders, buses out of the way.[25] This was an extraordinary overreaction,
even by the increasingly authoritarian Thatcher government's standards: the net
for the enemies within was being cast wider and wider. Flak-jaketed Defence
Secretary Michael Heseltine sternly placed himself in the front line for the vital
twin battles for British democracy and camera time.

Living spaces

In a sense the problem with writing about the Peace Convoy or the Rainbow
Village on the road is that you end up offering a series of spectacular
confrontations, which may only replicate the narrative approach of many erron-
eous media reports about the events: Nostell Priory in 1984, Rainbow Fields
Village at Molesworth in 1985, the Beanfield in 1985, Stoney Cross in 1986 –
these are the rural riots, largely the result of insensitive police and military
operations around Britain. This is the version of traveller as outcast, as pirate
even, offered by the Levellers, a leading crustie band (though some see *crustie* as
a pejorative term), and originally travellers themselves. A logo on their album
Levelling the Land shows crossed spanners beneath a rough 'folk devil' face in
imitation of the skull and crossbones. The spanners signal the affinity with buses
and trucks, another example of the self-reliant DIY ethos of resistance culture.[26]
The band's name, of course, itself recalls the radical republicans of the English
Civil War, and clearly (and self-heroically) locates the band within a historical
context of social uprising – which if you're a Rainbow Villager feeling you're in
'the middle of a civil war' at Molesworth may be accurate rather than dramatic.
But what's life like on a mundane level for New Age travellers, when they're not
locked in battle with various authorities?

Life is dirty. Everyone says so. In fact, dirt is a signifier of difference, of
outsiderness for travellers and other marginal groups. For the Amsterdam
squatters of the 1980s, according to ADILKNO (the anarcho-style Foundation for
the Advancement of Illegal Knowledge), 'The clothing of the movement looked
sexless (thus "masculine"), dirty and ripped. There were hard patches and nasty
stains on it. . . . Squat dress resembled the work clothes of miners, chimney
sweeps and tank cleaners. It looked at least as rough and filthy, only it couldn't
be traded in after work for a designer sweat suit.' Dirt as a focus both of
(political) identity and of revulsion is seen also in earlier press reports about
Greenham Common women. In *Femininity in Dissent*, Alison Young focuses on
the grotesque emphasis of media representations, 'a theme of startling intensity
and violence': 'It is a preoccupation to which the papers constantly return: dirt,
disease, disgust positioned against cleanliness, health, approval. . . . The grotesque [body] becomes synonymous with the marginal and the deviant, the cast-
out.'[27] While some may be easily identifiable by their clothes and behaviour,
New Age travellers as a larger community are more diverse and inventive in their
attitudes, motivations and living spaces. Describing a convoy going to an East
Anglian fair as early as 1978, Derek de Gale recalls a pause in the 'progress of

the long and ungainly snake. Then, green under the street lighting, its windows curtained, was a double-decker bus. It seemed so English and yet the language was not of Suffolk or England.'[28] This is a good point: as familiar icons like double-decker buses are transformed, so Britain is being rewritten around its margins. Those travellers who have vehicles want to tell you about the ways in which they've transformed them into homes. (Such a blurring of distinction between vehicle and home has apparently been overlooked by police: destroying homes at the Beanfield, impounding them *en masse* at Stoney Cross, the authorities contribute to rather than resolve social problems of homelessness. A darker reading – taking into account Thatcher's remark about being 'only too delighted to make life difficult for hippy convoys' – suggests that homelessness is the state's punishment for the travellers' transgressions.)

Exteriors of vehicles are decorated in a variety of beautiful, provocative or deliberately unassuming ways.[29] Slogans are offered: PEACE CONVOY on the side of a double decker, FIGHT TRUTH DECAY on the side of a coach. Destination windows of former public transport coaches read STONEHENGE GO FOR IT! and MYSTERY TOUR. Other coaches have signs saving CLOSED or PRIVATE on their windscreens. Rainbows, mandalas, suns, anarchy symbols – a coach entirely covered in a painting of a romantic landscape, all trees and mountains, with a huge tarpaulin running the length of its dodgy roof. Others are more practical: CAUTION! BABY & DOGS IN TRANSIT! – a plea to majority culture rather than a provocation?

Not all travellers sail the 'black tar rivers' of the Levellers' road mythology in particulate-puffing diesel trucks and buses. The nomadic nature of the lifestyle can be seen in other, more flimsy, living spaces. There's even a hierarchy of transitoriness: from a tipi to a bender, for instance, is a shift to ever-greater ephemerality. A sense of the sheer variety of sites that are occupied is evident in this description of the places the Travellers' Skool Bus parked up on in 1990:

6 areas of common land
6 green lanes
4 disused airfields
3 areas awaiting development
1 verge area
1 commercial festival
1 council-owned recreation area
1 farmer's field with permission
1 farmer's field without permission
1 national conference
1 field under disputed ownership.[30]

Peter Gardner notes that travellers in the South-west of England 'see thousands of acres set aside for military purposes and question the validity of the argument that there is no room for them. Surely there must be a small patch of land available somewhere for our peaceful coming together? – they ask.'[31] This question echoes the cheeky demand in the Albion Free State manifesto of 1974 for 'permanent free festival sites, collectives, and cities of Life and Love, maybe one every fifty miles or so' up and down the land.[32] Walking onto a lay-by, a piece of industrial waste ground, a disused quarry, some council-owned or common land, you may miss the resemblance to 'cities of Life and Love', but locating sites both temporary and permanent (or at least longer-term) is the constant struggle of New Age travellers. The Friends, Families and Travellers Support Group (FFT) is an organization that seeks to protect the rights of all travellers, whether gypsy or New Age. It too voices the need for identifiable sites, as well as for the maintenance of traditional stopping places. With the Criminal Justice and Public Order Act 1994 the prospect is not encouraging, as it actually repeals the obligation of local councils in the Caravan Sites Act 1968 to supply sites for travellers.

It's possible that travellers contribute to their own demonization: the brew crew, the younger crustie element that hangs around the edges of Tipi Valley, or begs for drugs in Bristol, represents the current popular image of travellers. Some travellers themselves might seem to revel in their alienation. (Why wouldn't you, if you're alienated?) Yet why should brew crews be seen as a problem, and one created by travellers? Alan Dearling suggests rather that they and their sites reflect the wider social situation: 'Sure, some Travellers are dirty; some get busted for dealing; some are suffering from mental health problems and drink- and drug-related problems. Unfortunately, the same is true for many other members of society.'[33] Even if hard drugs aren't present, life on site is increasingly punctuated by eviction orders, which are the standard way of moving the 'problem' on elsewhere – preferably, in the case of police and many county councils, over the border to another county's responsibility. Traveller Rabbit can see the positive side of this though: 'Suddenly everyone you haven't seen for ages is outside changing wheels and standing around with their engines in bits and everything gets going.'[34] This can be viewed as a sign of inertia: the travelling impulse as a response to external actions rather than initiated by travellers themselves. Or, more positively, it's a sign of New Age travellers' ingenuity: even the deliberate and systematic disruption of their lifestyle can be turned from a source of anger or despair into a routine of pleasure and energy.

Writing about the living spaces of New Age travellers, I've been struck by the extent to which they *are* living spaces, vibrant and imaginative. I'm most struck by this astute remark from Shannon, a traveller in the early 1990s, when the 'politics of envy' was a phrase much-favoured by the unimaginative political classes. Shannon reverses the typical poles of desire and disgust between travellers and the inhabitants of the majority culture: '. . . a lot of the hatred of travellers is out of envy'.[35]

Notes

1 Don Aitken, 'Twenty years of Festivals', *Festival Eye*, Summer 1990, p. 19.
2 See for example Lionel Rose, *'Rogues and Vagabonds': Vagrant Underworld in Britain 1815–1985*, London 1988. Oubridge's short piece 'Convoy Myth' also offers this line.
3 Which, of course, contributed greatly to the problem of homelessness in the first place – by placing restrictions on the building of council housing, the imposition of increasingly Draconian benefits rules, the introduction of the sick joke of care in the

community. . . . In Chapter 6 we'll see how the Criminal Justice Act 1994 extends such unsympathetic policies further still.

4 On the other hand, Tory policies could also insist on unemployed youth living nomadically. In December 1984, DHSS board-and-lodging allowance to unemployed people under twenty-six years of age was to be limited to periods of between two and eight weeks in any one district. 'Thereafter the claimant must move to another district to re-qualify. The object was to keep youngsters "on the move" looking for work. . . . To critics [these new regulations] smacked of a return in modern form to the old harassing approach to vagrants' (Rose, pp. 184–5).

5 Quoted in Richard Lowe and William Shaw, *Travellers: Voices of the New Age Nomads*, London 1993, a collection of first-person accounts of the lives of travellers, p. 56. In this chapter I refer frequently to the experiences of the interviewees in *Travellers*; all short quotations, unless otherwise annotated, are from this book.

6 Kevin Hetherington, 'Stonehenge and Its Festival: Spaces of Consumption', in Rob Shields, ed., *Lifestyle Shopping*, London 1992, p. 85. The experiences of earlier British homeless people are told in Steve Humphries and Pamela Gordon, *Forbidden Britain: Our Secret Past 1900–1960*, London 1994, pp. 190–1.

7 All three are quoted in Lowe and Shaw, pp. 240, 66, 162 respectively.

8 The narrator of the poem (published in 1853) exclaims, 'O life unlike to ours!' during his rapt description of the Oxford University dropout who 'roamed the world with that wild brotherhood'. Borrow's earlier *Lavengro* (1851), a novel partly about the Norfolk gypsy Ambrose Smith, takes its title from a gypsy term meaning 'philologist, lover of words, language, literature'. I suggested a link between New Age travellers and an alternative set of Victorian values first in a review of Lowe and Shaw, *Travellers* in *Anarchist Studies* vol. 2, no. 2, Autumn 1994, pp. 175–7.

9 For another look at this question see Fiona Earle et al., *A Time to Travel? An Introduction to Britain's Newer Travellers*, Lyme Regis, Dorset 1994, especially Chapter 6.

10 Quotations and information are from NCCL, *Stonehenge: A Report into the Civil Liberties Implications of the Events Relating to the Convoys of Summer 1985 and 1986*, London 1986, p. 35, emphasis added.

11 NCCL, p. 43.

12 Hetherington, p. 91.

13 Hetherington, p. 92; my emphasis.

14 Quoted in Lowe and Shaw, pp. 240–1.

15 Quoted in Ann Morgan and Bruce Garrard, *Travellers in Glastonbury*, Glastonbury 1989, p. 7.

16 See, for instance, *The Harmonic Convergence: Press Cuttings, August 1987*, Glastonbury 1987; Andrew Ross, *Strange Weather: Culture, Science and Technology in the Age of Limits*, London 1991, especially Chapter 1, 'New Age – a Kinder, Gentler Science?'

17 Stuart Hall, 'The Meaning of New Times', in Stuart Hall and Martin Jacques, eds, *New Times: The Changing Face of Politics in the 1990s*, London 1989, pp. 116–17; emphasis added.

18 Pat Kane, 'In Thrall to New Age Thrills', *Guardian*, section 2, 4 January 1995, p. 13.

19 Occult derives from the Latin verb *celere*, to conceal; culture – like cult – derives from *colere*, to cultivate.

20 Nigel Fountain, *Underground: The London Alternative Press, 1966–74*, London 1988, p. 215.

21 Derek Wall, 'Tepee Valley Left at Mercy of Big Chief Redwood', *Independent on Sunday*, 12 February 1995, p. 20.

22 Traveller Alex, quoted in Lowe and Shaw, pp. 23–4.

23 Bruce Garrard, ed., *The Green Collective: The Best from the Mailing, 1984*, Glastonbury 1986, pp. 27–8.

24 Bruce Garrard, *The Last Night of Rainbow Fields Village at Molesworth: An Account of the Eviction, February 5th–6th 1985*, Glastonbury 1986, revised edition, p. 8. Subsequent quotations from newspapers come from clippings reproduced in this booklet.

25 Prompted by Bruce Garrard, I'm happy to pass on this information: '(1) it took the MoD *3 months* to plan it, and (2) they spent more in that one night than CND's *entire annual budget!*' Personal letter, 28 June 1995.

26 Truck logos as images didn't start with the Levellers: earlier free festival favourites, based in a communal farm in Suffolk and travelling extensively during the early seventies, were the rock band/collective Global Village Trucking Company. Their eponymous album from 1976 features truck images on both cover and lyric sheet, signalling their affinity with travellers and festivals. More recently, at the bottom of every page of the book *A Time to Travel?* are line drawings of travellers and their vehicles, displaying the variety, the ingenuity of the lifestyle. The Levellers' DIY style extends to instrumentation: the violin 'was typically unconventional and ecologically pure, "recycled" from three different broken violin bodies', while the guitarist used 'an old record player arm acting as pick-ups, as well as an amplifier acquired from the Revillos' (Colin Larkin, ed., *The Guinness Who's Who of Indie and New Wave Music*, Enfield 1992, p. 161).

27 Young, pp. 55, 57.

28 Quoted in Richard Barnes, p. 100.

29 See Peter Gardner's collection of photographs in *Medieval Brigands* and the photographs throughout Earle et al.'s *A Time to Travel?* for illustrations.

30 Earle et al., pp. 67–8.

31 Gardner, p. 7.

32 Alan Beam, *Rehearsal for the Year 2000 (Drugs, Religions, Madness, Crime, Communes, Love, Visions, Festivals and Lunar Energy): The Rebirth of the Albion Free State (known in the Dark Ages as England): Memoirs of a Male Midwife (1966–1976)*, London 1976, p. 160.

33 Alan Dearling, in Earle et al., p. iv.

34 Both quotations are from Lowe and Shaw, pp. 96–7, 4.

35 Quoted in Lowe and Shaw, p. 239.

13 Sallie Westwood
Racism, Black Masculinity and the Politics of Space

Excerpts from: J. Hearn and D. Morgan (eds), *Men, masculinities and social theory*, pp. 55–71. London: Unwin Hyman (1990)

This chapter is an exploratory and tentative attempt to grapple with some of the issues raised by the current discussions of masculinity. I did not set out to research masculinity in a substantive way nor to provide an ethnography of the lives of some black men in a provincial city. The research grew out of an earlier research project which reconstructed the history of a black youth project, a local *cause célèbre* in the mid-1980s. The men who are the subjects of this paper were politicized through the project and I came to know them through the research. The men who have contributed towards this paper have been involved with the process throughout. I am conscious, however, that it has, ultimately, been constructed out of an ongoing process of negotiation in which I, as a white female researcher with a university background, have sought to interpret the lives of black working-class men. I am aware of the limits of this encounter and the ways in which racism and sexism intervene in the research process.

There is no innocent ethnography and I am mindful of the critiques of white sociology generated by black critics like Lawrence (1982). Black people have been poorly served by a sociology and an anthropology which has contributed to the definition of black people as 'Other' through the silences on black people's lives and the racism of British society, or through the presentation of black people's lives as a form of exotica, or through the use of stereotypes which have framed the representation of black people through the fixity of the stereotypes (Bhabha 1983). That this has happened partly relates to racism within the social sciences which is suffused with the commonsense of the day and equally it relates, I would argue, to the essentialism of the subject in sociological discourses. My hope is that, by developing an alternative, non-essentialist account which places racism centrally in the analysis, previous errors will not be reproduced within this paper.

Multiple selves

My task in this paper is to de-centre the subject and to de-essentialize black masculinity. The essentialism of the constructions that surround the black man and of black masculinity have given plenty of scope for racist accounts through stereotyping and the construction of black men as 'the Other'. For black men of African descent the stereotypes have been fixed on the body, on physicality, physical strength, and as a site for European fantasies about black male sexuality. Orientalism has generated a different picture for men of Asian descent. The colonial designation of the 'martial races' of northern India produced an account of men of vigour and strength, fighting men who were at some distance from the wily oriental who, by being tied to conceptions of manipulation and wiliness, became feminized in the eyes of the white men of the colonial era. The fixity of these stereotypes places 'races', genders, motivations, and behaviours in such a way that they become naturalized and a substitute for the complex realities they seek to describe. It does not help to try and replace one set of stereotypes by another; the point is to dismantle all stereotypes and to move to a much more shifting terrain in which identities are not seen as fixed and cannot, therefore, become stereotypical.

No one understands this better than the men who are the subjects of this paper. They understand themselves simultaneously as Gujaratis or Punjabis, as Hindus, Muslims, or Sikhs, as Afro-Caribbeans or Jamaicans, and as black men, a politically forged identity in post-colonial Britain. They understand that they are legally British but not culturally British or English because, as Desh commented, 'British means white' and as Kuldip said with feeling 'Black people will never be allowed to be British'. Their identies have been forged in relation to the specificities of their cultural and historical backgrounds and yet they recognize commonalities with white working-class men because of their class position which generates both political and cultural overlaps related to them as men. But these identities are not separable; to be one does not deny the other. They are lived and experienced simultaneously which is why in any analysis we have to make sense of this. Identities are not finished products like the old socialization model would have us believe, rather they are constantly produced and thus a shifting terrain. However, shifting identities are not just freefloating; they are positioned within histories, cultures, language, community, and class. This suggests a lack of fixity and thus a politically uncomfortable position in which contingency matters. There are contradictions and coalescences between the cultural and the social, identity and difference.

Public spaces

In my own account I want to try to unravel the cross-cutting discourses on black masculinity in relation to the way in which they are played out in different spaces that are privileged. Discourses are constructed and elaborated and they too cross-cut the spaces. They are not mutually exclusive and are played out through the discursive practices that surround lived experience in specific spaces. There is, however, a materiality to the politics of these spaces. We find mutuality and contradiction, dominant and subordinate discourses, and the ways in which black masculinities are made, remade, and reworked as an essential part of the cultures of resistance of black people in Britain against the racism of British society.

Discourses as registers of masculinity are worked through in a variety of spaces and this chapter concentrates attention upon the public realm and only two sites within this, inner-city streets and football, the most public of public spaces. Like the world of family and household these are arenas of contested terrain, areas of struggle in which masculinities are called up as part of the fight against racism. To consider black masculinities as part of the cultures of resistance to racism is not to suggest that they are not in themselves contradictory. These areas, the streets and football, are important to consider because it is precisely in these areas that current stereotypes about black men have been generated and have become part of the commonsense racism of today.

Although this paper concentrates upon the public sphere the reader should be aware that this does not exhaust the fields for black masculinities, or the discourses on black masculinities, and does not suggest that these fields have pre-eminence and are in themselves privileged over the worlds of work and, most importantly, family and the home. On the contrary, the world of home and family is vitally important especially as a site of resistance to racism in British society. It is, however, an arena in which women have power and this is acknowledged by the men with whom I spent time. Although as black men they had power in the home as brothers, sons, husbands, and fathers they acknowledged that the domestic world was a power base for women in which they often had to defer to the power of women to define this realm. It is also important to emphasize that the private–public dichotomy is not a real opposition; the public intervenes in the private world of the family and none more so than in the case of black families subjected to specific forms of state intervention (Westwood & Bhachu 1988) which often break up black families. Black families have therefore sought strength and unity as a form of resistance to racism in Britain (Mercer & Julien 1988).

For the black working-class men who are the subjects of this paper there is a powerful distinction to be drawn between the world of home and the world of the streets and the football pitch. I want to concentrate upon the latter because I am interested in the ways that masculinities are played out and validated not for home and family but for other men. The subjects of this paper are black working-class men not middle-class black men for whom the nuances of masculinity will be different but who will, I would suggest, still require a male audience for the validation of certain forms of masculinity. Women are not entirely absent from this public male world. They enter it through specific discourses on women which place women in relation to the home and domesticity and to romance and sexuality. The men who are the subjects of the lived realities presented here are also makers of discourses which is part of an engagement involved in the production of black, male, working-class identities.

In concentrating attention upon the streets and upon football in this account I am calling up not the masculinities of physical prowess, although this is part of the discussion, but the cultural politics of 'race' in which popular culture is a crucial site. Football is a vital part of this and while it has since its inception been tied to 'manliness and vigour' it can also be analysed in relation to cultures of resistance.

Racism

Staples (1985, p. 20), writing on black masculinity, comments: 'In the case of black men, their subordination as a racial minority has more than cancelled out their advantages as males in the larger society'; and he continues: 'The issues of masculinity and race are too interwoven to separate at this time.' The men in this paper are acutely aware of the articulation between masculinity and 'race' and they see in racism an affront to black masculinity. As a counter to racism black masculinity is called up as part of the cultures of resistance developed by black men in Britain.

Racism in Britain is, as Cohen (1988, p. 63) notes, 'not something "tacked on" to English history, by virtue of its imperialist phase, one of its aberrant moments; it is constitutive of what has become known as "the British Way of Life"' (emphasis in original) or as Gilroy points out: 'Racism is not a unitary event based on psychological aberration nor some historical antipathy to blacks . . . It must be understood as a process.' (Gilroy 1987, p. 27) The process is elaborated in 19th-century discourses that have become part of the commonsense racism of English life (Cohen 1988). Generally, the process to which Gilroy refers is characterized by prejudice and discrimination against groups of people understood to share common racial heritages. Racism rests, in part, upon the mistaken belief that there are biologically distinct races in nature and therefore people have natural attributes determined by their racial origins. Implicit in racism is the idea that one race is culturally superior to another and is, therefore, justified in exercising political and economic hegemony.

To take this one step further we have to consider the forms of racism and again Cohen's work is useful (Cohen 1988). He seeks to differentiate the forms of working-class racism by concentrating attention upon a reworking of the rough/respectable divide and by introducing gender as a crucial determinant. Cohen emphasizes 'the nationalism of the neighbourhood', that 'imagined community' to which loyalty is displayed and which invokes the politics of space as crucial to an understanding of the articulations between territoriality, masculinity, and racism. The defence of defined space as 'ours' is itself exclusionary, and the definitions of imagined communities are themselves exclusionary acts; they exclude black people. Tolson (1977, p. 34) also emphasizes the importance of space and neighbourhood when he writes: 'The intense localism and aggressive style shape the whole experience of working-class masculinity.' He is writing of white working-class men but the intense localism is something shared by the black working-class men whose lives are discussed in this paper.

It is clear that the 'nationalism of the neighbourhood' is not confined to white men. In black areas it is appropriated and the inner-city space becomes 'our place' for black people and 'our streets' become the space to be defended against the incursions of white men and most especially the police. This takes me into the discussion of privileged spaces which begins with life on the streets.

The local context

The men live in a provincial city which has undergone major restructuring of its manufacturing base in the 1980s with high unemployment rates as a consequence. On current estimates 25% of the city population are black British with a large Asian population now settled in the city and a much smaller Afro-Caribbean presence. The men belong to these diverse communities and refer to themselves in relation to specific cultural and religious identities but also, as I have emphasized, as black men. They often remarked that Asians and Afro-Caribbeans were closer in this city than in London or Birmingham. The men have grown up in the same inner-city neighbourhood, attended the same schools, and as young men were all involved with the politics of a youth project in their locality which had a crucial role in forging black political identities. They also share experiences of racism and encounters with the police. Most of the men had had brushes with the law in their younger days and continued to be wary of the police.

Among a wide network of contacts there was, during the time of the research, a core group of fifteen Asian and Afro-Caribbean men who lived locally, met regularly, and were in close and fairly constant contact. They were a predominantly Asian group but Asian and Afro-Caribbean men were close friends. Half of this group were unemployed, several worked in manual jobs in local factories or for the city council, one man worked on a community project, and another for a trade union. They were in the 20–30 age group and a third were married or had longstanding partners; some had young children.

Their world is an intensely local world and their social life relates to the inner-city area which is their home. Moves outside this world might be made for weddings in nearby cities, for football tournaments and matches, or for the occasional visit to clubs in a town thirty miles away.

The streets

There is a long history of young working-class men being presented as 'dangerous' in official discourse and in the media (Pearson 1983). But this has of late become specifically racialized with young black men positioned within street life in particular ways, from the moral panics surrounding mugging and criminality explored by Hall *et al.* (1978) and Gilroy (1987) to the more recent accounts of the 'riots' in Britain. These latter accounts, both from the media and within official discourse, articulate a moral panic with the current crises in British society and the British state (Benyon & Solomos 1987, Solomos 1988). Black men are highly visible in relation to street life which is expressed through visual representations in the media and in the current discourses on law and order. What it means for their lives is that they are subject to 'the Empire of the gaze' or as Foucault (1979, p. 187) elaborates: 'It is the fact of being constantly seen, of being able always to be seen that maintains the disciplined individual in his subjection.' The 'disciplined individual', however, has many ways of resisting 'the Empire of the gaze' and the streets of the 1980s have borne witness to this in spectacular ways.

There are, however, many less spectacular strategies available with which to resist surveillance. Thus, life on the streets generates a series of discourses that relate to the importance of being streetwise, but these are not easily articulated because they are part of the commonsense of urban life for black working-class men. What is called up in relation to being streetwise is the ability to handle the dangers of street life, and this links masculinity, defence, and manly behaviour. Contrary to the popular views, however, that being streetwise privileges physical prowess and fighting acumen, it is essentially an intellectual, cerebral attribute. What is required as a context for being streetwise and being able to operate safely on the streets is an intimate knowledge of locality which all the men shared. Thus the links between territoriality and white working-class masculinity underlined by the work of Tolson (1977) and Cohen (1988) can also be seen in the lives of black working-class men.

The area with which I am familiar is one of contested terrain in the material sense of terrain. There are no longer white gangs like the 'ketchup gang' (whose 'trademark' was to beat up Asian men and then douse them in ketchup) or the National Front; latterly these have been routed from the area. But the state intervenes in the form of policing adopted for the inner city. The area is very close to the main city police station and the police can arrive immediately, as they did following a pub shooting. In seconds there were armed police on every corner, long before the ambulance arrived to attend to the injured man. However, this was an extraordinary event. What affects the lives of black men on a day-to-day basis is the level of routine harassment where they are stopped, searched, intimidated, and subjected to racially abusive name-calling and physical duress. The hostel, one outcome of the youth project with which the men were involved, is regularly raided, the doors smashed, and the residents taken from their rooms. These encounters with the police undermine the collective strength of black people and are seen by the men as an affront to black men as men.

The politics of the street as a clash between men in blue and men without uniforms is an old story of urban life, but the documented racism of the police (*Broadwater Farm Inquiry* 1986; Institute of Race Relations 1987) is a crucial element for black people generally and black men in particular. In addition, the police now have available an armoury of weapons with which to police the streets. The men were acutely aware of both the changes in tactics and technology and the ways that these were used. 'They try to wind us up so they can show off their hardware', said Amrit, and continued: 'but we're not stupid. We know their games.' Wiliness, not confrontation, was seen to be the best way to resist. The point was to outwit the police. 'You've got to use your brain and act fast', said Mark with a grin, 'the coppers don't have much brain or pace.'

Being streetwise is not only about negotiating the city, it is also about satisfying needs and demands through a very local network of goods and services. Thus the men could organize vans and drivers to take them to football matches or acquire some of the latest fashion gear cheaply, or get some assistance with welfare rights or housing. They could get their houses painted cheaply or electrical work done or a car fixed through a complex process of reciprocal relations which enabled them to earn some money, raise their standard of living, and have fun. When I lent some of the men my car for a football weekend it was returned the following week with a new MOT, my trade for the loan of the car. Being able to summon up and control access to resources directly is a mark of power and status and an acknowledgement of male control over the immediate environment. In this way it has a direct bearing upon masculinity and the power that a man can exercise in the group. Favours are traded and power is demonstrated by the range of favours owed to any one man. Within the group one man was pre-eminent and his advice and guidance was sought with legal and economic problems and this was traded against the skills he did not possess like driving. Thus, there was a micro-politics to the reciprocity set up within the economic system of the inner city.

Football

In a favourite pub in the locality a huge silver cup is passed around the bar. It is filled with Barcardi and coke, a gift from the Asian landlord, in celebration of the triumph of the Saints in the Sunday league. The cheer and high spirits are for football, not just any football but amateur football and the dominance of a black team of Afro-Caribbean and Asian players over the white world of soccer – a heady moment, relived by recalling the final and other memorable matches when black teams have carried off the cups. For despite the glamour of John Barnes and Ruud Gullit and the media coverage of black players, English football is well known for its banana skins and racist chants. The racism and the masculinity of white working-class men are welded into one by the articulations between a white working-class masculinity socially and ideologically constructed out of the post-imperial chauvinism of British society. Modern football, growing up as it did in the Victorian and Edwardian eras, is redolent with the politics of 'race'.

Racism, however, does not go uncontested in Britain, on or off the soccer pitch. Black men play soccer because they love the game but their entry into the game means that they are immediately involved in the cultural politics of 'race'. So much of the impact of this politics is missed if local amateur and semi-professional football is ignored. The lads, as they sometimes called themselves, were not city supporters and part of the reason for their lack of support was their view that the city team had not tried to recruit black players, despite the numbers of young Asian and Afro-Caribbeans who went for trials at the club. The team does have one black player but, unlike Arsenal, young black players are not filtered through the youth scheme so there are no local black players. Manjit's view was clear. 'A city with so many black people and no black footballers. Ask the manager why is that? We know why, it's racism, that's why.' The absence of Asian men from the national football scene is discussed by Rajan Datar. Writing in *The Guardian* (28 June 1989) he points to the passion for football among Bengali boys and comments: 'Frustrated by the lack of opportunities to play organised football, many young Asians are now involved in their own leagues. From Huddersfield to Hayes in west London new competitions are sprouting up to accommodate demand.'

The response from black footballers and supporters is a collective one, generated and sustained through the local level in the amateur sphere of FA football. Like cities up and down the country the town has a thriving amateur section with a proliferation of leagues and knockout competitions in which black teams of Afro-Caribbeans, Asians, and those (which I know best) bringing Asians and Afro-Caribbeans together, are very prominent. Some of the black teams have grown out of the youth project in which the men were involved, others from community organizations and some from the Gurdwaras. The latter is an interesting way in which ethnicity, masculinity, and popular culture are fused. Some of the most fiercely contested tournaments in black football are those organized around the Sikh festival of Vaisakhi. It is celebrated in the temples and homes of Sikh people but also on the football pitch.

The teams and the prowess of black players are a collective statement about the power of black men to dominate in a sport where white men consider themselves superior. Consequently, matches between black and white teams are fought on and sometimes off the pitch and old rivalries between city teams of black men and county teams of white men are renewed. It is a fiercely contested terrain because, as the black players and supporters say: 'White teams don't like being beaten by a black team.' It is, in effect, an injury to white masculine pride and a source of power and celebration to black masculine pride when white teams are beaten.

Romance and rivalry

There is a romance to football. It offers an arena in which men can work together and invoke loyalty and camaraderie while it also offers a space for the drama of performance where individuals can shine, and have status and acclaim. Equally men can exercise power as 'nifty' players through their mastery of the ball and the pitch which requires both physical dexterity and tactical prowess. Football represents for the men I know their own specific history. They can recount the events of football matches from one year to the next and time is measured in terms of who won what in any specific year. It allows a space in which aspects of masculinity can be elaborated through the nuances of football: loyalty, brotherhood, collective responsibilities, status, and, of course, when things don't go well a man has to live with his own and his team's failures. It can be an uncomfortable and fraught space.

Football offers status through performance validated by other men and the chance to win and carry off the prizes in one very important area of life. It is part of the cultural politics of 'race' in Britain today and it is a politics deeply rooted in masculinity. The discourses on masculinity developed through the discussion of football and its history are also displayed when the men appear in the pub limping or on crutches with a variety of knee and leg injuries which are shown off with some pride like battle wounds. Injuries are recognized as the consequences of an active involvement with the game, as part of some daring play the effects of which can now be displayed and even celebrated.

What is important to the politics of football in the teams I know, however, is the relations between Asians and Afro-Caribbeans who play together in these teams, a legacy of the youth project teams. The men were very proud of their multi-ethnic teams, contrasting their teams with specifically Asian or Afro-Caribbean teams. In fact, their 'black' teams have created problems for them within black football. The Sikh festival tournaments declared that they were Indian and Afro-Caribbean players and at one tournament in Coventry the officials wanted to disqualify the city team on the grounds that it was not an Indian team. The team would not compromise and eventually took the case to the FA who ruled that they could play a joint team which is what they now do. Their loyalty to each other and to the politics of forging a black identity from Asians and Afro-Caribbeans did not waver and they now play as they have always done, as a team of Asians and Afro-Caribbeans.

Football is about a politics rooted in the cultures of resistance that have developed as part of the urban black experience in Britain. There is a collective mobilization through football that calls up black masculinities as part of the resistances that black men generate against the racisms of British society and by which they validate each other. It is a male space; women are not involved but the world of family and home is often present on warm days when the men bring babies and toddlers to the matches, caring for them while they watch the game and demonstrating as they do that even within the male world of football men too can call up very different black masculinities from those associated with the machismo world of the football pitch. The juxtaposition is an important visual reminder of the varieties of black masculinities and our understanding of them as shifting terrain.

References

Benyon, J. & J. Solomos (eds) 1987: *The Roots of Urban Unrest*. London: Pergamon.

Bhabha, H. K. 1983: The other question. *Screen* 24(6), 18–36.

Broadwater Farm Inquiry: Report of the Independent Inquiry into Disturbances of October 1985 at the Broadwater Farm Estate, Tottenham. Chaired by Lord Gifford, QC (1986). London: KARIA.

Cohen, P. 1988: The perversions of inheritance: studies in the making of multi-racist Britain. In *Multi-Racist Britain*, P. Cohen & H. Bains (eds). London: Macmillan.

Foucault, M. 1979: *Discipline and Punish*. Harmondsworth: Penguin.

Gilroy, P. 1987: *There Ain't No Black in the Union Jack: The Cultural Politics of Race and Racism*. London: Hutchinson.

Hall, S., C. Critcher, T. Jefferson, J. Clarke & B. Roberts 1978: *Policing the Crisis: Mugging, the State and Law and Order*. London: Macmillan.

Institute of Race Relations 1987: *Policing against Black People*. London: Institute of Race Relations.

Lawrence, E. 1982: Just plain commonsense: the 'roots' of racism. In *The Empire Strikes Back: Race and Racism in 70's Britain*, Centre for Contemporary Cultural Studies (ed.). London: Hutchinson.

Mercer, K. & I. Julien 1988: Race, sexual politics and black masculinity: a dossier. In *Male Order: Unwrapping Masculinity*, R. Chapman & J. Rutherford (eds), 97–164. London: Lawrence & Wishart.

Pearson, G. 1983: *Hooligan: A History of Respectable Fears*. London: Macmillan.

Solomos, J. 1988: *Black Youth, Racism and the State: The Politics of Ideology and Policy*. Cambridge: Cambridge University Press.

Staples, R. 1985: *Black Masculinity: The Black Male's Role in American Society*. London: Black Scholar Press.

Tolson, A. 1977. *The Limits of Masculinity*. London: Tavistock.

Westwood, S. & P. Bhachu 1988: Images and realities. *New Society*, 6 May.

14 Janet Wolff

On the Road Again: Metaphors of Travel in Cultural Criticism

Excerpts from: *Cultural Studies* 7, 224–39 (1993)

> Theory is a product of displacement, comparison, a certain distance. To theorize, one leaves home (Clifford, 1989: 177).

> A model of political culture appropriate to our own situation will necessarily have to raise spatial issues as its fundamental organizing concern. I will therefore provisionally define the aesthetic of such new (and hypothetical) cultural form as an aesthetic of *cognitive mapping* (Jameson, 1984: 89).

Vocabularies of travel seem to have been proliferating in cultural criticism recently: nomadic criticism, traveling theory, critic-as-tourist (and vice versa), maps, billboards, hotels and motels. There are good reasons why these particular metaphors are in play in current critical thought, and I'll review some of these in a moment. Mainly, I want to suggest that these metaphors are gendered, in a way which is for the most part not acknowledged. That is, they come to critical discourse encumbered with a range of gender connotations which have implications for what we do with them *in* cultural studies. My argument is that just as the practices and ideologies of *actual* travel operate to exclude or pathologize women, so the use of that vocabulary as metaphor necessarily produces andro-centric tendencies in theory. I think it therefore follows that it will not do to modify this vocabulary in the attempt to take account of women, as some critics have suggested we might do. Some discourses are too heavily compromised by the history of their usage, and it may be that the discourse of travel (or at least certain discourses of travel) should be understood in this way.

Gender is not, of course, the only dimension involved in travel. Disparities of wealth and cultural capital, and class difference generally, have always ensured real disparities in access to and modes of travel. In addition, it is clear that the *ways* in which people travel are very diverse, ranging from tourism, exploring and other voluntary activity to the forced mobility of immigrant workers and 'guest workers' in many countries, and to the extremes of political and economic exile. In examining here a single notion of travel, which for the most part rests on a Western, middle-class idea of the chosen and leisured journey, I am merely taking as my subject that metaphor which is in play in these particular discourses. (The fact that it *is* this idea of travel which is operating here is another important question which deserves examination, though it is not my focus here.)

Theory and travel

Quite apart from the increasing use of travel metaphors in critical theory, it is worth noting that a number of cultural analysts have been writing about travel itself.[1] Dean MacCannell's book, *The Tourist*, first published in 1976, was reissued in 1989 with a new introduction in which the author responds to more recent work on theory and travel.[2] Other studies of travel include John Urry's *The Tourist Gaze* (1990), and essays on the semioties of travel by Jonathan Culler (1988[3]) and John Frow (1991). In addition, the work of travel writers is more prominent, with (at least on an impressionistic rather than statistical count) more review space in newspapers and journals.[4] In some cases, the work of travel writers is cited by theorists of travel (and theorists of travel-theory), as for example in James Clifford's reference to Bruce Chatwin (Clifford, 1989: 183). Clearly this is a restless moment in cultural history.

In MacCannell's account of the nature of tourism, we are first presented with the idea that the tourist is typical of the modern person and, in particular, of the social theorist. In all three cases, it is a question of reacting to the increased differentiation of the contemporary world, and the consequent loss of sense or meaning. Social and cultural theory are then reconceptualized as a kind of tourism, or sightseeing, founded on the search for authenticity and the attempt to make sense of the social. (In the introduction to the second edition of the book, however, MacCannell firmly distances himself from postmodern theories which take the more radical view that there *is* no social, or that there is no fundamental ['real'] structure below the play of signifiers. In this his notion of sightseeing is something like Jameson's concept of 'mapping', which is also based on the need to negotiate the lost [but still existing] totality.)

A different link between travel and theory is made in Edward Said's influential essay, 'Traveling theory' (1983). Somewhat strangely, the use made of this notion hasn't always had much to do with Said's argument in that piece, in which he is interested in the question of what happens to theory when it does travel – for example, the transformations of a theory in passing from Lukács to Goldmann to Raymond Williams, and its location/interpretation in very different historical and political moments. (Said also refers to this as 'borrowed' theory: 241.) This is not the same thing as arguing that there is something mobile in the *nature* of theory, which is the way the notion of 'traveling theory' has been interpreted. In other words, the fact that theories sometimes travel (and therefore mutate) does not mean that theory (transported or not) is essentially itinerant. Actually, both senses of 'traveling theory' are in currency in cultural criticism.

Postcolonial criticism and travel

The quotation with which I began, from James Clifford, has to be seen in the context of important developments in postcolonial criticism. Here, the metaphors of mobility operate to destabilize the fixed, and ethnocentric, categories of traditional anthropology. Clifford is one of a number of cultural theorists who have recently revolutionized the methodologies and conceptual frameworks of cross-cultural study, at the same time demonstrating and deconstructing the entrenched ideologies of self and other on which such study has been based. For Clifford, the metaphor of 'travel' assists in the project of de-essentializing both researcher and subject of research, and of beginning to transform the unacknowledged relationship of power and control which characterized postcolonial encounters. Here, the notion of 'travel' operates in two ways. It is both *literal* – the ethnographer *does* leave home to do research – and *epistemological* – it describes knowledge in a different way, as contingent and partial. Related 'travel' vocabulary in this particular discourse includes Clifford's invocation of the hotel as 'a site of travel encounters', rather than either a fixed residence or a tent in a village (1992: 101). It is a notion (or, as he puts it, following Bakhtin, a 'chronotope') which registers both location and its provisional nature.

Postmodern theory and the need for maps

The quotation from Fredric Jameson is from his important essay on postmodernism as 'the cultural logic of late capitalism' (1984). The motivation behind a travel vocabulary here (and in the next case I shall take) has something in common with its location in postcolonial criticism: namely the response to, and attempt to negotiate, a crisis in both the social and the representational in the late twentieth century. Nevertheless, we should not equate postcoloniality, postmodern theory and poststructuralism, though it is important to keep in mind that there are intellectual and political links between them.

Jameson's notion of 'cognitive mapping' (spelled out in more detail in a subsequent essay: 1988) is offered as a metaphor which captures the nature of theory in the postmodern age. As is well known, Jameson's argument here is that in the era of late capitalism, it is no longer possible to perceive the social totality. At the level of the economy, multinational capitalism is not 'visible' in the way that entrepreneurial capitalism, and even monopoly capitalism, were. At the level of technology, steam and electric power (characterizing respectively the two earlier stages of capitalism) have given way to the hidden processes of nuclear power and electronic knowledge. The social subject (and *a fortiori* the sociologist and cultural critic) must therefore resort to new strategies of orientation and analysis. Already immersed in the chaotic and disorganized flow of late capitalist society, the only strategy is to 'map' the social from within. As I said earlier, like MacCannell (and unlike other theorists of the postmodern) Jameson has not given up on totality himself. His argument is that we need new ways of grasping and understanding the fundamental social structures and processes in which we live.[5]

Poststructuralism and nomadic subjects

The third theoretical origin of travel vocabularies, and the last I shall discuss here, is the poststructuralist theory of the subject. The product of radical semiotics, Lacanian psychonalytic theory, and deconstruction, this critique demonstrates the fluid and provisional nature of the subject, who must now be seen as decentred. In the context of media studies and reception theory, Deleuze's notion of the 'nomadic subject' has been found to be a useful way of acknowledging the television viewer's (or the reader's) complex ability to engage with a text both from a position of identity and in an encounter which also (potentially) *changes* that identity. Lawrence Grossberg puts it like this:

> The nomadic subject is amoeba-like, struggling to win some space for itself in its local context. While its shape is always determined by its nomadic articulations, it always has a shape which is itself effective. (1987: 39. See also Grossberg, 1982 and 1988)

Similarly, Janice Radway (1988) has employed the notion of the nomadic subject to provide the necessary conception of readers/viewers as active producers of meaning in their engagement with texts. (See also Meaghan Morris, 1988)[6]

Related metaphors of travel here are the idea of the 'billboard' (Grossberg, 1987; Morris, 1988) as signposts which 'do not tell us where we are going but merely announce . . . the town we are passing through'. (Grossberg, 1987; 31); and the concept of the 'commuter', also suggested by Grossberg (1988: 384).[7]

Off the road: women and travel

So far, I have located the emergence of vocabularies of travel in three related major theorical developments; postcolonial criticism, postmodern theory and poststructuralist theories of the subject. There is no doubt that in each case the metaphors I have identified have proved useful and suggestive, as well as promising solutions to the ideological effects of dominant terminologies. In all three cases, it is easy to see why notions of mobility, fluidity, provisionality and process have been preferable to alternative notions of stasis and fixity. In cultural criticism in the late twentieth century we have had to realize that only ideologies and vested interests 'fix' meaning, and it is the job of cultural critics to destabilize those meanings.

Of course, this work has already been criticized on a number of counts. The radical relativism of some of these texts has proved unacceptable for those who are not prepared to abandon certain meta-narratives. In addition, there is a tension between what we might think of as the more and less radical versions of semiotics: in short on the question of what (if anything) lies behind the play of signifiers in our culture. From the point of view of engaged politics, and here specifically in relation to feminism, a certain postmodern stance is incompatible with the fundamental commitment to a critique which is premised on the existence of systematically structured, actual, inequalities (of gender). I don't want to rehearse the various critiques of 'post' theories here, but instead to focus on the narrower case of a possible feminist critique of travel metaphors.

I should start by explaining the *un*theoretical, and coincidental, origin of my unease with this vocabulary, which was twofold. Like many other people, I had read Bruce Chatwin's *The Songlines* (1987) a few years ago, when it first came out. Also like many of its readers, I found it a compelling journal of the author's travels in Australia. At the same time, I felt, in a somewhat unarticulated way, alienated from it as a 'masculine' text. In particular, in a long section of the book which consists entirely of quotations about travel (including from Chatwin's own diaries from other travels), I had the sense that women did not travel like this.[8] The fact that Clifford and others cite this particular text mobilized that reaction again.

Secondly, I had been doing some work on the 1950s – specifically on the fantasy of 'America' in Britain in that decade. In this connection, I was reading newly published accounts of the period by the women of the Beat generation: Joyce Johnson, Carolyn Cassady, Hettie Jones. This was the other side of the stories we already knew, which was until now unrecorded – as the title of Carolyn Cassady's book puts it, what it was like 'off the road'. Reading Johnson (one of Kerouac's women) and Cassady together was illuminating – one woman on each coast, both waiting for Jack or Neal (in Cassady's case, sometimes for both) to get back. Johnson once wondered if she could join Kerouac *on* the road:

> In 1957, Jack was still traveling on the basis of pure, naive faith that always seemed to renew itself for his next embarkation despite any previous disappointments. He would leave me very soon and go to Tangier . . . I'd listen to him with delight and pain, seeing all the pictures he painted so well for me, wanting to go with him. Could he ever include a woman in his journeys? I didn't altogether see why not. Whenever I tried to raise the question, he'd stop me by saying that what I really wanted were babies. That was what all women wanted and what I wanted too, even though I said I didn't . . . I said of course I wanted babies someday, but not for a long time, not now. Wisely, sadly, Jack shook his head. (Johnson, 1983: 126)[9]

Reading these texts, I was already sensitized to certain questions of gender and travel, and perhaps suspicious of the appearance of travel metaphors in cultural theory. But this, of course, only raises the question of whether such metaphors are gendered – it doesn't decide the issue.[10]

Histories of travel make it clear that women have never had the same access to the road as men:

> In many societies being feminine has been defined as sticking close to home. Masculinity, by contrast, has been the passport for travel. Feminist geographers and ethnographers have been amassing evidence revealing that a principal difference between women and men in countless societies has been the licence to travel away from a place thought of as 'home'. (Cynthia Enloe, 1989: 21)

Of course, women do have a place in travel, as also in tourism. Often that place is marginal and degraded. John Urry (1990) and Cynthia Enloe (1989) both discover women in the tourist industry in the role of hotel maids, or active in sex-tourism. I will come back to the question of women who do travel. For the moment, I am simply recording the limited access and problematic relationship women have generally had to varieties of travel. (Another example, from Judith Adler's history of tramping [1985], also confirms the tendency for such undirected mobility to be the preserve of men.)

I have yet to move from the recognition of travel as predominantly what men do to, first, an argument that there is something *intrinsically* masculine about travel, and, secondly, that therefore there are serious implications in employing travel *metaphors*. At this point, though, I should note that at least two of the cultural critics I have been discussing recognize that there is a problem. James Clifford acknowledges that travel, and therefore travel metaphors, are not gender-neutral:

> The marking of 'travel' by gender, class, race, and culture is all too clear . . . 'Good travel' (heroic, educational, scientific, adventurous, ennobling) is something men (should) do. Women are impeded from serious travel. (1992: 105)

Meaghan Morris, who, as I have said, has also employed travel metaphors in her recent work, makes the same point:

But, of course, there is a very powerful cultural link – one particularly dear to a masculinist tradition inscribing 'home' as the site both of frustrating containment (home as dull) and of truth to be rediscovered (home as real). The stifling home is the place from which the voyage begins and to which, in the end, it returns . . . The tourist leaving and returning to the blank space of the *domus* is, and will remain, a sexually in-different 'him' (1988: 12).

Her suggestion is that the metaphor of the 'motel' may prove more appropriate for a non-androcentric cultural theory.

With its peculiar function as a place of escape yet as a home-away-from . . . home, the motel can be rewritten as a transit-place for women able to use it . . . Motels have had liberating effects in the history of women's mobility.(2)

The question is: What is the link between women's exclusion from travel, and uses of notions of travel in cultural theory and analysis? (And then: Will modified metaphors of travel avoid the risk of androcentrism in theory?)

Masculinity and travel

If it is only a *contingent* fact that, as Eric Leed says, 'historically, men have traveled and women have not', there might be no reason to argue that the vocabulary of travel is irrevocably compromised and, hence, unacceptable to cultural criticism. Here, I want to explore the possibility that the connection isn't just contingent, but that there is an *intrinsic* relationship between masculinity and travel. (By 'intrinsic', though, I do *not* mean 'essential'; rather my interest is in the centrality of travel/mobility to *constructed* masculine identity.) Leed himself has a fairly straightforward view of what he calls the 'spermatic journey'. He argues that it is likely that 'much travel is stimulated by a male reproductive motive, a search for temporal extensions of self in children, only achievable through the agency of women' (1991: 114). On this view, women's identification with place is the result of reproductive necessities that require stability and protection by men.[11] Such an account, however, does not really get us very far in explaining the persistence of these arrangements in totally transformed circum-stances.

Mary Gordon has recently argued that men's journeys should be construed as a flight *from women*. In an essay on American fiction, she notes the centrality of the image of motion connected with the American hero (authors she discusses include Faulkner, Dreiser and Updike). At work here is 'a habit of association that connects females with stasis and death; males with movement and life' (1991: 17). Indeed, she notes how frequently in such fiction the females have to be killed to ensure the man's escape. According to Gordon, 'the woman is the centripetal force pulling [the hero] not only from natural happiness but from heroism as well' (6). I think it would be possible to pursue this suggestion in psychoanalytic terms (though I don't propose to do this here): for example, feminists have used the (very different) work of Nancy Chodorow and Julia Kristeva to explore the male investment in strong ego boundaries, and the consequent and continuing fear of engulfment (in the female) and loss of self.

In MacCannell's account of tourism, the search for 'authenticity' is fore-grounded. By this he means the attempt to overcome the sense of fragmentation and to achieve a 'unified experience', which is less to do with the 'authentic' *self* than a quest for an authentic *social* meaning. But tourism and sightseeing can be seen just as much to operate as productive for the 'postmodern self' so frequently diagnosed by sociologists, and to that extent I think the gender implications are equally clear. (If, that is, we do take the view that in our culture men have a different and exaggerated investment in a concept of a 'self'.) Some years ago, two British sociologists wrote a book entitled *Escape Attempts*, whose subtitle is 'The theory and practice of resistance to everyday life' (Cohen and Taylor, 1976).[12] I had always wondered why one would need to 'resist' everyday life (by which the authors mainly mean routine, meaninglessness, the domestic, repeti-tiveness). Certainly the particular account they give, both *of* that everyday life, and of the types and strategies of resistance discussed (including fantasy, hobbies, role-distance, holidays), are, to say the least, extraordinarily *male*. Could such a text be written from the point of view of women, or in a gender-neutral way? My sense is that it probably couldn't.

My suggestion that a connection of masculinity/travel/self can be made is not unproblematic. First, I could argue the opposite case, based on the same theories. For example, since (according to Chodorow and others) women are produced as gendered subjects at the expense of any clear sense of self (of definite ego boundaries) – the result of inadequate separation from the mother – one might think that women have *more* of an investment in discovering a 'self' and that, if travel is a mode of discovery, then this would have a strong attraction to women. (Indeed, Dea Birkett, in her study of Victorian lady travellers, has suggested that it operated this way for some women, whose fragile sense of identity collapsed on the death of parents in relation to whom such women defined themselves [Birkett, 1991: 71]). Here I think the important distinction is between the defence of a precarious but already constructed self (the masculine identity) and an unformed, and less crucial (*to* identity) sense of self. The investment in travel in relation to the former seems to me to be potentially far greater.

Secondly, the notion of feminine identity as relational, fluid, without clear boundaries seems more congruent with the perpetual mobility of travel than is the presumed solidity and objectivity of masculine identity. And thirdly, and related to this, it might be argued that women have an interest in destabilizing what is fixed in a patriarchal culture (as those who propose an alliance between feminism and postmodernism have suggested), and hence that methods and tactics of movement, including travel, seem appropriate.[13] I want to return to these last two points later, when I will suggest that such destabilizing has to be from *a location*, and that simple metaphors of unrestrained mobility are both risky and inap-propriate.

I haven't set myself the task of analysing in depth the gendered nature of travel and escape, and all I have done in this section is to indicate some of the ways in which this might be pursued. My interest in this paper is to explore how metaphors of travel work. So far, I hope to have shown that they are (in fact and perhaps in essence) androcentric. For although I do think it is interesting to pursue the possible connection between masculinity and travel, my real interest is in the discursive construction involved – in other words, in the ways in which narratives of travel, which are in play in the metaphoric use of the vocabulary, are gendered. As Georges van den Abbeele has recently argued (1992; xxv–xxvi), although there is nothing inherently or essentially masculine about travel, in the sense that women have certainly travelled, nevertheless 'Western ideas about travel and the concomitant corpus of voyage literature have generally – if not characteristically – transmitted, inculcated, and reinforced patriarchal values and ideology'. The discourse of travel, he argues, typically functions as a 'technology of gender'.

Women who travel

The major objection to my argument so far is that, in fact, women *do* travel.[14] The case of the 'Victorian lady travellers' is the prime example.[15] Isabella Bird, Isabelle Eberhardt, Mary Kingsley, Freya Stark, Marianne North, Edith Durham and many other redoubtable women at the end of the nineteenth and beginning of the twentieth century left homes that were often extremely constraining to travel the world in the most difficult and challenging circumstances. Their lives and their journeys have been well documented, both by themselves and by subsequent historians (Eberhardt, 1987; Birkett, 1991; Middleton, 1965; Russell, 1988). If indeed travel is gendered as male, and women's travel restricted, here is a case when at a moment of exaggerated gender ideologies of women's domestic mission the most dramatic exceptions occur. It is interesting, here, to consider how this travel was construed, and constructed, both by the travellers themselves and by the cultures they left and returned to.

Dea Birkett has suggested that in an important way these women inhabited the position of men. In fact, they rarely dressed as men (Hester Stanhope and Isabelle Eberhardt were among the few who did) though they did, like Isabella Bird, sometimes modify their dress in a practical way. Nor did they define themselves as anything other than 'feminine' (and in many cases, also anti-feminist). But the relationship of authority they unquestioningly entered with the natives of places they visited and traversed (they were often addressed as 'Sir': Birkett, 1991: 117), overrode considerations of gender:

> As women travellers frequently pointed to the continuities and similarities with earlier European male travellers, the supremacy of distinctions of race above those of sex allowed them to take little account of their one obvious difference from these forebears – the fact they were female. (125)

In addition, Birkett suggests that many women travellers had a strong identification with their fathers from an early age, and through that identification learned to value the prospect of escape and freedom, since several (Ella Christie and Mary Kingsley for example) had fathers who travelled widely:

> Their mothers', and sometimes sisters', domestic spheres were associated with cloistered, cramped ambition and human suffering. In response, they created their own sense of stability and belonging in exploring their paternal ancestors, thereby reinforcing their identity with their father and his lineage. (18)

When women do travel, then, their *mode* of negotiating the road is crucial. The responses to the lady travellers by those back home are also illuminating in their own contradictory negotiation of a threatening anomaly. Dea Birkett discusses the reactions of the press, other travellers and members of the Royal Geographical Society, which included minimizing the travels (in relation to those undertaken by male explorers), stressing the 'femininity' of these women (despite such masculine pursuits), and hinting that their conduct overseas might well have been improper. In other words, we do not need to discover that women travellers were in any straightforward sense 'masculine' to conclude that their activities positioned them in important ways as at least problematic with regard to gender identification.[16]

The gendering of travel is not premissed on any simple notion of public and private spheres – a categorization which feminist historians have shown was in any case more an ideology of place than the reality of the social world. What is in operation here, I think, *is* that ideology. The ideological construction of 'women's place' works to render invisible, problematic, and in some cases impossible, women 'out of place'. Lesley Harman, in her study of homeless women in Toronto, shows how the myth of home constructs homeless women in a very different way from homeless men. As she puts it, 'the very notion of "homelessness" among women cannot be invoked without noting the ideological climate in which this condition is framed as problematic, in which the deviant categories of "homeless woman" and "bag lady" are culturally produced' (1989: 10). It seems to me not entirely frivolous to consider the hysterical and violent responses to the film *Thelma and Louise* in the same way. As Janet Maslin has pointed out (1991), the activities of this travelling duo are as nothing compared with the destruction wrought in many male road movies. She writes in response to an unprecendented barrage of hostile reviews, of which one in *People Weekly* (10.6.91) is an example:

> Any movie that went as far out of its way to trash women as this female chauvinist sow of a film does to trash men would be universally, and justifiably, condemned. . . . The movie portrays Sarandon and Davis as sympathetic. . . . The music and the banter suggest a couple of good ole gals on a lark; the content suggests two self-absorbed, irresponsible, worthless people.

My argument is that the ideological gendering of travel (as male) both impedes female travel and renders problematic the self-definition of (and response to) women who *do* travel. As I have said, I don't claim to have offered an analysis of this gendering (though I have suggested that, for example, a psychoanalytic account would be worth pursuing). Nor, once again, am I arguing that women *don't* travel. I have been primarily interested in seeing how metaphors and ideologies of travel operate. In the final section, I will consider the implications of this for a cultural theory which relies on such metaphors.

Feminism, travel and place

By now, many feminists have made the point about poststructuralist theory that just as women are discovering their subjectivity and identity, theory tells us that we have to deconstruct and de-centre the subject. Susan Bordo has identified the somewhat suspicious timing by which 'gender' evaporates into 'genders' at the moment in which women gain some power in critical discourse and academic institutions (Bordo, 1990). In the same way, I'd like to suggest that just as women accede to theory, (male) theorists take to the road. Without claiming any conspiracy or even intention, we can see what are, in my view, exclusionary moves in the academy. The already-gendered language of mobility marginalizes women who want to participate in cultural criticism. For that reason, I believe there is no point in tinkering with the vocabulary of travel (motels instead of hotels) to accommodate women. Crucially, this is still *the wrong language*.

How is it that metaphors of movement and mobility, often invoked in the context of radical projects of destabilizing discourses of power, can have conservative effects? As I said earlier, one would think that feminism, like postcolonial criticism, could only benefit from participating in a critique of stasis. Here I think we confront the same paradox as in the proposed alliance between feminism and postmodernism. The appeal of postmodernism lies in its demolition of grand narratives (narratives which have silenced women and minorities). The problem with an overenthusiastic embrace of the postmodern is that that same critique undermines the very basis of feminism, itself necessarily a particular narrative. Feminists have only reached provisional conclusions here based on either a *relative* rejection of grand narratives, or a pragmatic retention of (less grand) theory.

In the same way, I think that destabilizing has to be situated, if the critic is not to self-destruct in the process. The problem with terms like 'nomad', 'maps' and 'travel' is that they are not usually located, and hence (and purposely) they suggest ungrounded and unbounded movement – since the whole point is to resist fixed selves/viewers/subjects. But the consequent suggestion of free and equal mobility is itself a deception, since we don't all have the same access to the road. Women's critique of the static, the dominant, has to acknowledge two important things: first, that what is to be criticized is (to retain the geographic metaphor) the dominant centre; and secondly, that the criticism, the destabilizing tactics, originate too from a place – the margins, the edges, the less visible spaces. There are other metaphors of space which I find very suggestive, and which may be less problematic, at least in this respect: 'borderlands', 'exile', 'margins' – all of which are premissed on the fact of dislocation from a given, and excluding, place. Elspeth Probyn (1990) recommends we start from the body – what Adrienne Rich has called 'the politics of location' (1986) – to insist on the situated nature of experience and political critique. Caren Kaplan's (1987) use of the notion of 'de-territorialization' similarly assumes a territory from which one is displaced, and which one negotiates, dismantles, perhaps returns to.

For all these metaphors, there *is* a centre. In a patriarchal culture we are not all, as cultural critics any more than social beings, 'on the road' together. We therefore have to think carefully about employing a vocabulary which, liberatory in many ways, also encourages the irresponsibility of flight and misleadingly implies a notion of universal and equal mobility. This involves challenging the exclusions of a metaphoric discourse of travel.

Metaphors, though, are not static.[17] My critique of the specific metaphors of travel in relation to gender should not, therefore, be read as either a ban on metaphors (which are inevitable in thought and writing, and which always import certain limits and ideologies), or as a definitive condemnation of travel metaphors, but rather as a provisional and situated analysis of the current working of discourse. In the end, too, a different critical strategy might be the *reappropriation*, not the avoidance, of such metaphors – a good postmodern practice which both exposes the implicit meanings in play, and produces the possibility of subverting those meanings by thinking against the grain.

Notes

1 The real proof of this is that in David Lodge's novel, *Paradise News* (Viking, 1992), there is a character who is a professor investigating the sightseeing tour as secular pilgrimage.

2 An early example of the current interest in travel and theory, however, was Georges van den Abbeele's (1980) review of MacCannell's book, four years after its publication.

3 In fact, first published in 1981 but, as far as I can see, provoking more of a response in the last few years.

4 For example, the full-length, front-page symposium in *The New York Times Book Review*, 18 August 1991, entitled 'Itchy feet and pencils', in which Jan Morris, Russell Banks, Robert Stone and William Styron discuss travel writing.

5 In a different use of the metaphor of a 'map', Iain Chambers, describing the intellectual as a 'humble detective', explicitly abandons the idea of social totality (1987).

6 My knowledge of Deleuze's work is mostly secondary, but it is worth pointing out that in his essay 'Nomad thought', Deleuze means something rather different: the nomad as someone who opposes centralized power (1977: 148–9). This doesn't seem to have anything to do with de-centred subjects, but more with the idea of displaced (groups of) people, able to contest authority from the outside.

7 This metaphor does not work so well, I think. Grossberg says this:

> Nomadic subjects are like 'commuters' moving between different sites of daily life . . . Like commuters, they are constantly shaped by their travels, by the roads they traverse . . . And like commuters, they take many different kinds of trips, beginning from different starting-points, punctuated by different interruptions and detours, and arriving at different stopping-points (1988: 384).

But I would have thought that the central characteristic of commuting is that you always start from the *same* starting-point and end at the *same* stopping-point: primarily, of course, home/work.

8 In going back to read that section, some forty pages of text, quotation and aphorism, I couldn't find very much to support my sense of it as 'masculine'. It even included a reference to nomadic cultures in which it is the women who initiated the move. One entry was clearly 'male' – even misogynistic:

> To the Arabian bedouin, Hell is a sunlit sky and the sun a strong, bony female – mean, old and jealous of life – who shrivels the pastures and the skin of humans.
>   The moon, by contrast, is a lithe and energetic young man, who guards the nomad while he sleeps, guides him on night journeys, brings rain and distils the dew on plants. He has the misfortune to be married to the sun. He grows thin and

> wasted after a single night with her. It takes him a month to recover. (Chatwin, 1987: 201)

Nevertheless, I retained the feeling that I was 'reading as a man' here – a feeling which partly motivated the initial attempt in this essay to analyse that feeling.

9 This is also cited by Alix Kates Shulman in her review of *Minor Characters* (1989).

10 For one thing, it may be that the fifties, and more particularly the so-called Beat Generation, constituted a very specific phenomenon, in which case any generalization would be totally misconceived. It's part of my project here to examine *how* general the gendering of travel might be. But here I might also mention a recent piece of journalism, which replays Kerouac/Cassady, if in ironic form. Nicolas Cage, the movie actor, wrote a piece for the magazine *Details* (July 1991) entitled 'On the road, again: retracing Kerouac's footsteps in the wild heart of the country', documenting his drive from LA to New Orleans and his experiences and reflections en route. To me, both the events and the recollections seem very much in the Kerouac mode, despite a certain self-awareness and irony (for example, in relating the fact that the first car he took broke down when he and his friend were 'still comfortably within the 213 area code').

11 This is an argument that was used, a few years ago, by feminist anthropologists concerned to explain the historical and apparently universal oppression and domestication of women. Here, as in its other manifestations, it is an argument that raises as many problems as it solves.

12 This is soon to be reissued, by Routledge, in a second edition.

13 Here Deleuze's sense of 'nomad criticism' (see note 6) is more appropriate than the usage I have taken up in this paper.

14 Often, the fact of women's travels has been obscured by historians and other narrators. Gordon DesBrisay has pointed out to me that recent historical research has shown that in early modern Scotland women were more mobile than has generally been thought. (Whyte, 1988)

15 I could, of course, have taken other examples – contemporary women explorers, round-the-world yachtswomen, female truckers, for instance. Many of the suggestions I make here about the Victorians would then be likely to be seen as specific to their case.

16 Box Car Bertha, who spent her life on the road, took advantage of similar ambiguities in gender identification, in this case in her unusual upbringing.

> My childhood was completely free and always mixed up with the men and women on the road. There weren't many dolls or toys in my life but plenty of excitement . . . We took for playthings all the grand miscellany to be found in a railroad yard. We built houses of railroad ties so big that it took four of us to lift one of them in place. We invented games that made us walk the tracks . . . We played with the men's shovels and picks and learned to use them . . . We girls dressed just like the boys, mostly in hand-me-down overalls. No one paid much attention to us. (in St Aubin de Teran, 1990: 48–9)

17 Indeed, van den Abbeele points out that the word 'metaphor' comes from the Greek word which means to transfer or transport (1992: xxii).

References

Abbeele, Georges van den (1980): 'Sightseers: the tourist as theorist', *Diacritics* December.

—— (1992): *Travel as Metaphor: From Montaigne to Rousseau*, University of Minnesota Press.

Adler, Judith (1985): 'Youth on the road. Reflections on the history of tramping', *Annals of Tourism Research* Vol. 12.

Birkett, Dea (1991): *Spinsters Abroad. Victorian Lady Explorers*, London: Victor Gollancz Ltd.

Bordo, Susan (1990): 'Feminism, postmodernism, and gender-scepticism', in Nicholson (1990).

Cage, Nicolas (1991): 'On the road, again', *Details* July.

Cassady, Carolyn (1990): *Off the Road. My Years with Cassady, Kerouac, and Ginsberg*, New York: Wm Morrow & Co.

Chatwin, Bruce (1987): *The Songlines*, Harmondsworth: Penguin.

Clifford, James (1989): 'Notes on travel and theory', *Inscriptions* 5.

—— (1992) 'Travelling cultures', in Grossberg, Nelson and Treichler (1992).

Cohen, Stanley and Taylor, Laurie (1976): *Escape Attempts. The Theory and Practice of Resistance to Everyday Life*, Allen Lane. New edition to be published by Routledge, 1992.

Culler, Jonathan (1988): 'The semiotics of tourism', in *Framing the Sign. Criticism and its Institutions*, University of Oklahoma Press.

Deleuze, Gilles (1977): 'Nomad thought', *The New Nietzsche*, David Allison (ed.) New York: Delta.

Eberhardt, Isabelle (1987): *The Passionate Nomad. The Diary of Isabelle Eberhadt*, London: Virago.

Enloe, Cynthia (1989): *Bananas, Beaches and Bases. Making Feminist Sense of International Politics*, London: Pandora Press; (University of California Press, 1990).

Frow, John (1991): 'Tourism and the semiotics of nostalgia', *October* 57, Summer.

Gordon, Mary (1991): 'Good boys and dead girls', in *Good Boys and Dead Girls and Other Essays*, Viking.

Grossberg, Lawrence (1982): 'Experience, signification, and reality: the boundaries of cultural semiotics', *Semiotica* 41–1//4.

—— (1987): 'The in-difference of television', *Screen* 28, 2

—— (1988): 'Wandering audiences, nomadics critics', *Cultural Studies* Vol. 2, No. 3, October.

Grossberg, Lawrence, Nelson, Cary and Treichler, Paula (editors) (1992): *Cultural Studies*, London: Routledge.

Harman, Lesley D. (1989): *When a Hostel becomes a Home. Experiences of Women*, Toronto: Garamond Press.

Jameson, Fredric (1984): 'Postmodernism, or the cultural logic of late capitalism', *New Left Review* 146.

—— (1988): 'Cognitive mapping', in Nelson and Grossberg (1988).

Johnson, Joyce (1983): *Minor Characters*, London: Picador.

Kaplan, Caren (1987): 'Deterritorializations: the rewriting of home and exile in western feminist discourse', *Cultural Critique* No. 6.

Leed, Eric J. (1991): *The Mind of the Traveller, From Gilgamesh to Global Tourism*, New York: Basic Books.

MacCannell, Dean (1989): *The Tourist. A New Theory of the Leisure Class*, New York: Schocken Books [1976].

Maslin, Janet (1991): 'Lay off "Thelma and Louise",' *The New York Times* 16 June.

Middleton, Dorothy (1965): *Victorian Lady Travellers*, Chicago: Academy Chicago Books.

Morris, Meaghan (1988): 'At Henry Parkes Motel', *Cultural Studies* Vol. 2, No. 1, January.

Nelson, Cary and Grossberg, Lawrence (editors) (1988): *Marxism and the Interpretation of Culture*, London: Macmillan.

Nicholson, Linda J. (editor) (1990): *Feminism/Postmodernism*, London: Routledge.

Probyn, Elspeth (1990): 'Travels in the postmodern: making sense of the local', in Nicholson (1990).

Radway, Janice (1988): 'Reception study: ethnography and the problems of dispersed audiences and nomadic subjects', *Cultural Studies* Vol. 2, No. 3, October.

Rich, Adrienne (1986): 'Notes towards a politics of location (1984)', in *Blood, Bread and Poetry, Selected Prose 1979–1985*, New York: W. W. Norton & Company.

Russell, Mary (1988): *The Blessings of a Good Thick Skirt. Women Travellers and their World*, London: Collins.

Said, Edward W. (1983): 'Traveling theory', in *The World, The Text, and the Critic*, Harvard University Press.

Shulman, Alix Kates (1989): 'The Beat Queens. Boho chicks stand by their men', *Voice Literary Supplement* June.

St Aubin de Teran, Lisa (editor) (1990): *Indiscreet Journeys. Stories of Women on the Road*, London: Faber & Faber.

Urry, John (1990): *The Tourist Gaze, Leisure and Travel in Contemporary Societies*, London: Sage.

Whyte, Ian D. (1988): 'The geographical mobility of women in early modern Scotland', in *Perspectives in Social History: Essays in Honour of Rosalind Mitchison*, Leah Leneman (ed.), Aberdeen University Press.

SECTION FOUR

NO PLACE LIKE HOME: THE REST IN THE WEST

Editorial introduction

In Section Four I want to turn explicitly to the question of 'race' and ethnicity which has already been partly addressed in Sections Two and Three. 'Race' is a key dimension of socio-spatial divisions in contemporary western societies. It is common practice to refer to 'race' in inverted commas to acknowledge the acceptance that 'race' is a social construct rather than a distinctive biological difference between human beings. The term is, however, useful to retain, especially as it allows the ideology of 'race' and hence discrimination on its supposed 'natural' basis to be captured in the term 'racism'.

Here, in Section Four, instead of focusing exclusively on community at the local scale, on place as a bounded location constructed through mechanisms of exclusion, I want to examine the links between 'race' and ethnicity, culture, nationality and identity. I want to look at movement and fluidity, changing places, at the development and theorisation of 'creolisation, metissage, mestizaje, and hybridity' (Gilroy, 1993a), what in defining the last of these four terms Homi Bhabha has termed 'cultural seepage'. (As I wrote this sentence I was struck by the often perjorative connotations of the terms laid out by Gilroy and the different associations of a term that has perhaps had a longer currency: that of cosmopolitanism.) But whatever terms we use, it is clear that the relationship between ethnicity, identity and territory is one of the most important questions at the end of the twentieth century as associations of identity and territory, at a range of scales, are torn asunder and reconstructed through war and migration, through the vilely named ethnic cleansing in Europe and the tribal wars in Africa, as well as through the collapse of communism in Europe and new forms of political organisation that either break up or transcend the nation state. Thus the connections to a 'homeland' are unclear for more and more people, wrenched from their place of birth by global proletarianisation and famine, as well as war, and living in what Said (1985) called a 'generalized condition of homelessness' (p. 18).

In the 'old' world, the nations of western Europe with well-developed discourses of nationality, of what it means to be French or to be English for example, in-migrants, who may or may not be literally homeless in the sense of without a place/home to call their own, find themselves 'homeless' in a more general sense, and at a larger spatial scale. They are excluded from belonging to the nation by a narrative or discourse that defines them as Other, as outsiders, or strangers. As Gilroy bitterly argued, 'there ain't no black in the Union Jack'. Ethnic difference is used to distinguish people one from another and construct an exclusive definition of nationality in the same way as communities are defined on the basis of exclusionary tactics. Through a set of national myths, traditions, associations (between the state and the Church of England, for example) the idea of Englishness as ethnically homogeneous, as white and Anglo-Saxon is dominant. A brief look at centuries of 'English' history, marked by invasion from north and south as well as a liberal reputation as a place of refuge for those persecuted on the grounds of religious or political beliefs (from the Huguenots to the Jews, but no longer under the xenophobic and restrictive legislation of the right-wing Governments of the last years), makes it immediately clear the extent to which this myth of Englishness is exactly that: a myth. Indeed, an earlier counter narrative was that of the dogged English mongrel, proud of her/his mixed origins.

Chambers (1989) in a paper examining the 'narratives of nationalism' traces the origins of contemporary notions of Englishness to the nineteenth century.

> The construction of an 'English' history in the nineteenth century, both in the world of academia and, more effectively, in the literary milieu of the period, impregnated the whole debate on culture with this national mythology. In its neo-Gothic architecture, Pre-Raphaelite paintings, chivalric poetry, socialist utopias, and its insistence on the earlier harmonies of rural life and artistic production, Victorian intellectuals of the most varied political persuasion sealed a pact between a timeless and mythical vision of the nation and their selection and installation of an acceptable cultural heritage. (p. 89)

As Chambers goes on to point out, this heritage 'has subsequently bequeathed that deep-seated intellectual and moral ambiguity towards modernism, mass culture, mass democracy, commerce, and urban life that has so frequently characterised English intellectual thought and official culture in the twentieth century' (p. 89).

The first extract here is taken from the book by **Paul Gilroy** mentioned above, in which he explains how English nationalism excludes Black Britons. It is a particularly interesting chapter for geographers as he is extremely critical of the work of both Benedict Anderson and Raymond Williams: work that has had a significant influence on geographic scholarship. (David Harvey, for example, in his recent book *Justice, Nature and the Geography of Difference* (Harvey, 1996) finds a powerful voice in Williams' fictional discussions of border identity and places on the margin with no comment on how his work ignores questions of 'race' and racism.) Williams, of course, is exactly in that tradition of the English intellectuals referred to by Chambers in the quotation above. Anderson's arguments about the construction of the nation as an 'imagined community' in which disparate individuals with no personal ties to each other are bound together by common myths and symbolic acts, whatever their social position, have also been applied by geographers at a range of spatial scales. But Gilroy disputes his conclusions and demonstrates the ways in which the politics of 'race' as well as institutional and legal frameworks exclude black people.

The history of migration, of ethnic difference and 'race' is different, of course, in the UK and the USA. Although there has been a long history of migrations, a significant 'black' presence was a post-Second World War phenomenon in Britain which coincided with the end of empire. Unlike the US experience, British slave owners held land and people 'off shore' as it were rather than in the same territory. The post-war movement of African-American people was from the south to the northern cities rather than an overseas migration. These movements and migrations of people from the periphery to the metropolitan centres, that Stuart Hall has captured in his marvellous reversal of the phrase 'the West and the rest' ('the rest in the West'), disrupt that Victorian narrative outlined by Chambers, unsettling the 'native' associations of place and identity. The idealised narrative of Englishness can no longer be contained within, for example, John Major's nostalgic set of associations, 'the thwack of the willow on the wood, spinsters cycling to church in the morning mist, warm beer', however much the Conservative party try to deny social change in contemporary Britain. In the post-imperial world, older nationalist rhetorics have been disrupted by multicultural symbols, meanings and sets of beliefs and behaviours. Gilroy's analysis of the 1980s raises interesting questions for a newly elected Labour Government reclaiming the rhetoric of 'One Nation' at the end of the 1990s. England is now a hybrid mixture of 'Roast Beef and Reggae Music'. The difficult questions raised for white liberals who do not want to claim the traditional conservative identity of Englishness but prefer to identify with 'Black' struggles and cultural movements are addressed in a personal piece of writing by **Diana Jeater** (Chapter 16). It is interesting to note that Kerouac, whose nomadism was discussed in Section Three, sometimes referred to himself as a White Negro in the context of his passion for jazz and his visits to predominantly Black clubs in New York and other cities. Mercer (1992) has suggested that in the USA Black culture, especially music, influenced 'white' culture in the 1960s and often led to a 'White' jealousy of, or desire for, blackness.

Jeater ends on a pessimistic note (and perhaps she might feel even more pessimistic if she were writing now rather than in the early 1990s). She concludes 'the fact that we missed that "moment of hybridity" frightens me. I wonder what we will produce in its place'. Stuart Hall too has argued that the reactions to the development of what he terms 'new ethnicities' (1992) might lead either to negative or more positive changes. As we saw in Section Three, ethnicity may be the basis of exclusionary organisation and actions, sometimes based on violence. The long-standing struggles in Ireland between 'nationalists' and 'loyalists' are one example which has been repeated too many times in contemporary Europe. And, as we also saw in Section Three, the repressed and the excluded often also develop a politics of resistance based on a solidaristic reaction to their exclusion. In the case of ethnicity (and sexuality) this may take the form of a positive claiming of the very attributes that construct excluded groups as Other – the black pride slogan of 'black is beautiful'; gay pride marches and queer politics or in the politics of Black masculinity in the Nation of Islam movement. The response to the assertion of ethnic identity may also be negative, in the resurgence of politics of racism in Europe, for example, based in the UK on a white working class 'Little Englander' myth. Further, as bell hooks pointed out in the context of race and gender politics, there are inherent dangers in the tendency of reversal, as it entails claiming the very characteristics by which inferiority is asserted. Hall also sees these dangers and instead longs for what, after Homi Bhabha, he terms 'translation': the melding and mixing that produces a 'third space' or 'third identity' that is neither One nor the Other but replaces both and out of which vibrant new cultural policies may emerge.

In the extract from **Stuart Hall's** work that I have included here, he turns specifically to questions of cultural production and the ways in which they both reflect and affect ethnic identities. As he argues, Black Caribbean identities have to be negotiated in the tension between poles of similarity or continuity and of difference and rupture. He draws on the work of Derrida to capture the complexity of this doubled difference in a way that is more nuanced than Otherness. The concept of doubleness has also been productively explored in Paul Gilroy's recent work. In his scholarly book *The Black Atlantic: Modernity and Double Consciousness* (Gilroy, 1993a), he suggests that African Americans and British peoples stand between two great cultural assemblages that have formed them: their origins in Africa and their history in the West, both of which cultural formations have been transformed in the modern period. Gilroy argues that a new transformative and progressive position in the space between these poles might be possible. This argument is similar to Hall's conception of cultural translation.

In the final chapter in this part, **Kevin Robins** picks up the idea of translation but examines it in a rather different sense and with a specific spatial emphasis. He is interested in the ways in which global migration and cultural hybridity, as well as economic restructuring and the marketing of 'ethnicity' and heritage might be transforming local difference, leading to a more globally homogeneous world. This argument returns us in a way to the fears about postmodernism outlined in the introduction. The overall emphasis in the papers here in Section Four on narratives, myths and imaginary histories and geographies, however, takes us straight on into Section Five where the significance of imagined places is the focus.

References and further reading

Anderson, B. 1983: *Imagined communities*. London: Verso.

Bhabha, H. 1994: *The location of culture*. London: Routledge.

Chambers, I. 1989: Narratives of nationalism: being 'British'. *New Formations* **7**, 88–103.

Chambers, I. 1994: *Migrancy, culture identity*, London: Routledge.

Frankenburg, R. 1993: Growing up white: feminism, racism and the social geography of childhood. *Feminist Review* **45**.

Frankenburg, R. 1994: *White women, race matters*. London: Routledge.

Gilroy, P. 1993a: *The Black Atlantic: Modernity and Double Consciousness.* London: Verso.

Gilroy, P. 1993b: *Small acts: thoughts on the politics of black cultures.* London: Serpent's Tail.

Hall, S. 1991: Old and new identities: old and new ethnicities. In King, A. (ed.) *Culture, globalisation and the world system.* London: Routledge.

Hall, S. 1995: Negotiating Caribbean identities. *New Left Review* **209**, 3–14.

Harvey, D. 1996: *Justice, nature and the geography of difference.* Oxford: Blackwell.

hooks, b. 1992: Representing whiteness in the black imagination. In *Black Looks: race and representation.* Boston: South End Press.

Jackson, P. 1989: *Maps of meaning.* London: Unwin Hyman (reprinted 1992 by Routledge).

James, W. 1992: Migration, racism and identity. *New Left Review* **193**, 15–55.

Kureishi, H. 1991: *The Buddha of suburbia.* London: Faber.

Maitland, S. 1988. *Very heaven: looking back at the 1960s.* London: Virago.

Mercer, K. 1992: '1968' periodizing politics and identity. In Grossberg, L., Nelson, C. and Treichler, P. (eds), *Cultural Studies.* London: Routledge, 424–39a.

Mohanty, C. T. 1991: Cartographies of struggle: Third World women and the politics of feminism. In the introduction to Mohanty, C. T. et al. *Third World women and the politics of feminism.* Bloomington, Indiana: Indiana University Press.

Pratt, M. B. 1988: Identity: skin blood heart. In Bulkin, E. et al. (eds) *Yours in struggle: three feminist perspectives on anti-semitism and racism.* Ithaca, New York: Firebrand Books (see also the comment on this paper by Martin, B. and Mohanty, C. T. What's home got to do with it? In De Lauretis, T. (ed.) 1986: *Feminist studies/critical studies.* Bloomington, Indiana: Indiana University Press).

Rowe, M. 1991: *So very English.* London: Serpent's Tail.

Rutkovsky-Ruskin, M. 1991: Land of the free, home of the brave. In Rieder, I. (ed.), *Cosmopolis: urban stories by women.* Dublin: Attic Press, 36–49.

Said, E. 1985: *Orientalism.* Harmondsworth: Penguin.

Segal, L. 1990: *Slow motion: changing masculinities, changing men.* London: Virago.

Western, J. 1992: *A passage to England: Barbadian Londoners speak of home.* London: UCL Press.

15 Paul Gilroy

'The Whisper Wakes, the Shudder Plays': 'Race', Nation and Ethnic Absolutism

Excerpts from: *There ain't no black in the Union Jack*, pp. 43–71. London: Routledge (1987)

Much of this criticizing the Left for its (lack of / troubling) response to racism in Britain.

> The Queen Mother swayed in a gentle dance when one of three steel bands began playing a lilting reggae tune. Five yards away, swaying with her, were a group of Rastafarians wearing the red, yellow and green tea cosy hats which are the badge of their pot-smoking set (*Daily Mail*, 21.4.83).

> The nation has been and is still being, eroded and hollowed out from within by implantation of unassimilated and unassimilable populations ... alien wedges in the heartland of the state (Enoch Powell, 9.4.76).

> Methinks I see in my mind a noble and puissant Nation rousing herself like a strong man after sleep, and shaking her invincible locks (Milton).

Racism has been described as a discontinuous and unevenly developed process. It exists in plural form, and I have suggested that it can change, assuming different shapes and articulating different political relations. Racist ideologies and practices have distinct meanings bounded by historical circumstances and determined in struggle. This chapter focuses on the distinctive characteristics of the racism which currently runs through life in Britain.

This particular form, which Martin Barker (1981) and others[1] have labelled 'the new racism', will be examined below with a view to focusing on the nature of its newness – the job it does in rendering our national crisis intelligible. It will be argued that its novelty lies in the capacity to link discourses of patriotism, nationalism, xenophobia, Englishness, Britishness, militarism and gender difference into a complex system which gives 'race' its contemporary meaning. These themes combine to provide a definition of 'race' in terms of culture and identity. What new right philosopher John Casey has called 'The whole life of the people'.[2] 'Race' differences are displayed in culture which is reproduced in educational institutions and, above all, in family life. Families are therefore not only the nation in microcosm, its key components, but act as the means to turn social processes into natural, instinctive ones.

These ideas have hosted an extraordinary convergence between left and right, between liberals and conservatives and between racists and some avowed anti-racists. These politically opposed groups have come together around an agreed definition of what 'race' adds up to. Their agreement can itself be understood as marking the newness of a new racism which confounds the traditional distinctions between left and right. Conservative thinkers, whether or not they follow the *Salisbury Review* in arguing that 'the consciousness of nationhood is the highest form of political consciousness',[3] are forced by the nature of their beliefs to be open about their philosophies of race and national belonging. The British left, caught between a formal declaration of internationalism and the lure of a pragmatic, popular patriotism (Seabrook, 1986), is less explicit and has been confounded by the shifting relationship between national sentiment, 'race' and class politics.[4] This difficulty is encountered in acute form by English socialist writers but it can be traced back into the writings of Marx and Engels (1973).

'Race', nation and the rhetoric of order

In his thoughtful study of nationalism, *Imagined Communities*, Benedict Anderson seeks to clarify the relationship between racism and nationalism by challenging Tom Nairn's (1977) argument that these two forms of ideology are fundamentally related in that the former derives from the latter. Anderson's conclusion is worth stating at length:

> The fact of the matter is that nationalism thinks in terms of historical destinies, while racism dreams of eternal contaminations transmitted from the origins of time through an endless sequence of loathsome copulations . . . The dreams of racism actually have their origins in the ideologies of class, rather than those of nation: above all in claims to divinity among rulers and to blue or white blood and breeding among aristocracies. No surprise then that . . . on the whole, racism and anti-semitism manifest themselves, not across national boundaries but within them. In other words they justify not so much foreign wars as domestic repression and domination (Anderson, 1983, p. 136).

Anderson's theory claims that racism is essentially antithetical to nationalism because nations are made possible in and through print languages rather than notions of biological difference and kinship. Thus, he argues that anyone can in theory learn the languages of the nation they seek to join and through the process of naturalization become a citizen enjoying formal equality under its laws. Whatever objections can be made to Anderson's general argument, his privileging of the written word over the spoken word for example, it simply does not apply in the English/British case. The politics of 'race' in this country is fired by conceptions of national belonging and homogeneity which not only blur the distinction between 'race' and nation, but rely on that very ambiguity for their effect. Phrases like 'the Island Race' and 'the Bulldog Breed' vividly convey the manner in which this nation is represented in terms which are simultaneously biological and cultural. It is important to recognize that the legal concept of partiality, introduced by the Immigration Act of 1968, codified this cultural biology of 'race' into statute law as part of a strategy for the exclusion of black settlers (WING, 1984). This act specified that immigration controls would not apply to any would-be settler who could claim national membership on the basis that one of their grandparents had been born in the UK. The Nationality Act of 1981 rationalized the legal vocabulary involved so that patrials are now known as British citizens.

A further objection to Anderson's position emerges from consideration of how the process of black settlement has been continually described in military metaphors which offer war and conquest as the central analogies for immigration. The enemy within, the unarmed invasion, alien encampments, alien territory and new commonwealth occupation have all been used to describe the black presence in this way. Enoch Powell, whose careful choice of symbols and metaphors suggests precise calculation, typifies this ideological strand:

> It is . . . truly when he looks into the eyes of Asia that the Englishman comes face to face with those who would dispute with him the possession of his native land.[5]

This language of war and invasion is the clearest illustration of the way in which the discourses which together constitute 'race' direct attention to national boundaries, focusing attention on the entry and exit of blacks. The new racism is primarily concerned with mechanisms of inclusion and exclusion. It specifies who may legitimately belong to the national community and simultaneously advances reasons for the segregation or banishment of those whose 'origin, sentiment or citizenship' assigns them elsewhere. The excluded are not always conceived as a cohesive rival nation. West Indians, for example, are seen as a bastard people occupying an indeterminate space between the Britishness which is their colonial legacy and an amorphous, ahistorical relationship with the dark continent and those parts of the new world where they have been able to reconstitute it. Asians on the other hand, as the Powell quote above suggests, are understood to be bound by cultural and biological ties which merit the status of a fully formed, alternative national identity. They pose a threat to the British way of life by virtue of their strength and cohesion. For different reasons, both groups are judged to be incompatible with authentic forms of Englishness (Lawrence, 1982). The obviousness of the differences they manifest in their cultural lives underlines the need to maintain strong and effective controls on who may enter Britain. The process of national decline is presented as coinciding with the dilution of once homogeneous and continuous national stock by alien strains. Alien cultures come to embody a threat which, in turn, invites the conclusion that national decline and weakness have been precipitated by the arrival of blacks. The operation of banishing blacks, repatriating them to the places which are congruent with their ethnicity and culture, becomes doubly desirable. It assists in the process of making Britain great again and restores an ethnic symmetry to a world distorted by imperial adventure and migration.

What must be explained, then, is how the limits of 'race' have come to coincide so precisely with national frontiers. This is a central achievement of the new racism. 'Race' is bounded on all sides by the sea. The effect of this ideological operation is visible in the way that the word 'immigrant' became synonymous with the word 'black' during the 1970s. It is still felt today as black settlers and their British-born children are denied authentic national membership on the basis of their 'race' and, at the same time, prevented from aligning themselves with the 'British race' on the grounds that their national allegiance inevitably lies elsewhere. This racist logic has pinpointed obstacles to genuine belonging in the culture and identity of the alien interlopers. Both are central to the theories of 'race' and nation which have emerged from the political and philosophical work of writers associated with Britain's 'new right' (Gamble, 1974; Levitas, 1986).

As part of their lament that the national heart no longer beats as one, Peregrine Worsthorne has pointed out that 'though Britain is a multi-racial society, it is still a long way from being a multi-racial nation'.[6] This is an important distinction which was also made fourteen years earlier by Enoch Powell. He drew attention to the difference between the merely formal membership of the national community provided by its laws, and the more substantive membership which derives from the historic ties of language, custom and 'race'. Parliament, suggested Powell, can change the law, but national sentiment transcends such narrow considerations: 'the West Indian does not by being born in England, become an Englishman. In law, he becomes a United Kingdom citizen by birth; in fact he is a West Indian or an Asian still'.[7]

It has been revealed that, at the suggestion of Churchill, a Conservative cabinet discussed the possibility of using 'Keep Britain White' as an electoral slogan as early as 1955 (MacMillan, 1973). Yet it is in the period between Powell's and Worsthorne's statements above that a truly popular politics of 'race' and nation flowered in Britain. Its growth, emanating mainly though not exclusively from the right (Rex, 1968; see also *Guardian*, 30.9.65) marks the divergence of what can loosely be called the patrician and populist orientations in the modern Conservative Party.

The 'metaphysics of Britishness' (Carter, 1983) which links patriotism, xenophobia, militarism, and nationalism into a series of statements on 'race' was a key element in the challenge to the old leadership between 1964 and 1970 and in the reconstitution of the party under Thatcher. These themes have been fundamental to the popularity of the party and conservative intellectuals have not concealed their instrumental use. The language of one nation provides a link between the populist effect of 'race' and a more general project which has attempted to align the British People with an anti-statism and in particular with Conservative criticism of the 'guilty public schoolboys' of the liberal intelligentsia who have wrecked the country with their consensual approach to politics. They are the men who imposed mass immigration on a reluctant populace.

The themes of national culture and identity have long histories inside the Conservative political tradition (Bennett, 1962). Yet the populist[8] form in which they emerge as Powellism breaks decisively with its predecessors even if its object, the conception of a 'unity of national sentiment transcending classes' (Cowling, 1978) remains the same. The reconstitution of Powellism as Thatcherism (Barnett, 1984; and Worsthorne, *Sunday Telegraph*, 12.6.83) points to the consolidation of a new political language which has became progressively more dominant within the representation of the present crisis and which, more specifically, has solved a profound political problem for the right. The 'one nation' message has been a means to escape from the shadows of paternalism which were the undoing of Alec Douglas Home. The national symbols of Powellism/Thatcherism are significant not only because a populist orientation is their primary characteristic. Conservative intellectuals have been candid about the role these ideas have played in the rebirth of their party which, in the period after Wilson, lacked a language adequate to its social and political vision. This problem was described vividly in an editorial which looked at the 1970s in the first issue of the new right journal *Salisbury Review*: 'never before had it seemed so hard to recreate the verbal symbols, the images and axioms, through which the concept of authority could be renewed'. The solution to it involved making 'race' and nation the framework for a rhetoric of order through which modern conservatism could voice populist protest against Britain's post-imperial plight and marshal its historic bloc. Enoch Powell's superficially simple question 'what kind of people are we?' summoned those very images and axioms and answered itself powerfully in the negative. 'We' were not muggers, 'we' were not illegal immigrants, 'we' were not criminals, Rastafarians, aliens or purveyors of arranged marriages. 'We' were the lonely old lady taunted by 'wide-grinning piccaninnies'. 'We' were the only white child in a class full of blacks. 'We' were the white man, frightened that in fifteen to twenty years, 'the black man would have the whip hand over us'. The black presence is thus constructed as a problem or threat against which a homogeneous, white, national 'we' could be unified. To put this operation into perspective, it must be emphasized that these were not the only images and definitions of nationhood which were mobilized during this period. Other voices from the left and from the black communities themselves were to be heard. Even within the right there were alternative conceptualizations of the relationship between 'race' and nation which were more in keeping with a patrician reading of imperial history.

I have already introduced the idea that the new racism's newness can be gauged by its capacity to operate across the broad range of political opinion. This claim can be pursued further. The distinction which Powell and Worsthorne make between authentic and inauthentic types of national belonging, appears in an almost identical form in the work of Raymond Williams (Williams, 1983; Mulhern, 1984). It provides a striking example of the way in which the cultural dimensions of the new racism confound the left/right distinction.

Williams combines a discussion of 'race' with comments on patriotism and nationalism. However, his understanding of 'race' is restricted to the social and cultural tensions surrounding the arrival of 'new peoples'. For him, as with the right, 'race' problems begin with immigration. Resentment of 'unfamiliar neighbours' is seen as the beginning of a process which ends in ideological specifications of 'race' and 'superiority'. Williams, working his way towards a 'new and substantial kind of socialism', draws precisely the same picture of the relationship between 'race', national identity and citizenship as Powell:

> ... it is a serious misunderstanding ... to suppose that the problems of social identity are resolved by formal (merely legal) definitions. For unevenly and at times precariously, but always through long experience substantially, an effective awareness of social identity depends on actual and sustained social relationships. To reduce social identity to formal legal definitions, at the level of the state, is to collude in the alienated superficialities of 'the nation' which are limited functional terms of the modern ruling class (Williams, 1983, p. 195).

These remarks are part of Williams's response to anti-racists who would answer the denial that blacks can be British by saying 'They are as British as you are.' He dismisses this reply as 'the standard liberal' variety. His alternative conception stresses that social identity is a product of 'long experience'. But this prompts the question – how long is long enough to become a genuine Brit? His insistence that the origins of racial conflicts lie in the hostility between strangers in the city makes little sense given the effects of the 1971 Immigration Act in halting primary black settlement. More disturbingly, these arguments effectively deny that blacks can share a significant 'social identity' with their white neighbours who, in contrast to more recent arrivals, inhabit what Williams calls 'rooted settlements' articulated by 'lived and formed identities'. He describes the emergence of racial conflict where

> an English working man (English in the terms of sustained modern integration) protests at the arrival or presence of 'foreigners' or 'aliens' and now goes on to specify them as 'blacks'.

Williams does not appear to recognize black as anything other than the subordinate moment in an ideology of racial supremacy. His use of the term 'social identity' is both significant and misleading. It minimizes the specificities of nationalism and ideologies of national identity and diverts attention from analysis of the political processes by which national and social identities have been aligned. Several questions which are absolutely central to contemporary 'race' politics are thus obscured. Under what conditions is national identity able to displace or dominate the equally 'lived and formed' identities which are based on age, gender, region, neighbourhood or ethnicity? How has it come to be expressed in racially exclusive forms? What happens when 'social identities' become expressed in conflicting political organizations and movements and when they appeal to the authority of nature and biology to rationalize the relations of domination and subordination which exist between them? How these social identities relate to the conspicuous differences of language and culture is unclear except where Williams points out that the forms of identity fostered by the 'artificial order' of the nation state are incomplete and empty when compared to 'full social identities in their real diversity'. This does not, of course, make them any the less vicious. Where racism demands repatriation and pivots on the exclusion of certain groups from the imagined community of the nation, the contradictions around citizenship that Williams dismisses as 'alienated superficialities' remain important constituents of the political field. They provide an important point of entry into the nation's sense of itself. Where racial oppression is practised with the connivance of legal institutions – the police and the courts – national and legal subjectivity will also become the focus of political antagonism. Williams's discussion of 'race' and nation does not address these issues and is notable for its refusal to examine the concept of racism which has its own historic relationship with ideologies of Englishnes, Britishness and national belonging.

Quite apart from Williams's apparent endorsement of the presuppositions of the new racism, the strategic silences in his work contribute directly to its strength and resilience. The image Williams has chosen to convey his grasp of 'race' and nation, that of a resentful English working man, intimidated by the alterity of his alien neighbours is, as we shall see below, redolent of other aspects of modern Conservative racism and nationalism.

The national community in peace and war

The imagery of black settlement as an invasion and the close association between racism and nationalism make it impossible to discuss the contemporary

Here: Gilroy reviews responses to the Falklands War and patriotism around it.

politics of 'race' without reference to the war with Argentina during 1982. The war analogy of black settlement laid the discursive foundations on which connections could be made between conflict abroad and the subversive activities of the fifth column within. The supreme expression of this theme was Margaret Thatcher's speech at Cheltenham on 3 July 1982. This defined the 'Falklands Factor' so as to link the struggle against the 'Argies' with the battle against British workers: the NUR and ASLEF (the rail unions) whose industrial actions were to be undone by the fact that such activities did not 'match the spirit' of the reborn Britain.

> What has indeed happened is that now once again Britain is not prepared to be pushed around. We have ceased to be a nation in retreat. We have instead a new-found confidence – born in the economic battles at home and tested and found true 8,000 miles away. That confidence comes from the rediscovery of ourselves and grows with the recovery of our self respect.

This speech, like Powell's critique of the royal household, made no open references to the issue of 'race'. Other new right thinkers were less circumspect. Wars, it was argued, are key moments in the process of national self-realization – the willingness to lay down one's life being the definition of a true patriot. The great distance involved and the tenuous constitutional connection between Britain and Port Stanley led commentators to speculate about the nature of the ties which could bind our national destiny to the fate of our distant 'kith and kin' in the South Atlantic. Peregrine Worsthorne went straight to the point:

> If the Falkland Islanders were British citizens with black or brown skins, spoke with strange accents or worshipped different Gods it is doubtful whether the Royal Navy and Marines would today be fighting for their liberation.[9]

The Falklands episode celebrated the cultural and spiritual continuity which could transcend 8000 miles and call the nation to arms in defence of its own distant people. Images of the nation at war were also used to draw attention to problems inherent in 'multi-racialism' at home. There was a rich irony discovered in the contrast between the intimacy of the 'natural' if long-distance relationship with the Falklanders and the more difficult task of relating to alien intruders who persisted in disrupting life in Britain and were not seen to be laying down their lives for the greater good. Again Worsthorne was the first to point this out: 'Most Britons today identify more easily with those of the same stock 8000 miles away . . . than they do with West Indian or Asian immigrants living next door.'[10] His potent image draws directly on Powell and emphasizes the strength of the cultural ties which mark the boundaries of 'race' as well as the exclusion of blacks from the definitions of nationality which matter most. The article from which it comes was as important for its recognition of the interrelationship of black and white life in the urban context as for the invocation of a mystic nationhood which would only be revealed on the battlefield. In the heat of combat, the nation would discover, or rather remember, what truly 'turned it on'.

The implicit need to recognize and devalue the quality of 'transracial' relation-ships between neighbours contains a tacit acknowledgement that such relation-ships do exist, even if the white Britons involved relinquish their membership of Worsthorne's nation at the point at which these friendships are conceived.

The popular power of patriotism revealed by the Falklands episode was not lost on commentators of the left. Their responses to it have significantly been characterized by a reluctance to challenge the model of national greatness and the metaphysical order of belonging on which it rests. In an influential piece which noted that the Falklands had 'stirred up an ugly nationalist sediment which would cloud our cultural and political life', E. P. Thompson argued that we would pay for the war 'for a long time in rapes and racism in our cities' (Thompson, 1982). Eric Hobsbawm, on the other hand, advanced a resolute polemic in favour of a 'left patriotism'. Patriotism, he wrote, would only spill over into undesirable xenophobia, racism and jingoism if the left allowed it to be 'falsely' separated from the sentiments and aspirations of the working class. It is dangerous, he continued 'to leave patriotism exclusively to the right' and, in language reminis-cent of Thompson's own frequent invocations of the popular traditions of British radicalism, he made a plea for a political orientation which could demonstrate radical patriotism and class consciousness could be yoked together in front of the socialist cart:

> The dangers of . . . patriotism always were and still are obvious, not least because it was and is enormously vulnerable to ruling-class jingoism, to anti-foreign nationalism and of course in our days to racism. . . . The reason why nobody pays much attention to the, let's call it, jingoism of the chartists is that it was combined with and masked by an enormous militant class consciousness. It's when the two are separated, that the dangers are particularly obvious. Conversely when the two go together in harness, they multiply not only the force of the working class but its capacity to place itself at the head of a broad coalition for social change and they even give it the possibility of wrestling hegemony from the class enemy.[11]

In a similar article, principally notable for its blank refusal to use the words 'race' or racism, Robert Gray of the Communist Party's 'Theory and Ideology Committee' attacked 'national nihilism in the name of abstract internationalism' and argued that what was required was a 'redefinition of the interests of the nation around the alternative leadership of the working class' (Gray, 1982). None of these contributions, even those which concede the unfortunate ambiguities in nationalist ideology, make any attempt to show how this valuable redefinition might be achieved. Apart from pointing out the conspicuous success of national-ist sentiment in renovating the Tory project, few arguments are made which justify the need to make the nation state a primary focus of radical political consciousness. It is as if the only problem with nationalism is that the Tories have secured a near exclusive monopoly of it.

The possibility of politically significant connections between nationalism and contemporary racism is either unseen or felt to be unworthy of detailed discussion. More importantly, the types of subjectivity which nationalisms bring into being and put to work pass unquestioned. The problem has become how socialists can (re)possess them from the right.

Why does a rhetoric of nation dominate Brit. socialism?

Two anachronistic images of Britain lurk behind these omissions. The first depicts the nation as a homogeneous and cohesive formation in which an even and consensual cultural field provides the context for hegemonic struggle. The second is attached to the idea that this country is, and must continue to be, a major world power. Patriotism, even in its combative proto-socialist form, is empty without a filling of national pride. British socialists have so far remained silent about how this misplaced pride can be detached from the vestigial desire for imperial greatness which has so disfigured recent political life (Barnett, 1983; The *Sun*, 25.5.82).

The frequency with which Labour's senior spokespeople invoke the national interest as a verbal bludgeon introduces another note of caution. What is this interest? How is it created? How can it be identified? And where does it reside? If appealing to it is nothing more than a rhetorical motif, why has it become necessary at this moment in time? What needs does it address in those who respond to it? Michael Ignatieff's discussion of the 1984–5 coal dispute illustrates some of these difficulties.[12] Throughout this article, the national interest is taken for granted as a meaningful analytic category. 'No one lives apart from the national community', writes Ignatieff. Governments which fail to uphold it lose their electoral support. The police and the courts are its own institutions.

Like many socialists in the post-Falklands period, Ignatieff argues that 'The left crucially overestimates Mrs Thatcher's electoral appeal if it believes that she has succeeded in monopolising the language of "one nation".' He opts to ignore the regional conflicts which were also at the heart of the coal dispute and which, I would argue, call into question the viability and desirability of the appeal to national unity to which he aspires. There is no reason why a political language based on the invocation of national identity should be the most effective where people recognize and define themselves primarily in terms of *regional* or *local* tradition. No coherent argument is provided as to why, for example, socialists should answer the voices of Wales, Yorkshire or Tyneside – all places where regional traditions are a key axis of political organization – with a language of the British national interest. 'Geordie', 'Hinny', 'Brummie' or 'Scouse' may all be political identities which are more in harmony with the advancement of socialist politics in this country than those conjured into being by the phrase 'fellow Britons' or even by the word citizen, given the way in which citizenship is allocated and withheld on racial grounds. These regional or local subjectivities simply do not articulate with 'race' in quite the same way as their national equivalent.

Colin Mercer's discussion of nation[13] parallels Ignatieff's and arrives at a similar image of British socialism: held to ransom by its national culture. Mercer 'owns up' to a 'sneaking admiration for Enoch Powell's prose' and thus recognizes his complicity in 'certain pleasures of Englishness'. He describes himself and by extension his audience of left cultural politicians, as unable either to 'interrupt' or 'stand outside' the complex combination of discursive effects that provide pleasure at the very moment in which 'Englishness' is constituted. The possibility that this particular brand of 'Englishness' may also enjoy a class character is left unexplored.

When it comes to their patriotism, it would appear that England's left intellectuals become so many radical rabbits transfixed and immobile in the path of an onrushing populist nationalism. How does the language of public good they propose, a necessary addition to radical speech if ever there was one, become the language of a nation so cohesive that 'no one lives outside the national community'? Indeed, the suggestion that no one lives outside the national community is only plausible if the issue of racism is excluded. What is being described by these writers is a national community, not imagined in the way that Benedict Anderson has suggested, but actual. The construction of that community is overlooked. It is accepted a priori as the structure around which the struggle to gain hegemony must take place.

The work of other socialist thinkers can be used to show that the images of paralysis which emerge from the work of Mercer and Ignatieff are only mild cases of this patriotic English disease. Where these two are simply inert in the face of national identity and its pleasures, E. P. Thompson, for example, is positively enthusiastic. He begins his pamphlet analysing the 1983 election by declaring 'whatever doubts we have, we can all think of things in the British way of life which we like, and we would want to protect these from attack' (1983). Thompson laments the fact that 'a large part of our free press has been bought, over our heads by money (some of it foreign money)'. For him, the activities of the women's peace movement are 'characteristically British', their mass action at the Greenham Common airbase on 12 December 1982 was 'a very untidy, low-key, British sort of do'. Thompson's version of Britishness locates the lingering greatness of the nation in the inheritance of popular resistance as well as in cultural achievements:

> This has not only been a nation of bullies. It has been a nation of poets and inventors, of thinkers and of scientists, held in some regard in the world. It has been, for a time, no less than ancient Greece before us, a place of innovation in human culture (Thompson, 1983, p. 34).

It is tempting to dispute the special status which Thompson accords to British culture and in particular his suggestion that modern Britain and ancient Greece are the primary innovators in human culture. However, these claims are not the main issue. What is more important is the way in which the preferred elements of English/British culture and society are described as if their existence somehow invalidated the side of our national heritage from which socialists are inclined to disassociate themselves.

Nationhood is not an empty receptacle which can be simply and spontaneously filled with alternative concepts according to the dictates of political pragmatism. The ideological theme of national belonging may be malleable to some extent but its links with the discourses of classes and 'races' and the organizational realities of these groups are not arbitrary. They are confined by historical and political factors which limit the extent to which nationalism becomes socialist at the moment that its litany is repeated by socialists. The intention may be radical but the effects are unpredictable, particularly where culture is also conceived within discrete, separable, national units coterminous with the boundaries of the nation state.

Having said this, it is impossible to deny that the language of the nation offers British socialists a rare opportunity. Through it, they can, like Thompson, begin to say 'we' and 'our' rather than 'I' and 'my'. It encourages them to speak beyond the margins of sectional interest to which they are confined by party and ideology. But there is a problem in these plural forms: who do they include, or, more precisely for our purposes, do they help to reproduce blackness and Englishness as mutually exclusive categories? Why is the racial inflection in the language of nation continually overlooked? And why are contemporary appeals to 'the people' in danger of transmitting themselves as appeals to the white people?

An answer to these questions can only begin from recognition of the way that Britain's languages of 'race' and nation have been articulated together. The effect of their combination can be registered even where 'race' is not overtly referred to, or where it is discussed outside of crude notions of superiority and inferiority. The discourses of nation and people are saturated with racial connotations. Attempts to constitute the poor or the working class as a class across racial lines are thus disrupted. This problem will have to be acknowledged directly if socialists are to move beyond puzzling over why black Britons (who as a disproportionately underprivileged group, ought to be their stalwart supporters) remain suspicious and distant from the political institutions of the working-class movement (Fitzgerald, 1984; Studlar, 1983; 1984; 1985).

Labour's occasional attempts to address nationalist sentiment have, as Anthony Barnett has demonstrated (1982), been a site of further difficulties. However, the concept of 'Churchillism' with which Barnett has tried to pin down the patriotic junction of Labourism and 'Thatcherism' is not adequate to all the permutations of Labour's failure. It plays down the specific attributes and appeal of socialist nationalism and suggests that the 'fatal dementia of national pride' has been injected into British socialism from the outside. However, left nationalism is a more organic, historic property of British socialism. Michael Foot's benign view of Enoch Powell's 'rivers of blood' intervention as a 'tragic irony' and 'pathos' (Foot, 1986) and Neil Kinnock's claim to working-class patriotism:

> . . . a confident and generous patriotism of freedom and fairness, not one of prejudice, vanity, or the patriotism of the 'presidential puppet' . . . a patriotism that is forgotten when the chequebook is waved . . . a patriotism which holds that our values are not for sale to anyone at any price at any time[14]

suggest a longer pedigree for 'Churchillism' than the association of 'Tory belligerents, Labour reformists, revolutionary anti-fascists, and the liberal intelligentsia' which Barnett (1982) proposes as its genealogy. If the writings of left intellectuals are anything to go by, it is born from something altogether more cultural than political; something rooted, not in the end of empire, but in the imperial experience itself (Mackenzie, 1984; 1986). A more complex illustration of these problems can be found in Tony Benn's (1982) attempt to define the British crisis in terms of a descent into colonial status. *New Socialist*, the Labour Party journal introduced his piece thus:

... the British establishment has opted for survival as the colonial administrators of a subject country. Tony Benn calls for the Labour movement to lead a national liberation struggle and restore to us our democratic rights.

Benn's description of the socialist struggle against the Thatcher government as a national liberation struggle was certainly imaginative. It was a clear attempt to harness for the left the yearning for a return to national greatness which has been used effectively by the right. It addressed the British inability to accept the end of empire and the national discomfort at the loss of world pre-eminence. It substitutes a stark image of reduced national status for the metaphysical yearning for greatness amplified by both Conservatives and the Alliance parties in their 1983 election manifestos (Barnett, 1983a). Yet the bloodshed and ruthless mass violence characteristic of decolonization have not, other than in the six counties of Northern Ireland, been evident in recent British politics. The effect of Benn's words on black citizens for whom decolonization is a memory rather than a metaphor is hard to estimate. It is, however, difficult to resist the conclusion that his choice of imagery trivialized the bitter complexities of anti-colonial struggle.

I am not suggesting that the differences between Labour and Conservative languages of nation and patriotism are insignificant, but rather that these languages overlap significantly. In contemporary Britain, statements about nation are invariably also statements about 'race'. The Conservatives appear to recognize this and seek to play with the ambiguities which this situation creates. Their recent statements on the theme of Britishness betray a sophisticated grasp of the interface between 'race' and nation created in the post-'rivers of blood' era. During the coal dispute, for example, in a speech on the enduring power of the national constitution entitled 'Why Democracy Will Last', Mrs Thatcher invoked the memory of the Somerset case of 1772. Lord Mansfield's famous judgment in this case declared that British slaveholders could no longer compel their slaves to leave the country against their will (Fryer, 1984; Shyllon, 1977; 1974). It matters little that Mrs Thatcher quoted the case wrongly, suggesting that it brought slavery in this country to an end. With no trace of irony, her speech boldly articulates an apparently anti-racist position at the heart of a nationalist and authoritarian statement in which the mining communities were identified as 'enemies within'.[15]

The Conservatives' ethnic election poster of 1983 provides further insight into the right's grasp of these complexities. The poster was presumably intended to exploit ambiguities between 'race' and nation and to salve the sense of exclusion experienced by the blacks who were its target. The poster appeared in the ethnic minority press during May 1983 and was attacked by black spokespeople for suggesting that the categories black and British were mutually exclusive. It set an image of a young black man, smartly dressed in a suit with wide lapels and flared trousers, above the caption 'Labour says he's black. Tories say he's British'. The text which followed set out to reassure readers that 'with Conservatives there are no "blacks" or "whites", just people'. A variant on the one nation theme emerged, entwined with criticism of Labour for treating blacks 'as a "special case", as a

With the Conservatives, there are no 'blacks', no 'whites', just people.

Conservatives believe that treating minorities as equals encourages the majority to treat them as equals.

Yet the Labour Party aim to treat you as a 'special case', as a group all on your own.

Is setting you apart from the rest of society a sensible way to overcome racial prejudice and social inequality?

The question is, should we really divide the British people instead of uniting them?

WHOSE PROMISES ARE YOU TO BELIEVE?

When Labour were in government, they promised to repeal Immigration Acts passed in 1962 and 1971. Both promises were broken.

This time, they are promising to throw out the British Nationality Act, which gives full and equal citizenship to everyone permanently settled in Britain.

But how do the Conservatives' promises compare?

We said that we'd abolish the 'SUS' law.

We kept our promise.

We said we'd recruit more coloured policemen, get the police back into the community, and train them for a better understanding of your needs.

We kept our promise.

PUTTING THE ECONOMY BACK ON ITS FEET

The Conservatives have always said that the only long term answer to our economic problems was to conquer inflation.

Inflation is now lower than it's been for over a decade, keeping all prices stable, with the price of food now hardly rising at all.

Meanwhile, many businesses throughout Britain are recovering, leading to thousands of new jobs.

Firstly, in our traditional industries, but just as importantly in new technology areas such as micro-electronics.

In other words, the medicine is working.

Yet Labour want to change everything, and put us back to square one.

They intend to increase taxation. They intend to increase the National Debt.

They promise import and export controls.

Cast your mind back to the last Labour government. Labour's methods didn't work then.

They won't work now.

A BETTER BRITAIN FOR ALL OF US.

The Conservatives believe that everyone wants to work hard and be rewarded for it.

Those rewards will only come about by creating a mood of equal opportunity for everyone in Britain, regardless of their race, creed or colour.

The difference you're voting for is this:

To the Labour Party, you're a black person.

To the Conservatives, you're a British Citizen.

Vote Conservative, and you vote for a more equal, more prosperous Britain.

LABOUR SAYS HE'S BLACK.
TORIES SAY HE'S BRITISH.

CONSERVATIVE ☒

Figure 15.1   Conservative Party election poster, 1983.

group all on your own'. At one level, the poster states that the category of citizen and the formal belonging which it bestows on its black holders are essentially colourless, or at least colour-blind. Yet as the writings of Powell and Worsthorne above illustrate, populist racism does not recognize the legal membership of the national community conferred by its legislation as a substantive guarantee of Britishness. 'Race' is, therefore, despite the text, being defined beyond these legal definitions in the sphere of culture. There is more to Britishness than a passport. Nationhood, as Alfred Sherman pointed out in 1976,

> remains ... man's main focus of identity, his link with the wider world, with past and future, 'a partnership with those who are living, those who are dead and those who are to be born'. . . . It includes national character reflected in the way of life . . . a passport or residence permit does not automatically implant national values or patriotism.[16]

His reading of the poster ↓

At this point the slightly too large suit worn by the young man, with its unfashionable cut and connotations of a job interview, becomes a key signifier. It conveys what is being asked of the black readers as the price of admission to the colour-blind form of citizenship promised by the text. Blacks are being invited to forsake all that marks them out as culturally distinct before real Britishness can be guaranteed. National culture is present in the young man's clothing. Isolated and shorn of the mugger's key icons – a tea-cosy hat and the dreadlocks of Rastafari – he is redeemed by his suit, the signifier of British civilization. The image of black youth as a problem is thus contained and rendered assimilable. The wolf is transformed by his sheep's clothing. The solitary maleness of the figure is also highly significant. It avoids the hidden threat of excessive fertility which is a constant presence in the representation of black women (Parmar, 1985). This lone young man is incapable of swamping 'us'. He is alone because the logics of racist discourse militate against the possibility of making British blackness visible in a family or an inter-generational group.[17] The black family is presented as incomplete, deviant and ruptured.

Conclusion

This chapter has sought to identify the links between the discourses of 'race' and nation and to use their proximity as an argument against the pre-eminent place which the idea of nationhood continues to enjoy in the work of English socialists and the practice of black cultural nationalists. Apart from the overlap with the concerns and premises of the right, there are other reasons why the language of nation has become an inappropriate one for the black movement and the socialist movement in Britain. The uneven development of national crisis has, for example, exacerbated regional differences to the point that we can speak routinely of two nations – north and south. These nations, which sometimes appear to co-exist with difficulty, are more than simply competing metaphors of Englishness (Weiner, 1981). The difference between them is rooted in the mode of production and is expressed not least in the different relationships each enjoys to national decline and de-industrialization. The industrial geography of the present crisis and changes in the geographies of class hierarchy and gender patterns of employment have been identified by Doreen Massey (1984) who has concluded that a new spatial division of labour is being created. This must be taken into account in discussions of the political language and concepts which radicals will require if they are to end the dominance of authoritarian populist nationalism.

Quite apart from 'racial' and gender considerations, the steady fragmentation of national unity and its recomposition along new economic and regional axes, militates against the language of patriotism and people retaining its wide popular appeal. Changes in Britain are matched by changes outside, in the character and composition of capital which has begun to organize itself into productive structures and operations which transcend the limits of the nation state and cannot therefore be combated by workers' organizations trapped and confused within national borders. The need to develop international dialogues and means of organization which can connect locality and immediacy across the international division of labour is perhaps more readily apparent to black populations who define themselves as part of a diaspora, and who have recent experience of migration as well as acute memories of slavery and international indenture.

I have not intended to suggest that the attempt to turn these insights into political practice necessitates the abandonment of any idea of Englishness or Britishness. We are all, no doubt, fond of things which appear unique to our national culture – queueing perhaps, or the sound of leather on willow. What must be sacrificed is the language of British nationalism which is stained with the memory of imperial greatness. What must be challenged is the way that these apparently unique customs and practices are understood as expressions of a pure and homogeneous nationality. British socialists often interpret the things they like or wish to encourage as repositories and emblems of national sentiment. Socialists from Orwell to Thompson have tried to find the answer to their marginalization in the creation of a popular patriotism, and described the oppression of the working class in nationalist images as diverse as 'the Norman yoke' (Hobsbawm 1978), and Tony Benn's colonial analogy discussed above. Their output owes its nationalist dimensions to several sources. It has been forged not only in the peculiarities of the route which brought the English proletariat into being within national limits (Linebaugh 1982; 1984) marked by the early liquidation of the peasant class and the protracted dominance of the aristocracy, but also by the concepts and methods of historical materialism itself. These have been shown to play a role in the reproduction of a blind spot around nationalism as far as Marxists are concerned (Gellner, 1983; Kitching, 1985; Nimni, 1985; Nairn, 1977). Marx and Engels' assertion that the workers have no fatherland sits uncomfortably beside their practice as German nationalists and its accompanying theory of historic and non-historic peoples which differentiated between 'the large viable European nations' and the 'ruins of peoples' (*Volkerabfalle*) which are found here and there and which are no longer capable of national existence (Robinson, 1983). Their dismissal of the nationalist movements which might arise from among 'historyless' peoples (*Geschitlossen Volker*) whose national communities did not conform to the precise equation of state and language which could guarantee them historical being, is one of the fundamental moments in which the Eurocentrism and statism of Marxism are brought into being. It illustrates the limitations that history has placed on the value of Marxian insights which may not be appropriate to the analysis of the relationship between 'race', nation and class in the post-industrial era. This legacy must be re-examined and dealt with if the hold of nationalism on today's socialists is to be broken. Its inversion in the form of black cultural nationalism simply replicates the problem.

Notes

1 Barker's concept has been usefully developed by Mark Duffield (1984) and Errol Lawrence (1982).
2 'One Nation: The Politics of Race' *Salisbury Review*, no. 1. See also Casey's 'Tradition and Authority', in Cowling (ed.) (1978), *Conservative Essays*.
3 Editorial *Salisbury Review*, no. 1., 1983.
4 The works of Anthony Barnett and Patrick Wright are honourable and notable exceptions here. Wright, in particular, tries to break the deadlock between these tendencies in a bold and innovative manner. The success of his attempt is, however, qualified by the fact that it is secured on the terrain of history. Barnett has consistently pointed to the profound relationship between Labourism and nationalism as well as to the contemporary dangers inherent in a populist nationalism.
5 Speech at Southall, 4.11.71.
6 *Sunday Telegraph*, 27.6.82.
7 Speech at Eastbourne, 16.11.68.
8 Jose Nun has produced a definitive discussion of this concept in his exchange with Ernesto Laclau in *Latin American Research Unit Studies*, 3, nos. 2–3, 1980.
9 *Sunday Telegraph*, 23.5.82.
10 *Sunday Telegraph*, 27.6.82.
11 *Guardian*, 20.12.82, reprinted from *Marxism Today*, January 1983.
12 *New Statesman*, 14.12.84.
13 *Formations*, 1, no. 1 (1983), p. 99.
14 *Guardian*, 8.3.86; see also reply by Hugo Young 'The love that dares to speak its name too often', *Guardian*, 11.3.86.
15 Speech at the Carlton Club, 26.11.1984.
16 *Sunday Telegraph*, 8.9.76.
17 The best example of this is the contrast between television situation comedies featuring blacks and whites. It is notable that none of the series featuring blacks seem able to portray inter-generational relations between black characters or show their experiences over time, in a diachronic dimension.

   The BBC series 'Frontline' (1985) about the relationship between two black brothers, one a 'Rasta', the other a policeman, began significantly with the death of their mother. An equivalent programme centred on a fractured white family in which notions of locality and 'ethnicity' play a similar role – 'Only Fools and Horses' – builds its humour out of the tension between generations.

References

Anderson, B. (1983), *Imagined Communities: Reflections on the Origin and Spread of Nationalism*, Verso, London.
Barker, M. (1981), *The New Racism*, Junction Books, London.
Barnett, A. (1984), 'Fortress Thatcher' in Ayrton, Englehart and Ware (eds.), *World View 1985*, Pluto Press, London.
Barnett, A. (1983), 'Getting it wrong and making it right', *New Socialist*, Sept./Oct..
Barnett, A. (1982), 'Iron Britannia', *New Left Review*, no. 134, July/August.
Benn, Tony (1981), 'Britain as a colony', *New Socialist*, no. 1., Sept./Oct..

Bennett, G. (1962), *The Concept of Empire from Burke to Attlee* (2nd edn), Adam and Charles Black, London.

Carter, A. (1983), 'Masochism for the Masses', *New Statesman*, 3.6.83.

Cowling, M. (ed.) (1978), *Conservative Essays*, Cassell, London.

Duffield, M. (1984), 'New Racism ... New Realism: Two Sides of The Same Coin', *Radical Philosophy*, **37**, summer.

Fitzgerald, M. (1984), *Political Parties and Black People: participation, representation and exploitation*, Runneymede/GLC.

Foot, M. (1986), *Loyalists and Loners*, Collins, London. See also 'The Lives of Enoch', an extract from this book, *Guardian*, 12.3.86.

Gamble, A. (1974), *The Conservative Nation*, RKP, London.

Gellner, E. (1983), *Nations and Nationalism*, Basil Blackwell, Oxford.

Hobsbawm, E. (1978), 'The Historians' Group of the Communist Party', in Cornforth (ed.), *Rebels and Their Causes: Essays in Honour of A. L. Morton*, Lawrence and Wishart, London.

Kitching, G. (1985), 'Nationalism: the instrumental passion', *Capital and Class*, no. 25, spring.

Lawrence, E. (1982), 'In the Abundance of Water the Fool is Thirsty: Sociology and Black Pathology', in CCCS (eds.), *The Empire Strikes Back*, Hutchinson, London.

Levitas, R. (1986), *The Ideology of the New Right*, Polity Press, Oxford.

Linebaugh, P. (1984), 'Reply To Sweeny', *Labour/Le Travail*, no. 14, autumn.

Linebaugh, P. (1982), 'All The Atlantic Mountains Shook', *Labour/Le Travail*, no. 10, autumn.

Mackenzie, J. M. (1984), *Propaganda and Empire*, Manchester University Press.

Mackenzie, J. M. (1986), *Imperialism and Popular Culture*, Manchester University Press.

MacMillan, H. (1973), *At The End of the Day*, Macmillan, London.

Marx, K. and Engels, F. (1973), D. Fernbach (ed.), *The Revolutions of 1848*, Penguin, Harmondsworth.

Massey, D. (1984), *Spatial Divisions of Labour*, Macmillan, London.

Mulhern, F. (1984), 'Towards 2000: News From You Know Where', *New Left Review*, no. 148, November/December.

Nairn, T. (1977), *The Break Up of Britain*, New Left Books, London.

Nimni, E. (1985), 'The great historical failure: Marxist theories of Nationalism', *Capital and Class*, no. 25, spring.

Parmar, P. (1984), 'Hateful Contraries', *Ten 8*, no. 16.

Rex, J. (1968), 'The Race Relations Catastrophe', in Burgess, T. (ed.), *Matters of Principle: Labour's Last Chance*, Penguin, Harmondsworth.

Robinson, C. (1982), *Black Marxism*, Zed, London.

Seabrook, J. (1986), 'The war against jingoism', *Guardian*, 31.3.86. See also the reply by Dafydd Elis Thomas, 'In a loveless state', *Guardian*, 14.4.86..

Shyllon, F. (1977), *Black People in Britain*, Oxford University Press.

Shyllon, F. (1974), *Black Slaves in Britain*, IRR/Oxford University Press.

Studlar, D. (1985), '"Waiting for the catastrophe": Race and the Political Agenda in Britain', *Patterns of Prejudice*, **19**, no. 1.

Studlar, D. (1984), 'Nonwhite Policy Preferences, Political Participation, and the Political Agenda in Britain', unpublished paper prepared for Conference on Race and Politics, St Hughs College, Oxford, 28–30 September.

Studlar, D. (1983), 'The Ethnic Vote: problems of analysis and interpretation', *New Community*, **11**.

Thompson, E. P. (1983), *The Defence of Britain*, END/CND, London.

Thompson, E. P. (1982), *Zero Option*, Merlin, London.
Weiner, M. J. (1981), *English Culture and the Decline of Industrial Spirit 1850–1980*, Cambridge University Press.
Williams, R. (1983), *Towards 2000*, Pelican, Harmondsworth.
WING (1985), *Worlds Apart: Women Under Immigration and Nationality Law*, Pluto Press, London.

16   Diana Jeater
Roast Beef and Reggae Music: The Passing of Whiteness

Excerpts from: *New Formations* **18**, 107–21 (1992)

I

'We are all . . .', Stuart Hall told us in 1988, 'ethnically *located and our ethnic identities are crucial to our subjective sense of who we are.'*[1] This may be true, but I find that white English people like me have a lot of difficulty in identifying what the parameters defining our ethnic identity might be, and even if we think we know, being very unsure about whether this is an identity we want. At a workshop on representations of black people in the media, during a *Marxism Today* conference in 1989, the (black) speakers seemed somewhat bemused by the predominantly white participants, who told them 'It's different for you. You can identify with your culture. We haven't got anything to identify with except roast beef and Yorkshire pudding.' At the time this struck me as rather facile; but the strength of emotion and the depth of confusion which that workshop revealed suggests that something more important might have been going on. Why were black 'cultural roots' seen as 'valid' in a way that Yorkshire pudding was not? There is one corner of this question that I want to nibble away at, and that is the crisis in white self-esteem that these issues seem to represent. I want to suggest that there is a significant body of young, or youngish, white people in Britain's urban centres who don't feel they have an 'ethnicity', or if they do, that it's not one they feel too good about. I want further to suggest that this feeling bad is not a very useful response to the issue of contemporary racism, or indeed to anything else, and that there must be more to being 'white' than this.

In discussing this issue of white identity, I am making no claims to be speaking for or about all white people. Indeed, one of the main points I wish to make is that, in contemporary urban conditions, there is no single, homogeneous category of 'white' in Britain's inner cities, any more than there is a single category of 'black'. This point is important because it seems to me that most of the best writing on 'race' over the past decade has investigated the complexity of a non-essentialist but nonetheless meaningful category 'black', but it has built into this analysis a fairly un-thought-out category of 'white'. I will return to this point later. For the moment, I simply wish to register that I don't want to be thought to be making claims on behalf of all white people.

It is perhaps unlikely that I might be so misinterpreted; white people have hardly been restricted in the range of experiences and viewpoints which they have expressed in the past. Nonetheless, the fact that I am writing about white people in the context of 'race' might encourage that binary mode of thought – black/white – which has dominated the way in which so much work on 'race' has been read in the past. If we are to develop a way of thinking about our identities which gives meaning to our lived experiences, then we need to be clear about what this entails. As Kobena Mercer has pointed out, contrasting recent films from black film-makers with earlier 'black' films:

> The critical difference in the contemporary situation . . . turns on the decision to speak *from* the specificity of one's circumstances and experiences, rather than the attempt, impossible in any case, to speak *for* the entire social category in which one's experience is constituted.[2]

This, then, is an attempt to speak from the specificity of my own circumstances, to attempt to understand why it feels so frustrating that there is no way of speaking of 'white' as an identity which refers to my everyday lived experience, comparable to the way that many of my friends, on an everyday level, derive some kind of sense of themselves and their identity from calling themselves 'black'.

It is only recently, it seems to me, that the issue of being 'white' has become problematic. At a recent conference in Oxford, Raphael Samuel pointed out that, as a young man growing up within the Communist Party tradition, the use of 'racialized' language was eschewed in favour of other solidarities. The language of 'black', 'Jewish' and 'white' was used by the racists, and was challenged by an appeal to class solidarity. White people did not worry about their identities as 'whites', or find it useful to think about the struggle against racism in terms of black interests opposing white. Of course, in retrospect, it seems clear that the subsuming of the struggle against racism into the struggle against capitalism effectively silenced much of what black people were experiencing. My point here is that the issue of being 'white' has to be seen as growing out of a particular recent history and politics.

So what is it about me, and my experiences, which makes this issue of white identity so important? It's largely because the issue of being white comes up for me every day. It's not a theoretical pondering; it's a day-to-day struggle to make sense of who I am. For reasons which I hope to explain, a large part of my life is spent engaging with 'black' issues. Under the influence of classic 1970s reggae, I studied African politics at university, and went on to write a doctorate about the construction of moral discourse in white settler-occupied Zimbabwe. From there, I was appointed to my current job specifically to design a course on black history. As well as my academic interests, there are other things about my life which raise for me the question of what it means to be 'white' in contemporary London. I live in Brixton; I occasionally read 'The Voice' and I listen to Choice FM; my lover is black; so is the woman I live with. I can cook sadza or curry at the drop of a hat, but I don't have a clue how to cook roast beef. Clearly my lifestyle is not entirely 'white' (whatever that might mean). Yet it is *my* lifestyle, and I am a white person. I'm not trying to be anyone other than me. Nonetheless, I find that when people discover these things about me, they accuse me (and it is an accusation) of trying to be some kind of 'honorary black'. It seems that there isn't a category to describe people like me.

When I talk about this, I find I'm fighting against waves of intense embarrassment which suddenly fill the room. On the one hand, the fact that many white people *have* wanted to be 'honorary blacks' seems to be a taboo subject, particularly for whites. On the other hand, my statement that that isn't my personal project could be interpreted as an arrogant and 'typically white' attempt to silence and deny the experience of black people who have suffered from the patronage of white people 'wanting to be black'. The reason why I, and other white people like me have ended up living lives which do not conform to the stereotype of 'white' is not necessarily because we want to appropriate some other existence which we can call 'black'. Rather, it is because of the historic moment at which we were born. It seems to me we need to find some way of talking about being white which recognizes that many white people of my generation grew up in a Britain where black people didn't appear as 'immigrants'; they were already visibly present as part of the world we were born into.

The period when my generation first developed a political awareness was, of course, the period of the breakdown of the postwar consensus. It was impossible to ignore the importance of 'race' in that process. As Stuart Hall expressed it: 'Race is the lens through which people come to perceive that a crisis is developing.'[3] I don't want to engage here with the process by which young black people developed political/cultural responses to this; Paul Gilroy and others have done that job elsewhere. I want to express rather what this centrality of race meant to me as I had to prepare myself for adulthood in a world in which the old certainties were disintegrating.

The disintegrating of old certainties can be experienced in many ways. For me, in a secure middle-class environment, it felt very exhilarating. The intense thrill I derived from seeing young black men chuck a hunk of rock through a police car windscreen and watch it reverse away in full retreat at the Notting Hill Carnival was rooted in seeing it as a kind of metaphor for my arguments with my parents. This is not as facile as it might at first appear. The actions of those black people on the television news seemed to express, very clearly, a range of things which mattered very much to me. It was no coincidence that it was black youth who were bearing the brunt of the crisis, and who were therefore the ones whose resistance was most visible. But this fact had an impact on us, their white contemporaries, who began to develop a sense of solidarity with that struggle, and to privilege the politics of 'race' over other issues.

The point that seems particularly important about this is that we were looking to the black community to provide the leadership and direction for what we, nonetheless, conceptualized as 'our' struggle. As long as the enemy was seen as 'the State', there was nothing problematic about seeing Darcus Howe, for example, as the spokesman for the anger felt by disaffected youth, whatever our skin colour. So we left the 1970s and entered the 1980s with a politics which had built into it a strong sense of affiliation to black politics.

The affiliation of black politics was reinforced by the music. Gilroy has described how Jamaican reggae – crying out against hunger and confinement in the Caribbean – was equally powerful as metaphor for the black experience in Britain.[4] But the metaphor could go further; white youth, too, read into 1970s reggae a metaphor for the growing sense of containment that state responses to the crisis were producing. 'Babylon' served as a metaphor for the encroaching state opposition to squatting, street demonstrations, radical politics and a vision of a better world for white kids at the same time that it was also being picked up by African-Caribbean kids experiencing parallel, and more vicious, oppressions. This was particularly important at a time when white music was fissured between 1960s rock, whose politics were directed towards older struggles against the Vietnamese war and the right to grow one's hair, and 1970s punk, where the politics were spot-on, but which had a very limited emotional range. I absorbed from 1970s reggae music a political vision, akin perhaps to what Gilroy has described as a 'politics of transfiguration',[5] which was spiritual and inspirational as well as utterly uncompromising. For someone growing up under the influence of the white anti-war movements in the United States, there was a great liberation in hearing someone sing 'I don't want no peace; I want equal rights and justice', and in a voice so sweet but so sure that it lifted up the soul and left it cleansed and refreshed to continue the fight.

It was the music which brought white and African-Caribbean youth together in the Rock Against Racism (RAR) movement, and the Anti-Nazi League (ANL). Even then, there was no sense of alienation or lack of identity for whites. Rather the opposite in fact: both movements had white leaderships. Punk bands were billed alongside reggae bands, and neither was considered to be more 'valid' or 'authentic' than the other. However, when The Clash issued 'White Man in Hammersmith Palais', it was a hint of some of the problems which were to emerge in the mid-1980s. In that song, this white punk band expressed a fairly unclear sense of disquiet at feeling excluded from the 'black' scene, and a sense of feeling undervalued because of their whiteness. The Clash, of course, were one of the punk bands who worked most closely with the Caribbean reggae artists in Britain, and were clearly finding that boundaries were being drawn to separate collaboration from appropriation. This drawing of boundaries was confusing for white people who knew they weren't 'racist' in the sense that they understood, but found that, nonetheless, their 'whiteness' meant that they weren't necessarily welcome in the black communities.

For me, at the time, the question of my 'whiteness' didn't arise. I just loved the music and the politics with which I associated it. Of course, this phenomenon, of white English youth finding inspiration in reggae music, and particularly in its politics, has been documented elsewhere. Considerable attention has been paid to the ways in which white urban youth picked up on reggae music and found ways to negotiate the kind of boundaries which The Clash hinted at in 'Hammersmith Palais'. However, for me the engagement with black struggles was always from a distance, cheering at the sidelines. Where I differ from the subjects of those studies, perhaps, is that the world in which I moved was wholly white. I didn't know any black people except the man who sold me my reggae records, and I couldn't really claim to know him. So I never tried to be a part of 'black culture'; conversely, I never experienced any overt exclusion from black communities.[6] I absorbed what seemed to matter to me, and my right to do this was never challenged.

It was in this context that black influences became significant in the development of my sense of self, along with all the other things that influenced me. I revised for my school exams playing Bob Marley's 'Exodus' over and over again, so that I can't hear it now without associating it with the English country garden where I grew up. I'm not sure what Marley would have felt about that, but to me there never seemed to be any contradiction about this. I was white and English, but much of the way that I thought about the world was structured by my contact with black music and my sense of connection with my black contemporaries. They were living in the same country as me and I felt that we were fighting the same battles, albeit on a battlefield 'chosen' by them: 'sus' law, anti-Nazism, the oppressive state.

II

That, of course, was a long time ago. When I look at London now, it seems that the music and the politics have really changed. Culturally, black styles are much more central than they were then. The novel experience that my generation had, of being inspired by black music but not being able to appropriate it, has become more of a norm. The increasing control that black people have gained over successful nightclubs, record labels and radio stations is matched by a self-confidence in the black communities that have moved from positions of defence to positions of assertion.[7]

These changes have also had an impact politically. In discussion about 'identity' in the 1980s, the issue of 'race' became less an issue of 'the state', and more an issue of 'black' versus 'white'. This was something of a shock to those of us in the white communities who'd always thought of the black politics of the 1970s as part of our history. While we had assumed an unproblematic solidarity with Notting Hill rioters in 1976, we found, by 1986, that black people didn't necessarily see things the same way. What I want to do in the rest of this piece is to investigate how significant that 1970s engagement with reggae music and black politics has been for white people of my generation, and how we can begin to formulate a way of thinking about 'race' politics that doesn't deny the specificity of black experiences but at the same time doesn't deny the value of interactions between white and black people in addressing such a politics.

Once the issue has been stated as an issue of 'white' versus 'black', then it would seem that we need to know what we mean when we talk about 'white'. However, this was a debate that never took place. The reason why my generation engaged as we did with black politics and culture was because, in Kobena Mercer's phrase, 'race' provided a 'privileged metaphor'[8] for our own experiences. The diaspora was about the absence of fixed identity, and about the constant construction of provisional identities, or identity as process. As the collapse of the postwar consensus exposed the cracks in the old certainties and opened up spaces for us to recognize our own fragmentation and relative marginalization, we too experienced the crisis through the 'lens' of race. Most of the time, this happened unconsciously; but it is very significant that for almost all the 'white Left' (although this concept itself was disintegrating) of my generation, our entry into politics was through the Anti-Nazi League. Those alliances were cemented both musically and politically in the summer of 1981, when The Specials' 'Ghost Town' was number one in the charts as the inner cities burned. But precisely because it was the black experience which created the possibility for those events, they became fixed in the public image as black.[9] This 'racialization' of events marginalized and obscured the extent of white involvement, from whites themselves as well as from blacks. In other words, we did not *see* ourselves in the crisis, although we *felt* and *acted* as if the 'black' struggle was also our struggle.

When we did begin to question what it meant to be 'white', we found that the reasons for asking the question confined the scope of the answer. It was in the context of the 'New Racism' that certain definitions of 'whiteness' came under scrutiny. Although 'race' was clearly vital to any understanding of what was happening to Britain from the mid-1970s, most white commentators did not really engage with the question of what 'racial' identity might mean for them. This is not because white intellectuals were not aware that 'race' was a very important part of what was going on, but they tended to concentrate on *racism*, which effectively meant a concentration on white constructions of black identity. The work of Martin Barker, for example, was immensely significant in identifying the development of a 'New Racism', which replaced a crude obsession with pigmentation with a much more subtle argument based on culture.[10] This analysis became the springboard for a burst of creative intellectual work examining the nature of 'black culture', and attempting to assess what it means to root an identity in culture. That work in its turn has provided the foundation for my analysis here. But the focus on racism has meant that discussions of white interactions with black people have tended to supply only one symbol of whiteness; that of the racist.

Discussion of the New Racism has necessarily concentrated on the ways in which romantic constructions of 'Englishness' have been used to efface, exclude and attack the black presence here. Clearly to be 'English' is a national identity, not a racial one. Nonetheless, the 'New Racism' was an attempt to construct 'English' in such a way that it excluded people who were not white. It arose as a response to a situation in which most black people in England were not immigrants, but born here. The language of anti-immigration was no longer appropriate to attack them. Instead, the New Racism attacked their putative 'culture'. 'English' culture was differentiated from 'other' cultures, with the implication that those who did not share this culture, which effectively meant those of 'New Commonwealth' descent, were not really 'English', regardless of their place of birth. Accusations of crude racism were thereby neatly sidestepped: culture was more important than 'race' (i.e., skin colour), and the 'problem' with the children of immigrants was not their skin but their culture. Children who had 'integrated' into the culture of their 'host' country (which was of course not a host at all, but their motherland) were not a 'problem'. So when Margaret

Thatcher said in 1978 that some people were worried about being 'swamped', she referred specifically to different ways of behaving, different 'cultures', rather than the number of black faces in a street. The political point about this was that access to 'Englishness' entailed access to privileged treatment by the state, particularly in the spheres of policing and education.

The weakness of the New Racist analysis was that culture is not static. I have the same skin colour as my parents, but my cultural references are very different. However, the extent of interaction between white and black youth was purposely overlooked. Whereas Bob Marley seemed to be an integral part of the garden where I grew up, he is explicitly excluded from Roger Scruton's. It is his garden, not mine, that is important for the 'New Racism'. And it was therefore his garden, not mine, that came under scrutiny from the white academics trying to respond to this 'cultural' racism. Consequently, in most of the work produced in the early 1980s dealing with discourses of 'race', only one construction of white identity is discussed, and that is the one which racists themselves had produced in response to the changing nature of the black presence in Britain.

Critique of the New Racism tended to challenge its celebration of 'English' culture, and to expose the roots of that culture in the necessary presence and exploitation of black people. The slave trade was revealed as the foundation upon which the country houses of Jane Austen's novels were built. However, the typification of that 'English' culture as the heritage of white people in Britain was largely unchallenged. This seems a serious oversight. I am *not* trying to challenge the importance of racist structures in English cultural heritage, or deny that I absorbed a part of that racism. What I do think we need to recognize, however, is that, for my generation at least, other things were going on as well, and that there are more aspects of white identity than are dreamed of in Roger Scruton's garden.

Nevertheless, the concentration upon 'culture' was the Achilles' heel of the New Racism, and it was a weakness which was magnificently exploited by a new group of thinkers, who were well attuned to the fluidity of any definition rooted in 'culture'. The work of black British intellectuals and film-makers appeared, in the mid-1980s, to offer a vision of 'racial identity' which was based upon experiences parallel to my own. They began to map out the parameters of a 'Third Space'[11] where black identity was not essentialist (confining them within some immutable 'black' culture), but neither was it integrationist (requiring a rejection of 'black' culture and an embracing of an 'English' culture which had been defined as synonymous with white').

In exploring these possibilities, the writers and film-makers focused on what it meant to be 'black', given that this was not an essentialist identity. Clearly, however, it was not a free-floating voluntarist application of any definition they liked to the signifier /black/: the shared history of black people, particularly the paradox of an identity rooted in the shared experience of having *lost* an identity through the diaspora, provided a framework within which it became meaningful to talk in terms of black ethnicity, black culture, and black political sensitivity.[12] This was not a 'blackness' inherited as part of a fixed and separate cultural tradition, but a 'blackness' which emerged out of *interactions* between cultures and histories which could be traced back to both Europe and Africa. It is because culture is not fixed and immutable that contemporary forms of 'black' culture appeared.

This was exciting and illuminating for me. However, there was no interest here in what it might mean to be 'white', if that too is not a fixed, unitary identity; but then, that was not the question being raised. It was precisely because of the specific nature of black history that this analysis became possible and meaningful. Moreover, the New Racism wasn't challenging white identity; it was challenging blacks. All the same, as a white person, speaking from the specificity of my own experiences, this analysis did raise for me the issue of what my identity was in the 'Third Space'. I could see my interaction with reggae music as part of that process of cultural interaction being discussed, but the work as it stood was still concerned specifically with definitions of 'blackness'.

The 1980s saw the institutionalization of 'identity politics' in the thinking of the Left. Everyone had to have an identity to anchor their politics. And the problem with being white was that it did not seem to bestow an identity which could be linked to any kind of oppositional politics. Indeed, we never even seemed to question what 'whiteness' was. Although the term was bandied about happily by all kinds of people – the phrase 'white middle-class heterosexual male' tripping easily off the tongue – we never really thought about what we meant by it, except when we stumbled over the issue almost by accident, as at the *Marxism Today* workshop mentioned above. So I began to look at the ways in which references to 'whiteness' were used, to see if this would help.

Primarily, it seemed to me, 'whiteness' was used to signify the centre which pushes out, excludes, appropriates and distorts the margins. Although there is detailed understanding of the process whereby this happens, there is little analysis of what exactly that centre is, and, importantly for me, where I fit into it. We all recognize that the centre is a composite entity, and that the signifier /white/ must be read to refer to a set of articulated constituencies whose composition is conditional and in flux. But it still appears that all these constituencies are, in some way or another, defined primarily by their opposition to black people, or, to put it crudely, their racism. The analysis is useful for understanding how 'race' came to be the privileged metaphor for the collapse of the postwar consensus. It is less useful as a way of understanding what it means to me to be 'white'.

The problem for me, then, was to find a way of talking about myself as 'white' which recognized that at least part of what I was had been influenced by my sense of being situated *alongside* my black contemporaries, in opposition to the centre. White identity, it seemed, could only be seen through the lens of racism; any other sense of 'white' seemed to be absent from the work of those writers whose concerns seemed closest to my own. Stuart Hall, perhaps, was trying to express what I felt, when he said:

> Now that, in the postmodern age, you all feel so dispersed, I become centred. What I've thought of as dispersed and fragmented comes, paradoxically, to be *the* representative modern experience! . . . welcome to migranthood.[13]

But the fact that he was welcoming the whites to migranthood, to becoming part of the *black* experience, demonstrates the very problem that I have. If the centre is always represented by the signifier /white/, and 'white' has no meaning except as a way of describing the centre, then anyone who falls outside that definition is left with no identity except 'not-white'. Stuart Hall went on to say:

> Envy is a very funny thing for the British to feel at this moment in time – to want to be black! Yet I feel some of you surreptitiously moving towards that marginal identity. I welcome you to that, too.

It's nice to be welcomed, of course, but I think it's important to understand who is being welcomed to what. If our sense of ourselves as 'white' has become problematic, because the cohesion of our polity has broken down, 'whiteness' has become synonymous with the state's preferred modes of citizenship, and many of our political and cultural references have come from groups defined by their exclusion from that polity and by extension from 'whiteness', does this mean that we are, or want to be, 'honorary blacks'? Or does it mean that there is something wrong with the way we think about 'whiteness'?

III

It seems, then, that the work of black intellectuals gives me no way to engage with black issues except as someone who wants to be black. So I need to find a way of redefining 'white', to carry out some kind of raid on the garden where the integrationists managed to keep the term penned in. This cannot be a voluntaristic project. It is precisely because 'racial' signifiers are not free-floating that the issue of white identity is problematic for us in the first place. As Stuart Hall himself argued:

> 'Black' could not be converted to 'black is beautiful' simply by wishing it so. It had to become part of an organized practice of struggles requiring the building up of black resistances as well as the development of new forms of black consciousness.

If we want, then, to find a way to convert 'white', if not to 'white is beautiful', at least to 'white is OK in some circumstances', then we need to do so in concrete praxis, in direct engagement with issues of so-called 'racial politics'.

There is an immediate problem in finding a meaning for 'whiteness' other than that defined by integrationism and multiculturalism (both of which either promote, or collude in, the conflation of 'white' with a supposed 'English cultural heritage'). It is a simple question of power. The last ten years have seen, first, the elision of the idea of 'whiteness' with conformity to a putative 'English' way of behaving; second, sanctions against those who do not behave in those ways; and last, the re-creation of all those among us who have accepted this mode of behaviour as 'citizens' who have a right to consume goods and services, but not to question the wares on offer. If white people want to open out the meaning of 'whiteness', we can only do so as part of an overall assault upon the system which has created the dominant definition. And unfortunately, the obvious allies in this assault, the politically active members of the black communities, are increasingly accepting and perpetuating the dichotomy between notions of 'black' and 'white'.

The identity politics of the 1980s achieved many useful things. However, it also served to drive wedges between people whose 'identities' were presumed to be in opposition: male/female; Jewish/Palestinian; black/white. The easy assumption of an identification of my struggles with those of my black contemporaries which characterized the heady days of RAR and ANL was clearly no longer possible. As soon as we tried to engage with struggles affecting black people, we were confronted by a powerful Afrocentric strand which didn't necessarily exclude alliances with us, but was not particularly interested in white people's struggles and, unsurprisingly, had a limited vision of what it means to be 'white'. In pointing this out, I'm not challenging the separatist nature of black struggles; the right to separate organization is a fundamental tenet of the politics I grew up in. What I'm interested in is the definition of 'white' that seems to go with it.

The political tradition that is most active in issues around 'race' in London at present is not the one that underlies most of the theoretical work on 'race'. There is a crack between the intellectual and political leaderships among new social movements as a whole, which is by no means confined to issues of 'race'. This fissure seems to me to have arisen during the early 1980s, when Left intellectuals fell back on writing about how things *might* or *ought to* be, while Left activists found themselves defending old victories. Meanwhile, the activists of the 'new social movements' got on with much more pragmatic and expansionary projects, which often absorbed many of the values of Thatcherism. The wedges driven between people by identity politics encouraged the privileging and defending of a sectional interest against all comers. Margaret Thatcher proclaimed the end of 'society', and it became easy to replace 'society' with 'identity', especially in the boom period of the mid-1980s when 'identities' could also be markets. The *organic* intellectuals of black activism in London today are not the *academic* intellectuals whose work has been so influential on my own structures of thinking. While the work of the intellectuals focused on ways of breaking down essentialist modes of thought, and demonstrated the *complexity* of our identities, the growing Afrocentricity of black political activism largely reinstated essentialist modes of thought. Darcus Howe was perhaps the last prominent black activist thinker whose perspective was not explicitly Afrocentric, and he seemed to disappear into a Channel 4 career some time in the early 1980s. So it is important to understand the specific constructions of 'white' which are found in activism, and to distinguish them from those found in theory.

The definition of 'white' which sees racism as an attribute of individuals rather than of social structures seems to me to have been much influenced by the feminist theories about male violence against women which gained wide currency in the late 1970s. The one-to-one private and personal nature of that violence, and the recognition that violent men looked and behaved much like all other men, encouraged an analysis which saw maleness itself as inherently oppressive. The glib but inappropriate transfer of this analysis to other forms of oppression was too easy to resist. As Jenny Bourne pointed out:

Feminism has diluted the meaning of racism itself by personalizing it. Racism has ceased to be seen as the primarily *structural* and *institutional* issue that it was shown to be in the 1960s and 1970s and has become, under the impact of the tendencies in the Woman's Movement, an internalized matter of prejudice.[14]

This way of understanding racism was never particularly productive. It was taken over almost wholesale by the corporate anti-racism of the GLC, where, as Paul Gilroy has definitively demonstrated, it was largely counter-productive.[15] Policy replaced politics and Londoners wcre told that they had only two possible identities when it came to 'race': perpetrators or victims of racism. The effect on white political activism was to kill it:

> Issues of racism and sexism – instead of being tackled on a institutional, societal basis – tended to be reduced to a personalized level, with individuals either deeply complacent or paralysed by guilt.[16]

The effect on black political activism was more complicated, however. For black para-professionals whose identity seemed to straddle an uneasy space between 'white' and 'black' worlds, the claims of a single black 'nation' not only provided a sense of belonging but also justified their claims to speak on behalf of all black people, and not only their own class.[17] There is, therefore, a strong political reason for black activism to tend towards essentialism, which will not be swayed by the presence of white people claiming to espouse the same cause.

Within this political tradition, the nature of 'whiteness' is ascribed, almost entirely, by reference to events within the black communities. What white people themselves do is not really important to these ascriptions, although, of course, the continuing activities of racists seem to provide ample confirmation of the inherently racist aspects of white identity. It is not just racism, however, which gets tied up in these debates about 'whiteness'. 'White' seems rather to be a general term to describe anything situated outside the Afrocentric tradition. In the row over sponsorship of last year's Notting Hill Carnival, for example, many white people were surprised to discover that the London-wide radio station, Kiss FM, a former pirate station, was a 'white' radio station. This surprise was shared by a lot of black listeners, who had supported Kiss' campaign to be granted a licence precisely because its mandate was to service the black community.

In the world of black activism, then, 'white' is a matter of black ascription, and what white people do is marginal to the construction of these ascriptions. Their marginality extends beyond the sphere of politics and into the rest of the community. So, for example, objections were raised earlier this year when a photographic exhibition of images of Brixton was fronted by a white woman. This was not, note, an exhibition of images of Brixton's black community, but this recognition that Brixton is a mixed community was felt to undermine the importance of Brixton as an icon of black struggle. White activism is not ignored; but the legacy of corporate anti-racism means that there is only one way to interpret this activism. If racism is a personal attribute, and all white people are necessarily racist, then any white engagement with issues of race must be motivated by guilt. The only role for a white person in black politics, or indeed as a member of a 'black community' such as Brixton, is that of the guilty (in the objective as well as subjective sense) liberal.

So it seems that there were three main representations of 'whiteness' which emerged in the late 1970s and 1980s, on which white people could found an 'identity'. We could embrace a cultural conception of 'whiteness' as bestowing a limited and conservative 'Englishness'; we could be guilty liberals; or we could accept that our marginalization from mainstream 'whiteness' made us in some sense 'honorary blacks'. It still doesn't seem to be much of a basis for founding a politics, or indeed a sense of what it means to be 'white'.

I have a feeling that there was a moment, some time in the mid-1980s, when other options seemed to be available. In the process of criticizing the facile nature of corporate 'racism awareness training' and the essentialism of multicultural education, we seemed to glimpse a promise that we could all develop a sense of how we inhabited the 'Third Space'. What this meant for me, in political terms, was something to do with celebrating the hybrid nature of all Britishness. There was an energy and excitement about the idea of hybrid 'ethnicities'. We would destroy the New Racism and construct a new kind of society, typified, in Homi Bhabha's words, by cultural *difference*, rather than by the liberal construct of cultural diversity.[18] In direct opposition to the New Racism's proposition that cultural identities defined the fundamental divisions between white people and black, we all began to celebrate the complexities and interdependencies of our cultural heritages. White people like myself, whose sense of ourselves had been forged through the lens of 'race', who grew up listening to reggae music and who perhaps took part in the urban uprisings against the state, were as much a part of this project as everyone else.

In the early stages of developing this idea, it seemed easy to imagine what such an identity might look like. In London, in particular, the cosmopolitanism and the dynamic interactions of cultural traditions[19] created a real sense that the world was there to be forged in a new way. We did not yet live in that world, but the seeds were already present. For me, the work of the black film collectives Sankofa and Black Audio, and the writing of the Zimbabwean novelist Dambudzo Marechera, stood out as prefigurative inspiration for the complex whole that Englishness might become.

Somewhere, however, I feel that this project began to lose its momentum. Whites seemed to be more and more excluded from it, and the sense that this was something we were all involved in together began to fade. In understanding this, I want to return to the issue of 'ethnicity' raised at the very start of this essay. For, as the sharp-eyed among you will have spotted, this piece has so far been about 'whiteness' and about 'Englishness', and the ways in which the two ideas have been linked in the politics of the Thatcher years. 'Ethnicity' is something rather different, and something which can easily be linked to a (so-called) white skin colour. Much recent writing on 'ethnicity', especially coming out of the USA, has not been about people of colour at all, but about Poles, Irish, Germans, etc. As Stuart Hall has suggested, ethnic identities are becoming increasingly significant as the disintegration of the old certainties and the continuing racism of British institutional life makes a sense of national identity harder and harder to sustain as meaningful:

The slow contradictory movement from 'nationalism' to 'ethnicity' as a source of identities is part of a new politics. It is also part of the 'decline of the West' – that immense process of historical relativization which is just beginning to make the British, at least, feel just marginally 'marginal'.[20]

'Ethnicity' seems to lie at the heart of modern urban culture; it's something the white kids have to engage with as much as their black classmates. For black youth, 'ethnicity' seems to be something which applies to them, which they can transform or adopt as they please. There is a feeling of energy, exuberance and creativity growing out of this sense of self and exploration of potential. In this sense, 'ethnicity' 'belongs' to black kids in a way it doesn't belong to white kids. However, this does not seem to me to be because there is any specific connection between 'ethnicity' and 'race': it is perfectly possible to be black and Scottish, for example. 'Ethnicity' in this context is about 'superficial' cultural markers such as clothes, speech patterns and music. According to its dominant usage, whiteness cannot be an 'ethnic' identity, because it provides the standard against which ethnicity is defined. For white youth, then, and specifically the ones who cannot claim to be Irish, or Polish and so on, 'ethnicity' seems to be something which always belongs to someone else, which they can purchase, in clothes shops or record stores, but never possess.

This 'consumption' of an 'ethnic' identity is a very different thing from the sense of a shared politics which my love of reggae music was rooted in. Like political identities, 'ethnic' identities could be marketed – and they were extensively, by the late-1980s. The 'ethnic' identities in question were purely based on external forms: clothes, music etc. They could be easily adopted by young people from 'ethnic' backgrounds, who wanted a sense of their place in the urban environment, but did not feel strongly rooted in any community. Multicultural programmes in schools contributed to this replacement of 'racial' identity with 'ethnic' identity, a concentration on the external forms. The link with politics, if any, was largely coincidental. The voracious pillaging of any 'ethnic' identity by white trendies who had no apparent 'ethnicity' of their own to claim, and who were not motivated by any real sense of solidarity with the peoples whose shirts they were wearing, finally severed the links between politics and culture which had seemed so central in the late-1970s.

This strikes me now as a tragedy. It was precisely at that moment when something very exciting and important seemed to be happening in Britain, just at the time when cultural interaction seemed to suggest a whole new definition of 'Englishness' which would undermine its appropriation by the state and open up a new form of anti-racist practice; a tragedy that the project was drowned, as it were, in a cacophony of 'ethnic' identities, none of which engaged seriously with the issues of how 'cultural' identity was used by the state. The lack of a sense of creativity or potential or new identity coming from young white people in London now strikes me as an indication of how the vision of hybridity has been lost. This doesn't mean that we might as well all give up and concede that 'whiteness' will always be limited and fundamentally racist. But I think the task of reinventing 'whiteness' to meet the challenge of racism has been made harder, because we let slip the moment when racist ideology itself was vulnerable as it was shifting from a biological to a cultural determinism.

I understand, I think, why we failed to make the most of that moment. The continuing power of racism made it difficult to see how whites could fit into the picture without returning blacks to the margin. As most whites were not particularly engaged with the project anyway, the attention turned primarily towards a specifically 'black' experience of hybridity. And then, as I have shown, white identity had to be conceptualized as a unity, to provide a sense of what hybridity was *not*.

I think that we need to rethink our metaphors if we are going to find a way to talk meaningfully about whiteness. 'Centre' and 'margin' are not useful metaphors to describe a 'Third Space'. Where there is a centre and a margin, there is a circle, or some kind of enclosed space. And in an enclosed space, there is no 'third space'; there is *only* the centre and the margin. The only way you can move from that to a third space is by an act of will; there is nothing in the structure of centre and margin itself which can be transformed into a multi-spatial structure. Homi Bhabha's use of the idea of 'translation' will perhaps be more fruitful here.[21] But we also need to rethink our politics and to be much more open to the possibility of a radical white identity that isn't guilty, doesn't eat roast beef, and isn't trying to be black.

I'm making this suggestion because I think that more than just our self-esteem depends upon it. The passing of that moment in which it was possible to imagine a white hybrid identity has been accompanied by a passing of the era in which it was possible to have a unified politics. Ten years ago, when we fought the police on the streets of Brixton, and six years ago, when we did it again, we knew who our enemy was. There was an understanding – not always clearly articulated, but always present – that the enemy was the state. Now I'm afraid that we more often think the enemy is each other. The fact that we missed that 'moment of hybridity' frightens me. I wonder what we will produce in its place.

Notes

1 Stuart Hall, 'New Ethnicities', in *ICA Documents*, no. 7, London 1988.
2 Kobena Mercer, 'Recoding Narratives of Race and Nation' in *ICA Documents*, no. 7, London 1988.
3 Stuart Hall, *Racism and Reaction*, CRE/BBC, London 1978.
4 Paul Gilroy, *There Ain't No Black in the Union Jack*, Macmillan, London 1987, p. 207.
5 Paul Gilroy, 'It's Not Where You're From, It's Where You're At', in *Third Text*, no. 13, Winter 1990/91.
6 I am thinking particularly here of Dick Hebdige's concept of 'white ethnicity', expressed in the punk phenomenon as a response to the exclusion of young white men from the politics of Rastafari. See also Paul Gilroy's comment that: 'The realization that black youth would insist on the particularity of their oppression no matter how much the whites identified with Rastafari, precisely because white interest was a threat to the autonomy of the movement, returned white youth to the drab horizons of the skinhead cult with particular bitterness.' 'Steppin' Out of Babylon – Race, Class and Autonomy', in CCCS, *The Empire Strikes Back*, Hutchinson, London 1982, p. 296.
7 Stuart Hall, 'Minimal Selves', in *ICA Documents*, no. 6, 1987.

8 Kobena Mercer, 'Welcome to the Jungle: Identity and Diversity in Postmodern Politics', in Jonathan Rutherford (ed.), *Identity: Community, Culture, Difference*, Lawrence & Wishart, London 1990, p. 61.

9 Paul Gilroy, *Union Jack*, *op. cit.* p. 32, points out: 'The riots of 1981 and 1985 are remembered as somehow racial events. Given that a minority – between 29% and 33% – of those arrested in 1981 were "non-white" it is essential to ask how this memory has been constructed.'

10 Martin Barker, *The New Racism*, Junction Books, London 1981.

11 The term 'Third Space' comes from a variety of work by Homi Bhabha: see, for example, 'The Third Space', interview with Jonathan Rutherford, in Rutherford (ed.), *op. cit.*

12 See, for example, Stuart Hall, 'Cultural Identity and Diaspora', in *ibid.*; and Gilroy, 'It's Not Where You're From', *op. cit.*

13 Hall, 'Minimal Selves', *op. cit.*

14 Jenny Bourne, 'Homelands of the Mind: Jewish Feminism and Identity Politics', in *Race and Class*, Vol. 29 no. 1, Summer 1987.

15 Gilroy, *Union Jack*, *op. cit.*, chapter 4.

16 Andrea Stuart, 'Feminism: Dead or Alive?', in Rutherford (ed.), *op. cit.*, p. 35.

17 Gilroy, *Union Jack*, *op. cit.*, pp64ff; Gilroy, 'It's Not Where You're From', *op. cit.*, p. 6.

18 Cf. Bhabha, 'Third Space', in Rutherford (ed.), *op. cit.*, p. 209.

19 I am using the word 'tradition' here in the sense in which we speak of artistic traditions, which may be constructed and communicated within particular 'ethnic' communities: I do not intend to imply an essentialist 'tradition'.

20 Hall, 'Minimal Selves', *op. cit.*

21 Bhabha, 'Third Space', in Rutherford (ed.), *op. cit.* pp. 209–11.

---

17 Stuart Hall
Cultural Identity and Diaspora

Excerpts from: J. Rutherford (ed.) *Identity: community, culture difference*, pp. 222–37. London: Lawrence and Wishart (1990)

A new cinema of the Caribbean is emerging, joining the company of the other 'Third Cinemas'. It is related to, but different from the vibrant film and other forms of visual representation of the Afro-Caribbean (and Asian) 'blacks' of the diasporas of the West – the new post-colonial subjects. All these cultural practices and forms of representation have the black subject at their centre, putting the issue of cultural identity in question. Who is this emergent, new subject of the cinema? From where does he/she speak? Practices of representation always implicate the positions from which we speak or write – the positions of *enunciation*. What recent theories of enunciation suggest is that, though we speak, so to say 'in our own name', of ourselves and from our own experience, nevertheless who speaks, and the subject who is spoken of, are never identical, never exactly in the same place. Identity is not as transparent or unproblematic as we think. Perhaps instead of thinking of identity as an already accomplished fact, which the new cultural practices then represent, we should think, instead, of identity as a 'production', which is never complete, always in process, and always constituted within, not outside, representation. This view problematises the very authority and authenticity to which the term, 'cultural identity', lays claim.

We seek, here, to open a dialogue, an investigation, on the subject of cultural identity and representation. Of course, the 'I' who writes here must also be thought of as, itself, 'enunciated'. We all write and speak from a particular place and time, from a history and a culture which is specific. What we say is always 'in context', *positioned*. I was born into and spent my childhood and adolescence in a lower-middle-class family in Jamaica. I have lived all my adult life in England, in the shadow of the black diaspora – 'in the belly of the beast'. I write against the background of a lifetime's work in cultural studies. If the paper seems preoccupied with the diaspora experience and its narratives of displacement, it is worth remembering that all discourse is 'placed', and the heart has its reasons.

There are at least two different ways of thinking about 'cultural identity'. The first position defines 'cultural identity' in terms of one, shared culture, a sort of collective 'one true self', hiding inside the many other, more superficial or artificially imposed 'selves', which people with a shared history and ancestry hold in common. Within the terms of this definition, our cultural identities reflect the common historical experiences and shared cultural codes which provide us, as 'one people', with stable, unchanging and continuous frames of reference and meaning, beneath the shifting divisions and vicissitudes of our actual history. This 'oneness', underlying all the other, more superficial differences, is the truth, the essence, of 'Caribbeanness', of the black experience. It is this identity which a Caribbean or black diaspora must discover, excavate, bring to light and express through cincmatic representation.

Such a conception of cultural identity played a critical role in all the post-colonial struggles which have so profoundly reshaped our world. It lay at the centre of the vision of the poets of 'Negritude', like Aimée Cesaire and Leopold Senghor, and of the Pan-African political project, earlier in the century. It continues to be a very powerful and creative force in emergent forms of representation amongst hitherto marginalised peoples. In post-colonial societies, the rediscovery of this identity is often the object of what Frantz Fanon once called a

> passionate research . . . directed by the secret hope of discovering beyond the misery of today, beyond self-contempt, resignation and abjuration, some very beautiful and splendid era whose existence rehabilitates us both in regard to ourselves and in regard to others.

New forms of cultural practice in these societies address themselves to this project for the very good reason that, as Fanon puts it, in the recent past,

Colonisation is not satisfied merely with holding a people in its grip and emptying the native's brain of all form and content. By a kind of perverted logic, it turns to the past of oppressed people, and distorts, disfigures and destroys it.[1]

The question which Fanon's observation poses is, what is the nature of this 'profound research' which drives the new forms of visual and cinematic representation? Is it only a matter of unearthing that which the colonial experience buried and overlaid, bringing to light the hidden continuities it suppressed? Or is a quite different practice entailed – not the rediscovery but the *production* of identity. Not an identity grounded in the archaeology, but in the *re-telling* of the past?

We should not, for a moment, underestimate or neglect the importance of the act of imaginative rediscovery which this conception of a rediscovered, essential identity entails. 'Hidden histories' have played a critical role in the emergence of many of the most important social movements of our time – feminist, anti-colonial and anti-racist. The photographic work of a generation of Jamaican and Rastafarian artists, or of a visual artist like Armet Francis (a Jamaican-born photographer who has lived in Britain since the age of eight) is a testimony to the continuing creative power of this conception of identity within the emerging practices of representation. Francis's photographs of the peoples of The Black Triangle, taken in Africa, the Caribbean, the USA and the UK, attempt to reconstruct in visual terms 'the underlying unity of the black people whom colonisation and slavery distributed across the African diaspora.' His text is an act of imaginary reunification.

Crucially, such images offer a way of imposing an imaginary coherence on the experience of dispersal and fragmentation, which is the history of all enforced diasporas. They do this by representing or 'figuring' Africa as the mother of these different civilisations. This Triangle is, after all, 'centred' in Africa. Africa is the name of the missing term, the great aporia, which lies at the centre of our cultural identity and gives it a meaning which, until recently, it lacked. No one who looks at these textural images now, in the light of the history of transportation, slavery and migration, can fail to understand how the rift of separation, the 'loss of identity', which has been integral to the Caribbean experience only begins to be healed when these forgotten connections are once more set in place. Such texts restore an imaginary fullness or plentitude, to set against the broken rubric of our past. They are resources of resistance and identity, with which to confront the fragmented and pathological ways in which that experience has been recon-structed within the dominant regimes of cinematic and visual representation of the West.

There is, however, a second, related but different view of cultural identity. This second position recognises that, as well as the many points of similarity, there are also critical points of deep and significant *difference* which constitute 'what we really are'; or rather – since history has intervened – 'what we have become'. We cannot speak for very long, with any exactness, about 'one experience, one identity', without acknowledging its other side – the ruptures and discontinuities which constitute, precisely, the Caribbean's uniqueness. Cultural identity, in this second sense, is a matter of 'becoming' as well as of 'being'. It belongs to the future as much as to the past. It is not something which already exists, transcending place, time, history and culture. Cultural identities come from somewhere, have histories. But, like everything which is historical, they undergo constant transformation. Far from being eternally fixed in some essentialised past, they are subject to the continuous 'play' of history, culture and power. Far from being grounded in a mere 'recovery' of the past, which is waiting to be found, and which, when found, will secure our sense of ourselves into eternity, identities are the names we give to the different ways we are positioned by, and position ourselves within, the narratives of the past.

It is only from this second position that we can properly understand the traumatic character of 'the colonial experience'. The ways in which black people, black experiences, were positioned and subjected in the dominant regimes of representation were the effects of a critical exercise of cultural power and normalisation. Not only, in Said's 'Orientalist' sense, were we constructed as different and other within the categories of knowledge of the West by those regimes. They had the power to make us see and experience *ourselves* as 'Other'. Every regime of representation is a regime of power formed, as Foucault reminds us, by the fatal couplet, 'power/knowledge'. But this kind of knowledge is internal, not external. It is one thing to position a subject or set of peoples as the Other of a dominant discourse. It is quite another thing to subject them to that 'knowledge', not only as a matter of imposed will and domination, by the power of inner compulsion and subjective conformation to the norm. That is the lesson – the sombre majesty – of Fanon's insight into the colonising experience in *Black Skin, White Masks*.

This inner expropriation of cultural identity cripples and deforms. If its silences are not resisted, they produce, in Fanon's vivid phrase, 'individuals without an anchor, without horizon, colourless, stateless, rootless – a race of angels'.[2] Nevertheless, this idea of otherness as an inner compulsion changes our conception of 'cultural identity'. In this perspective, cultural identity is not a fixed essence at all, lying unchanged outside history and culture. It is not some universal and transcendental spirit inside us on which history has made no fundamental mark. It is not once-and-for-all. It is not a fixed origin to which we can make some final and absolute Return. Of course, it is not a mere phantasm either. It is *something* – not a mere trick of the imagination. It has its histories – and histories have their real, material and symbolic effects. The past continues to speak to us. But it no longer addresses us as a simple, factual 'past', since our relation to it, like the child's relation to the mother, is always-already 'after the break'. It is always constructed through memory, fantasy, narrative and myth. Cultural identities are the points of identification, the unstable points of identi-fication or suture, which are made, within the discourses of history and culture. Not an essence but a *positioning*. Hence, there is always a politics of identity, a politics of position, which has no absolute guarantee in an unproblematic, transcendental 'law of origin'.

This second view of cultural identity is much less familiar, and more unsettling. If identify does not proceed, in a straight, unbroken line, from some fixed origin, how are we to understand its formation? We might think of black

Caribbean identities as 'framed' by two axes or vectors, simultaneously operative: the vector of similarity and continuity; and the vector of difference and
rupture. Caribbean identities always have to be thought of in terms of the dialogic
relationship between these two axes. The one gives us some grounding in, some
continuity with, the past. The second reminds us that what we share is precisely
the experience of a profound discontinuity: the peoples dragged into slavery,
transportation, colonisation, migration, came predominantly from Africa – and
when that supply ended, it was temporarily refreshed by indentured labour from
the Asian subcontinent. (This neglected fact explains why, when you visit
Guyana or Trinidad, you see, symbolically inscribed in the faces of their peoples,
the paradoxical 'truth' of Christopher Columbus's mistake: you *can* find 'Asia'
by sailing west, if you know where to look!) In the history of the modern world,
there are few more traumatic ruptures to match these enforced separations from
Africa – already figured, in the European imaginary, as 'the Dark Continent'. But
the slaves were also from different countries, tribal communities, villages,
languages and gods. African religion, which has been so profoundly formative in
Carribbean spiritual life, is precisely *different* from Christian monotheism in
believing that God is so powerful that he can only be known through a
proliferation of spiritual manifestations, present everywhere in the natural and
social world. These gods live on, in an underground existence, in the hybridised
religious universe of Haitian voodoo, pocomania, Native pentacostalism, Black
baptism, Rastafarianism and the black Saints Latin American Catholicism. The
paradox is that it was the uprooting of slavery and transportation and the insertion
into the plantation economy (as well as the symbolic economy) of the Western
world that 'unified' these peoples across their differences, in the same moment as
it cut them from direct access to their past.

Difference, therefore, persists – in and alongside continuity. To return to the
Carribbean after any long absence is to experience again the shock of the
'doubleness' of similarity and difference. Visiting the French Carribbean for the
first time, I also saw at once how different Martinique is from, say, Jamaica: and
this is no mere difference of topography or climate. It is a profound difference of
culture and history. And the difference *matters*. It positions Martiniquains and
Jamaicans as *both* the same and different. Moreover, the boundaries of difference
are continually repositioned in relation to different points of reference. Vis-à-vis
the developed West, we are very much 'the same'. We belong to the marginal,
the underdeveloped, the periphery, the 'Other'. We are at the outer edge, the
'rim', of the metropolitan world – always 'South' to someone else's *El Norte*.

At the same time, we do not stand in the same relation of the 'otherness' to the
metropolitan centres. Each has negotiated its economic, political and cultural
dependency differently. And this 'difference', whether we like it or not, is
already inscribed in our cultural identities. In turn, it is this negotiation of identity
which makes us, vis-à-vis other Latin American people, with a very similar
history, different – Caribbeans, *les Antilliennes* ('islanders' to their mainland).
And yet, vis-à-vis one another, Jamaican, Haitian, Cuban, Guadeloupean, Barbadian, etc . . .

How, then, to describe this play of 'difference' within identity? The common
history – transportation, slavery, colonisation – has been profoundly formative.

For all these societies, unifying us across our differences. But it does not constitute a common *origin*, since it was, metaphorically as well as literally, a translation. The inscription of difference is also specific and critical. I use the word 'play' because the double meaning of the metaphor is important. It suggests, on the one hand, the instability, the permanent unsettlement, the lack of any final resolution. On the other hand, it reminds us that the place where this 'doubleness' is most powerfully to be heard is 'playing' within the varieties of Caribbean musics. This cultural 'play' could not therefore be represented, cinematically, as a simple, binary opposition – 'past/present', 'them/us'. Its complexity exceeds this binary structure of representation. At different places, times, in relation to different questions, the boundaries are re-sited. They become, not only what they have, at times, certainly been – mutually excluding categories, but also what they sometimes are – differential points along a sliding scale.

One trivial example is the way Martinique both *is* and *is not* 'French'. It is, of course, a *department* of France, and this is reflected in its standard and style of life, Fort de France is a much richer, more 'fashionable' place than Kingston – which is not only visibly poorer, but itself at a point of transition between being 'in fashion' in an Anglo-African and Afro-American way – for those who can afford to be in any sort of fashion at all. Yet, what is distinctively 'Martiniquais' can only be described in terms of that special and peculiar supplement which the black and mulatto skin adds to the 'refinement' and sophistication of a Parisian-derived *haute couture*: that is, a sophistication which, because it is black, is always transgressive.

To capture this sense of difference which is not pure 'otherness', we need to deploy the play on words of a theorist like Jacques Derrida. Derrida uses the anomalous 'a' in his way of writing 'difference' – *differance* – as a marker which sets up a disturbance in our settled understanding or translation of the word/concept. It sets the word in motion to new meanings without erasing the *trace* of its other meanings. His sense of *differance*, as Christopher Norris puts it, thus

> remains suspended between the two French verbs 'to differ' and 'to defer' (postpone), both of which contribute to its textual force but neither of which can fully capture its meaning. Language depends on difference, as Saussure showed . . . the structure of distinctive propositions which make up its basic economy. Where Derrida breaks new ground . . . is in the extent to which 'differ' shades into 'defer' . . . the idea that meaning is always deferred, perhaps to this point of an endless supplementarity, by the play of signification.[3]

This second sense of difference challenges the fixed binaries which stabilise meaning and representation and show how meaning is never finished or completed, but keeps on moving to encompass other, additional or supplementary meanings, which, as Norris puts it elsewhere,[4] 'disturb the classical economy of language and representation'. Without relations of difference, no representation could occur. But what is then constituted within representation is always open to being deferred, staggered, serialised.

Where, then, does identity come in to this infinite postponement of meaning? Derrida does not help us as much as he might here, though the notion of the 'trace' goes some way towards it. This is where it sometimes seems as if Derrida has permitted his profound theoretical insights to be reappropriated by his disciples into a celebration of formal 'playfulness', which evacuates them of their political meaning. For if signification depends upon the endless repositioning of its differential terms, meaning, in any specific instance, depends on the contingent and arbitrary stop – the necessary and temporary 'break' in the infinite semiosis of language. This does not detract from the original insight. It only threatens to do so if we mistake this 'cut' of identity – this *positioning*, which makes meaning possible – as a natural and permanent, rather than an arbitrary and contingent 'ending' – whereas I understand every such position as 'strategic' and arbitrary, in the sense that there is no permanent equivalence between the particular sentence we close, and its true meaning, as such. Meaning continues to unfold, so to speak, beyond the arbitrary closure which makes it, at any moment, possible. It is always either over- or under-determined, either an excess or a supplement. There is always something 'left over'.

It is possible, with this conception of 'difference', to rethink the positionings and repositionings of Caribbean cultural identities in relation to at least three 'presences', to borrow Aimée Cesaire's and Leopold Senghor's metaphor: *Présence Africaine*, *Présence Européenne*, and the third, most ambiguous, presence of all – the sliding term, *Présence Americain*. Of course, I am collapsing, for the moment, the many other cultural 'presences' which constitute the complexity of Caribbean identity (Indian, Chinese, Lebanese etc). I mean America, here, not in its 'first-world' sense – the big cousin to the North whose 'rim' we occupy, but in the second, broader sense: America, the 'New World', *Terra Incognita*.

Présence Africaine is the site of the repressed. Apparently silenced beyond memory by the power of the experience of slavery, Africa was, in fact present everywhere: in the everyday life and customs of the slave quarters, in the languages and patois of the plantations, in names and words, often disconnected from their taxonomies, in the secret syntactical structures through which other languages were spoken, in the stories and tales told to children, in religious practices and beliefs, in the spiritual life, the arts, crafts, musics and rhythms of slave and post-emancipation society. Africa, the signified which could not be represented directly in slavery, remained and remains the unspoken, unspeakable 'presence' in Caribbean culture. It is 'hiding' behind every verbal inflection, every narrative twist of Caribbean cultural life. It is the secret code with which every Western text was 're-read'. It is the ground-bass of every rhythm and bodily movement. *This* was – is – the 'Africa' that 'is alive and well in the diaspora'.[5]

When I was growing up in the 1940s and 1950s as a child in Kingston, I was surrounded by the signs, music and rhythms of this Africa of the diaspora, which only existed as a result of a long and discontinuous series of transformations. But, although almost everyone around me was some shade of brown or black (Africa 'speaks'!), I never once heard a single person refer to themselves or to others as, in some way, or as having been at some time in the past, 'African'. It was only in the 1970s that this Afro-Caribbean identity became historically available to the great majority of Jamaican people, at home and abroad. In this historic moment,

Jamaicans discovered themselves to be 'black' – just as, in the same moment, they discovered themselves to be the sons and daughters of 'slavery'.

This profound cultural discovery, however, was not, and could not be, made directly, without 'mediation'. It could only be made *through* the impact on popular life of the post-colonial revolution, the civil rights struggles, the culture of Rastafarianism and the music of reggae – the metaphors, the figures or signifiers of a new construction of 'Jamaican-ness'. These signified a 'new' Africa of the New World, grounded in an 'old' Africa: – a spiritual journey of discovery that led, in the Caribbean, to an indigenous cultural revolution; this is Africa, as we might say, necessarily 'deferred' – as a spiritual, cultural and political metaphor.

It is the presence/absence of Africa, in this form, which has made it the privileged signifier of new conceptions of Caribbean identity. Everyone in the Caribbean, of whatever ethnic background, must sooner or later come to terms with this African presence. Black, brown, mulatto, white – all must look *Présence Africaine* in the face, speak its name. But whether it is, in this sense, an *origin* of our identities, unchanged by four hundred years of displacement, dismemberment, transportation, to which we could in any final or literal sense return, is more open to doubt. The original 'Africa' is no longer there. It too has been transformed. History is, in that sense, irreversible. We must not collude with the West which, precisely, normalises and appropriates Africa by freezing it into some timeless zone of the primitive, unchanging past. Africa must at last be reckoned with by Caribbean people, but it cannot in any simple sense by merely recovered.

It belongs irrevocably, for us, to what Edward Said once called an 'imaginative geography and history', which helps 'the mind to intensify its own sense of itself by dramatising the difference between what is close to it and what is far away'. It 'has acquired an imaginative or figurative value we can name and feel'.[6] Our belongingness to it constitutes what Benedict Anderson calls 'an imagined community'.[7] To *this* 'Africa', which is a necessary part of the Caribbean imaginary, we can't literally go home again.

The character of this displaced 'homeward' journey – its length and complexity – comes across vividly, in a variety of texts. Tony Sewell's documentary archival photographs, Garvey's Children: the Legacy of Marcus Garvey, tells the story of a 'return' to an African identity which went, necessarily, by the long route – through London and the United States. It 'ends', not in Ethiopia but with Garvey's statue in front of the St Ann Parish Library in Jamaica: not with a traditional tribal chant but with the music of Burning Spear and Bob Marley's Redemption Song. This is our 'long journey' home. Derek Bishton's courageous visual and written text, *Black Heart Man* – the story of the journey of a *white* photographer 'on the trail of the promised land' – starts in England, and goes through Shashemene, the place in Ethiopia to which many Jamaican people have found their way on their search for the Promised Land, and slavery; but it ends in Pinnacle, Jamaica, where the first Rastafarian settlement was established, and 'beyond' – among the dispossessed of 20th-century Kingston and the streets of Handsworth, where Bishton's voyage of discovery first began. These symbolic journeys are necessary for us all – and necessarily circular. This is the Africa we must return to – but 'by another route': what Africa has *become* in the New World, what we have made of 'Africa': 'Africa' – as we re-tell it through politics, memory and desire.

What of the second, troubling, term in the identity equation – the European presence? For many of us, this is a matter not of too little but of too much. Where Africa was a case of the unspoken, Europe was a case of that which is endlessly speaking – and endlessly speaking *us*. The European presence interrupts the innocence of the whole discourse of 'difference' in the Caribbean by introducing the question of power. 'Europe' belongs irrevocably to the 'play' of power, to the lines of force and consent, to the role of the *dominant*, in Caribbean culture. In terms of colonialism, underdevelopment, poverty and the racism of colour, the European presence is that which, in visual representation, has positioned the black subject within its dominant regimes of representation: the colonial discourse, the literatures of adventure and exploration, the romance of the exotic, the ethnographic and travelling eye, the tropical languages of tourism, travel brochure and Hollywood and the violent, pornographic languages of ganja and urban violence.

Because *Présence Européenne* is about exclusion, imposition and expropriation, we are often tempted to locate that power as wholly external to us – an extrinsic force, whose influence can be thrown off like the serpent sheds its skin. What Frantz Fanon reminds us, in *Black Skin, White Masks*, is how this power has become a constitutive element in our own identities.

> The movements, the attitudes, the glances of the other fixed me there, in the sense in which a chemical solution is fixed by a dye. I was indignant; I demanded an explanation. Nothing happened. I burst apart. Now the fragments have been put together again by another self.[8]

This 'look', from – so to speak – the place of the Other, fixes us, not only in its violence, hostility and aggression, but in the ambivalence of its desire. This brings us face to face, not simply with the dominating European presence as the site or 'scene' of integration where those other presences which it had actively disaggregated were recomposed – re-framed, put together in a new way; but as the site of a profound splitting and doubling – what Homi Bhaba has called 'the ambivalent identifications of the racist world ... the "otherness" of the self inscribed in the perverse palimpsest of colonial identity.'[9]

The dialogue of power and resistance, of refusal and recognition, with and against *Présence Européenne* is almost as complex as the 'dialogue' with Africa. In terms of popular cultural life, it is nowhere to be found in its pure, pristine state. It is always-already fused, syncretised, with other cultural elements. It is always-already creolised – not lost beyond the Middle Passage, but ever-present: from the harmonics in our musics to the ground-bass of Africa, traversing and intersecting our lives at every point. How can we stage this dialogue so that, finally, we can place it, without terror or violence, rather than being forever placed by it? Can we ever recognise its irreversible influence, whilst resisting its imperialising eye? The engima is impossible, so far, to resolve. It requires the most complex of cultural strategies. Think, for example, of the dialogue of every Caribbean filmmaker or writer, one way or another, with the dominant cinemas and literature of the West – the complex relationship of young black British filmmakers with the 'avant-gardes' of European and American filmmaking. Who could describe this tense and tortured dialogue as a 'one way trip'?

The Third, 'New World' presence, is not so much power, as ground, place, territory. It is the juncture-point where the many cultural tributaries meet, the 'empty' land (the European colonisers emptied it) where strangers from every other part of the globe collided. None of the people who now occupy the islands – black, brown, white, African, European, American, Spanish, French, East Indian, Chinese, Portugese, Jew, Dutch – originally 'belonged' there. It is the space where the creolisations and assimilations and syncretisms were negotiated. The New World is the third term – the primal scene – where the fateful/fatal encounter was staged between Africa and the West. It also has to be understood as the place of many, continuous displacements: of the original pre-Columbian inhabitants, the Arawaks, Caribs and Amerindians, permanently displaced from their homelands and decimated; of other peoples displaced in different ways from Africa, Asia and Europe; the displacements of slavery, colonisation and conquest. It stands for the endless ways in which Caribbean people have been destined to 'migrate'; it is the signifier of migration itself – of travelling, voyaging and return as fate, as destiny; of the Antillean as the prototype of the modern or postmodern New World nomad, continually moving between centre and periphery. This preoccupation with movement and migration Caribbean cinema shares with many other 'Third Cinemas', but it is one of our defining themes, and it is destined to cross the narrative of every film script or cinematic image.

Présence Americaine continues to have its silences, its suppressions. Peter Hulme, in his essay on 'Islands of Enchantment'[10] reminds us that the word 'Jamaica' is the Hispanic form of the indigenous Arawak name – 'land of wood and water' – which Columbus's re-naming ('Santiago') never replaced. The Arawak presence remains today a ghostly one, visible in the islands mainly in museums and archeological sites, part of the barely knowable or usable 'past'. Hulme notes that it is not represented in the emblem of the Jamaican National Heritage Trust, for example, which chose instead the figure of Diego Pimienta, 'an African who fought for his Spanish masters against the English invasion of the island in 1655' – a deferred, metonymic, sly and sliding representation of Jamaican identity if ever there was one! He recounts the story of how Prime Minister Edward Seaga tried to alter the Jamaican coat-of-arms, which consists of two Arawak figures holding a shield with five pineapples, surmounted by an alligator. 'Can the crushed and extinct Arawaks represent the dauntless character of Jamaicans? Does the low-slung, near extinct crocodile, a cold-blooded reptile, symbolise the warm, soaring spirit of Jamaicans?' Prime Minister Seaga asked rhetorically.[11] There can be few political statements which so eloquently testify to the complexities entailed in the process of trying to represent a diverse people with a diverse history through a single, hegemonic 'identity'. Fortunately, Mr Seaga's invitation to the Jamaican people, who are overwhelmingly of African descent, to start their 'remembering' by first 'forgetting' something else, got the comeuppance it so richly deserved.

The 'New World' presence – America, *Terra Incognita* – is therefore itself the beginning of diaspora, of diversity, of hybridity and difference, what makes Afro-

Caribbean people already people of a diaspora. I use this term here metaphorically, not literally: diaspora does not refer us to those scattered tribes whose identity can only be secured in relation to some sacred homeland to which they must at all costs return, even if it means pushing other people into the sea. This is the old, the imperialising, the hegemonising, form of 'ethnicity'. We have seen the fate of the people of Palestine at the hands of this backward-looking conception of diaspora – and the complicity of the West with it. The diaspora experience as I intend it here is defined, not by essence or purity, but by the recognition of a necessary heterogeneity and diversity; by a conception of 'identity' which lives with and through, not despite, difference; by *hybridity*. Diaspora identities are those which are constantly producing and reproducing themselves anew, through transformation and difference. One can only think here of what is uniquely – 'essentially' – Caribbean: precisely the mixes of colour, pigmentation, physiognomic type; the 'blends' of tastes that is Caribbean cuisine; the aesthetics of the 'cross-overs', of 'cut-and-mix', to borrow Dick Hebdige's telling phrase, which is the heart and soul of black music. Young black cultural practitioners and critics in Britain are increasingly coming to acknowledge and explore in their work this 'diaspora aesthetic' and its formations in the post-colonial experience:

> Across a whole range of cultural forms there is a 'syncretic' dynamic which critically appropriates elements from the master-codes of the dominant culture and 'creolises' them, disarticulating given signs and re-articulating their symbolic meaning. The subversive force of this hybridising tendency is most apparent at the level of language itself where creoles, patois and black English decentre, destabilise and carnivalise the linguistic domination of 'English' – the nation-language of master-discourse – through strategic inflections, re-accentuations and other performative moves in semantic, syntactic and lexical codes.[12]

It is because this New World is constituted for us as place, a narrative of displacement, that it gives rise so profoundly to a certain imaginary plenitude, recreating the endless desire to return to 'lost origins', to be one again with the mother, to go back to the beginning. Who can ever forget, when once seen rising up out of that blue-green Caribbean, those islands of enchantment. Who has not known, at this moment, the surge of an overwhelming nostalgia for lost origins, for 'times past'? And yet, this 'return to the beginning' is like the imaginary in Lacan – it can neither be fulfilled nor requited, and hence is the beginning of the symbolic, of representation, the infinitely renewable source of desire, memory, myth, search, discovery – in short, the reservoir of our cinematic narratives.

We have been trying, in a series of metaphors, to put in play a different sense of our relationship to the past, and thus a different way of thinking about cultural identity, which might constitute new points of recognition in the discourses of the emerging Caribbean cinema and black British cinemas. We have been trying to theorise identity as constituted, not outside but within representation; and hence of cinema, not as a second-order mirror held up to reflect what already exists, but as that form of representation which is able to constitute us as new kinds of subjects, and thereby enable us to discover places from which to speak. Communities, Benedict Anderson argues in *Imagined Communities* are to be distinguished, not by their falsity/genuineness, but by the style in which they are imagined.[13] This is the vocation of modern black cinemas: by allowing us to see and recognise the different parts and histories of ourselves, to construct those points of identification, those positionalities we call in retrospect our 'cultural identities'.

> We must not therefore be content with delving into the past of a people in order to find coherent elements which will counteract colonialism's attempts to falsify and harm . . . A national culture is not a folk-lore, nor an abstract populism that believes it can discover a people's true nature. A national culture is the whole body of efforts made by a people in the sphere of thought to describe, justify and praise the action through which that people has created itself and keeps itself in existence.[14]

Notes

1   Frantz Fanon, 'On National Culture', in *The Wretched of the Earth*, London 1963, p. 170.
2   *Ibid.*, p. 176.
3   Christopher Norris, *Deconstruction: Theory and Practice*, London 1982, p. 32.
4   Christopher Norris, *Jacques Derrida*, London 1987, p. 15.
5   Stuart Hall, *Resistance Through Rituals*, London 1976.
6   Edward Said, *Orientalism*, London 1985, p. 55.
7   Benedict Anderson, *Imagined Communities: Reflections on the Origin and Rise of Nationalism*, London 1982.
8   Frantz Fanon, *Black Skin, White Masks*, London 1986, p. 109.
9   Homi Bhabha, 'Foreword' to Fanon, *ibid.*, xv.
10   In *New Formations*, no. 3, Winter 1987.
11   *Jamaica Hansard*, vol. 9, 1983–4, p. 363. Quoted in Hulme, *ibid.*
12   Kobena Mercer, 'Diaspora Culture and the Dialogic Imagination', in M. Cham and C. Watkins (eds), *Blackframes: Critical Perspectives on Black Independent Cinema*, 1988, p. 57.
13   Anderson, *op. cit.*, p. 15.
14   Fanon, *op. cit.*, 1963, p. 188.

Kevin Robins

Tradition and Translation: National Culture in Its Global Context

Excerpts from: J. Corner and S. Harvey (eds), *Enterprise and heritage: crosscurrents of national culture*, pp. 21–44. London: Routledge (1991)

Where once we could believe in the comforts and continuities of Tradition, today we must face the responsibilities of cultural Translation.

Homi Bhabha

This chapter is about *changing geographies* – particularly the new forces of globalization that are now shaping our times – and what they mean for the economic and cultural life of contemporary Britain. It is in this global context, I think, that we can begin to understand the emergence, over the last decade or so, of both enterprise and heritage cultures. It is also in this context that the *problem of empire*, for so long at the heart of British national culture and identity, is now taking on a new significance.

Tradition and translation

Recent debate on the state of British culture and society has tended to concentrate on the power of Tradition. Accounts of the crisis of British (or English) national traditions and cultures have described the cultural survivalism and mutation that comes in the aftermath of an exploded empire. As Raphael Samuel argues in his account of the pathology of Tradition, the idea of nationality continues to have a powerful, if regressive, afterlife, and 'the sleeping images which spring to life in times of crisis – the fear, for instance, of being "swamped" by foreign invasion – testify to its continuing force.'[1] It is a concern with the past and future of British Tradition that has been central to Prince Charles's recent declamations on both enterprise and heritage. A 'new Renaissance for Britain' can be built, he suggests, upon a new culture of enterprise; a new business ethos, characterized by responsibility and vision, can rebuild the historical sense of community and once again make Britain a world actor. What is also called for, according to the Prince's 'personal vision', is the revival and re-enchantment of our rich national heritage. As Patrick Wright argues, the Prince of Wales has been sensitive to 'the deepest disruptions and disappointments in the nation's post-war experience,'[2] and his invocation of so-called traditional and spiritual values is again intended to restore the sense of British community and confidence that has collapsed in these modern or maybe postmodern times.

This prevailing concern with the comforts and continuities of historical tradition and identity reflects an insular and narcissistic response to the break-down of Britain. In the broader political and cultural sphere what is called for is our recognition or other worlds, the dis-illusioned acknowledgement of other cultures, other identities and ways of life.

This is what I take Homi Bhabha to mean by the responsibility of cultural Translation. It is about taking seriously 'the deep, the profoundly perturbed and perturbing question of our relationship to others – other cultures, other states, other histories, other experiences, traditions, peoples, and destinies'.[3] This responsibility demands that we come to terms with the 'geographical disposition' that has been so significant for what Edward Said calls the 'cultural structures of the West'. 'We could not have had empire itself,' he argues, 'as well as many forms of historiography, anthropology, sociology, and modern legal structures, without important philosophical and imaginative processes at work in the production as well as the acquisition, subordination, and settlement of space.'[4] Empire has long been at the heart of British culture and imagination, manifesting itself in more or less virulent forms, through insular nationalism and through racist paranoia. The relation of Britain to its 'Other' is one profoundly important context in which to consider the emergence of both enterprise and heritage cultures. The question is whether, in these supposedly post-imperial times, it is possible to meet the challenge of Translation; whether it is now possible for Britain to accept the world as a sufficiently benign place for its weakness not to be catastrophic. The challenge is not easy, as the Rushdie affair has made clear, for 'in the attempt to mediate between different cultures, languages and societies, there is always the threat of mis-translation, confusion and fear'.[5] There is also, and even more tragically, the danger of a fearful refusal to translate: the threat of a retreat into cultural autism and of a rearguard reinforcement of imperial illusions.

Global culture

The historical development of capitalist economies has always had profound implications for cultures, identities, and ways of life. The globalization of economic activity is now associated with a further wave of cultural transformation, with a process of cultural globalization. At one level, this is about the manufacture of universal cultural products – a process which has, of course, been developing for a long time. In the new cultural industries, there is a belief – to use Saatchi terminology – in 'world cultural convergence'; a belief in the convergence of lifestyle, culture, and behaviour among consumer segments across the world. This faith in the emergence of a 'shared culture' and a common 'world awareness' appears to be vindicated by the success of products like *Dallas* or *Batman* and by attractions like Disneyland. According to the president of the new Euro Disneyland, 'Disney's characters are universal'. 'You try and convince an Italian child', he challenges, 'that Topolino – the Italian name for Mickey Mouse – is American.'[6]

As in the wider economy, global standardization in the cultural industries reflects, of course, the drive to achieve ever greater economies of scale. More precisely, it is about achieving both scale and scope economies by targeting the shared habits and tastes of particular market segments at the global level, rather than by marketing, on the basis of geographical proximity, to different national audiences. The global cultural industries are increasingly driven to recover their escalating costs over the maximum market base, over pan-regional and world markets. They are driven by the very same globalizing logic that is reshaping the economy as a whole.

The new merchants of universal culture aspire towards a borderless world. Sky and BSB (which merged their activities in October 1990) beam out their products to a 'world without frontiers'; satellite footprints spill over the former integrity of national territories. With the globalization of culture, the link between culture and territory becomes significantly broken. A representative of Cable News Network (CNN) describes the phenomenon:

> There has been a cultural and social revolution as a consequence of the globalisation of the economy. A blue-collar worker in America is affected as much as a party boss in Moscow or an executive in Tokyo. This means that what we do for America has validity outside America. Our news is global news.[7]

What is being created is a new electronic cultural space, a 'placeless' geography of image and simulation. The formation of this global hyperspace is reflected in that strand of postmodernist thinking associated particularly with writers like Baudrillard and Virilio. Baudrillard, for example, invokes the vertigo, the disorientation, the delirium created by a world of flows and images and screens. This new global arena of culture is a world of instantaneous and depthless communication, a world in which space and time horizons have become compressed and collapsed.

The creators of this universal cultural space are the new global cultural corporations. In an environment of enormous opportunities and escalating costs, what is clearer than ever before is the relation between size and power. What we are seeing in the cultural industries is a recognition of the advantages of scale, and in this sphere too, it is giving rise to an explosion of mergers, acquisitions, and strategic alliances.[8] The most dynamic actors are rapidly restructuring to ensure strategic control of a range of cultural products across world markets. America's largest broadcasting company, NBC, is now, in the words of its vice-president, keenly 'developing global partnerships' and 'encouraging those companies in Europe and Japan who are interested in working in a partnered or allied way'.[9]

If the origination of world-standardized cultural products is one key strategy, the process of globalization is, in fact, more complex and diverse. In reality, it is not possible to eradicate or transcend difference. Here, too, the principle of equidistance prevails: the resourceful global conglomerate exploits local difference and particularity. Cultural products are assembled from all over the world and turned into commodities for a new 'cosmopolitan' market-place: world music and tourism; ethnic arts, fashion, and cuisine; Third World writing and cinema. The local and 'exotic' are torn out of place and time to be repackaged for the world bazaar. So-called world culture may reflect a new valuation of difference and particularity, but it is also very much about making a profit from it. Theodore Levitt explains this globalization of ethnicity. The global growth of ethnic markets, he suggests, is an example of the global standardization of segments:

Everywhere there is Chinese food, pitta bread, country and western music, pizza, and jazz. The global pervasiveness of ethnic forms represents the cosmopolitanisation of speciality. Again, globalisation does not mean the end of segments. It means, instead, their expansion to worldwide proportions.[10]

Now it is the turn of African music, Thai cuisine, Aboriginal painting, and so on, to be absorbed into the world market and to become cosmopolitan specialities.

Jean-Hubert Martin's recent exhibition at the Pompidou Centre, *Magiciens de la Terre*, is an interesting and significant barometer, in the world of high art, of this new climate of cultural globalization.[11] In this exhibition, Martin assembles original works by one hundred artists from all over the world: from the major artistic centres of Europe and America, but also from the 'margins' of Haiti, Nepal, Zaire and Madagascar. Here the discourse of high art converges with that of ethnography, the work of the Euro-American avant-garde is contiguous with that of Third World 'primitives'. Hubert's aim in developing 'the truly international exhibition of worldwide contemporary art' was to question the 'false distinction' between western cultures and other cultures, to 'show the real difference and the specificity of the different cultures', and to 'create a dialogue' between western and other cultures. *Magiciens de la Terre* brings 'world art' into being. Artistic texts and artifacts are pulled out of their original contexts and then reinserted and reinterpreted in a new global context. The global museum is a decentred space: Hubert cultivates an 'equidistance of perspective' in which each exhibit, in equal dialogue with all the rest, is valued for its difference and specificity.

Is *Magiciens de la Terre* about something more than simply absorbing new products into the international art market? 'What is it', in the words of Coco Fusco, 'that makes ethnicity attractive and marketable at a particular moment such as ours?'[12] Why does it resonate so much with the times? At one level, the project is genuinely exciting and challenging. This kind of cosmopolitanism is to be preferred to parochialism and insularity. There is indeed an immediate pleasure and exhilaration in seeing such a juxtaposition of diverse and vibrant cultures. But the exhibition touches deeper and darker chords. In its preoccupation with 'magic' and the 'spirituality' of Third World art, *Magiciens de la Terre* seeks to expose a certain emptiness, a spiritual vacuum, in western culture. There is, of course, something very suspect and problematical about this western idealization of 'primitiveness' and 'purity', this romance of the 'Other'. The exhibition in no way confronts or handles this inadequacy. None the less, even if there is no resolution, it does pose important questions about the nature of cultural identity, and about its relation to 'Otherness'. How do we now define ourselves as western? And how does this western identity relate to 'other', non-western, identities in the world?

If the global collection and circulation of artistic products has been responsible for new kinds of encounter and collision between cultures, there have also been more direct and immediate exchanges and confrontations. The long history of colonialism and imperialism has brought large populations of migrants and refugees from the Third to the First World. Whereas Europe once addressed African and Asian cultures across vast distances, now that 'Other' has installed itself within the very heart of the western metropolis. Through a kind of reverse invasion, the periphery has infiltrated the colonial core. The protective filters of time and space have disappeared, and the encounter with the 'alien' and 'exotic' is now instantaneous and immediate. The western city has become a crucible in which world cultures are brought into direct contact. As Neil Ascherson argues,

> the history of immigration into Europe over the past quarter century may seem like the history of increasing restrictions and smaller quotas. Seen in fast forward, though, it is the opposite: the beginning of a historic migration from the South into Europe which has gained its first decisive bridgehead.

It is a migration that is shaking up the 'little white "christian" Europe' of the past.[13] Through this irruption of empire, the certain and centred perspective of the old colonial order is confronted and confused.

Time and distance no longer mediate the encounter with 'other' cultures. This drama of globalization is symbolized perfectly in the collision between western 'liberalism' and Islamic 'fundamentalism' centred around the Rushdie affair. How do we cope with the shock of confrontation? This is perhaps the key political agenda in this era of space/time compression. One danger is that we retreat into fortress identities. Another is that, in the anxious search for secure and stable identities, we politicize those activities – religion, literature, philosophy – that should not be *directly* political. The responsibility of Translation means learning to listen to 'others' and learning to speak to rather than for or about, 'others'. That is easily said, of course, but not so easy to accomplish. Hierarchical orders of identity will not quickly disappear. Indeed, the very celebration and recognition of 'difference' and 'otherness' may itself conceal more subtle and insidious relations of power. When Martin turns world art into a spectacle in *Magiciens de la Terre*, might this not simply represent a new and enhanced form of western colonial appropriation and assimilation?

Global–local nexus

Globalization is about the compression of time and space horizons and the creation of a world of instantaneity and depthlessness. Global space is a space of flows, an electronic space, a decentred space, a space in which frontiers and boundaries have become permeable. Within this global arena, economies and cultures are thrown into intense and immediate contact with each other – with each 'Other' (an 'Other' that is no longer simply 'out there', but also within).

I have argued that this is the force shaping our times. Many commentators, however, suggest that something quite different is happening: that the new geographies are, in fact, about the renaissance of locality and region.[14] There has been a great surge of interest recently in local economies and local economic strategies. The case for the local or regional economy as the key unit of production has been forcefully made by the 'flexible specialization' thesis. Basing its arguments on the economic success of the 'Third Italy', this perspective stresses the central and prefigurative importance of localized production complexes. Crucial to their success, it is suggested, are strong local institutions and infrastructures: relations of trust based on face-to-face contact; a 'productive community' historically rooted in a particular place; a strong sense of local pride and attachment.[15] In Britain this localizing ethos, often directly influenced by the 'flexible specialization' thesis, was manifest in a number of local economic development strategies undertaken by local authorities (notably the Greater London Council, Sheffield City Council and West Midlands County Council).[16]

In the cultural sphere too, localism has come to play an important role. The 'struggle for place' is at the heart of much contemporary concern with urban regeneration and the built environment. Prince Charles's crusade on behalf of community architecture and classical revivalism is the most prominent and influential example. There is a strong sense that modernist planning was associated with universalizing and abstract tendencies, whilst postmodernism is about drawing upon the sense of place, about revalidating and revitalizing the local and the particular. A neo-Romantic fascination with traditional and vernacular motifs is supposedly about the re-enchantment of the city.[17] This cultural localism reflects, in turn, deeper feelings about the inscription of human lives and identities in space and time. There is a growing interest in the embeddedness of life histories within the boundaries of place, and with the continuities of identity and community through local memory and heritage. Witness the enormous popularity of the Catherine Cookson heritage trail in South Tyneside, of 'a whole day of nostalgia' at Beamish in County Durham, or of Wigan Pier's evocation of 'the way we were'. If modernity created an abstract and universal sense of self, then postmodernity will be about a sense of identity rooted in the particularity of place: 'it contains the possibility of a revived and creative human geography built around a newly informed synthesis of people and place'.[18]

Whilst globalization may be the prevailing force of our times, this does not mean that localism is without significance. If I have emphasized processes of de-localization, associated especially with the development of new information and communications networks, this should not be seen as an absolute tendency. The particularity of place and culture can never be done away with, can never be absolutely transcended.[19] Globalization is, in fact, also associated with new dynamics of *re*-localization. It is about the achievement of a new global–local nexus, about new and intricate relations between global space and local space.[20] Globalization is like putting together a jigsaw puzzle: it is a matter of inserting a multiplicity of localities into the overall picture of a new global system.

We should not idealize the local, however. We should not invest our hopes for the future in the redemptive qualities of local economies, local cultures, local identities. It is important to see the local as a relational, and relative, concept. If once it was significant in relation to the national sphere, now its meaning is being recast in the context of globalization. For the global corporation, the global–local nexus is of key and strategic importance. According to Olivetti's Carlo de Benedetti, in the face of ever-higher development costs, '*globalisation* is the only possible answer'. 'Marketers', he continues, 'must sell the latest product everywhere at once – and that means producing *locally*.'[21] Similarly, the mighty Sony describes its operational strategy as 'global localisation'.[22] NBC's vice-president, J. B. Holston III, is also resolutely 'for localism', and recognizes that globalization is 'not just about putting factories into countries, it's being part of that culture too'.[23]

What is being acknowledged is that globalization entails a corporate presence in, and understanding of, the 'local' arena. But the 'local' in this sense does not correspond to any specific territorial configuration. The global–local nexus is about the relation between globalizing and particularizing dynamics in the strategy of the global corporation, and the 'local' should be seen as a fluid and relational space, constituted only in and through its relation to the global. For the global corporation, the local might, in fact, correspond to a regional, national or even pan-regional sphere of activity.

This is to say that the 'local' should not be mistaken for the 'locality'. It is to emphasize that the global–local nexus does not create a privileged new role for the locality in the world economic arena. Of course local economies continue to matter. That is not the issue. We should, however, treat claims about new capacities for local autonomy and proactivity with scepticism. If it is, indeed, the case that localities do now increasingly bypass the national state to deal directly with global corporations, world bodies or foreign governments, they do not do so on equal terms. Whether it is to attract a new car factory or the Olympic Games, they go as supplicants. And, even as supplicants, they go in competition with each other: cities and localities are now fiercely struggling against each other to attract footloose and predatory investors to their particular patch. Of course, some localities are able successfully to 'switch' themselves in to the global networks, but others will remain 'unswitched' or even 'unplugged'. And, in a world characterized by the increasing mobility of capital and the rapid recycling of space even those that manage to become connected in to the global system are always vulnerable to the abrupt withdrawal of investment and to disconnection from the global system.

The global–local nexus is also not straightforwardly about a renaissance of local cultures. There are those who argue that the old and rigid hegemony of national cultures is now being eroded from below by burgeoning local and regional cultures. Modern times are characterized, it is suggested, by a process of cultural decentralization and by the sudden resurgence of place-bound traditions, languages and ways of life. It is important not to devalue the perceived and felt vitality of local cultures and identities. But again, their significance can only be understood in the context of a broader and encompassing process. Local cultures are over-shadowed by an emerging 'world culture' – and still, of course, by resilient national and nationalist cultures.

It may well be that, in some cases, the new global context is recreating sense of place and sense of community in very positive ways, giving rise to an energetic cosmopolitanism in certain localities. In others, however, local fragmentation may inspire a nostalgic, introverted and parochial sense of local attachment and identity. If globalization recontextualizes and reinterprets cultural localism, it does so in ways that are equivocal and ambiguous.

It is in the context of this global–local nexus that we can begin to understand the nature and significance of the enterprise and heritage cultures that have been developing in Britain over the past decade or so. I want now to explore two particular aspects of contemporary cultural transformation (each in its different way centred around the relationship between Tradition and Translation).

On not needing and needing Andy Capp

Why discuss enterprise and heritage together? Is there really any connection between the modernizing ambitions of enterprise culture and the retrospective nostalgia of heritage culture? The argument put forward in this section is that there is in fact a close and *necessary* relation between them. The nature of this relationship becomes clear, I suggest, when we see that each has developed as a response to the forces of globalization. Insight into this relational logic then helps us to understand the neurotic ambivalence that is, I think, at the heart of contemporary cultural transformation.

Enterprise culture is about responding to the new global conditions of accumulation. British capital must adapt to the new terms of global competition and learn to function in world markets. It must pursue strategic alliances and joint ventures with leading firms in Europe, North America, and Japan. In all key sectors – from pharmaceuticals to telecommunications, from automobiles to financial services – 'national champions' are being replaced by new flexible transnationals. In the cause of global efficiency, it is necessary to repudiate the old 'geography-driven' and home-centred ethos, and to conform to the new logic of placelessness and equidistance of perspective. The broadcasting industries are a good example. In the new climate, it is no longer viable to make programmes for British audiences alone. One way of understanding the debate around 'public service versus the market' is in terms of the displacement of nationally centred broadcasting services by a new generation of audio-visual corporations, like Crown Communications and Carlton Communications, operating in European and global markets. As the recent White Paper on broadcasting makes clear, television is 'becoming an increasingly international medium' centred around 'international trade in ideas, cultures and experiences'.[24] The consequence of these developments, across all sectors, is that the particularity of British identity is de-emphasized. In a world in which it is necessary to be 'local' everywhere – to be 'multidomestic' – certain older forms of national identity can actually be a liability. The logic of enterprise culture essentially pushes towards the 'modernization' of national culture. Indeed it is frequently driven by an explicit and virulent disdain for particular aspects of British culture and traditions. This scorn is directed against what the self-styled 'department for Enterprise' calls 'the past anti-enterprise bias of British culture'.[25] The spirit of enterprise is about eradicating what has been called the 'British disease': the 'pseudo-aristocratic' snobbery that has allegedly devalued entrepreneurial skills and technological prowess, and which has always undermined Britain's competitive position in world trade.[26]

If enterprise culture aims to refurbish and refine national culture and identity, there are, however, countervailing forces at work. Globalization is also underpinned by a quite contrary logic or development. As Scott Lash and John Urry argue, the enhanced mobility of transnational corporations across the world is, in fact, associated with an increased sensitivity to even quite small differences in the endowments of particular locations. 'The effect of heightened spatial difference', they suggest, 'has profound effects upon particular places ... contemporary developments may well be heightening the salience of localities.'[27] As global corporations scan the world for preferential locations, so are particular places forced into a competitive race to attract inward investors. Cities and localities must actively promote and advertise their attractions. What has been called the 'new urbanity'[28] is very much about enhancing the profile and image of places in a new global context. It is necessary to emphasize the national or regional distinctiveness of a location. As Margit Mayer points out, 'endogenous potentials' come to be cultivated: 'cities have come to emphasise, exploit and even produce (cultural and natural) local specificity and assets ... In this process, place-specific differences have become a tool in the competition over positional advantages.'[29] In this process, local, regional, or national cultures and heritage will be exploited to enhance the distinctive qualities of a city or locality.[30] Tradition and heritage are factors that enhance the 'quality of life' of particular places and make them attractive locations for investment. An emphasis on tradition and heritage is also, of course, important in the development of tourism as a major industry. Here, too, there is a premium on difference and particularity. In a world where differences are being erased, the commodification of place is about creating distinct place-identities in the eyes of global tourists. Even in the most disadvantaged places, heritage, or the simulacrum of heritage, can be mobilized to gain competitive advantage in the race between places. When Bradford's tourist officer, for example, talks about 'creating a product' – weekend holidays based around the themes of 'Industrial Heritage' and 'In the Steps of the Brontës',[31] he is underlining the importance of place-making in placeless times, the heightened importance of distinction in a world where differences are being effaced.

In the new global arena, it is necessary, then, simultaneously to minimize and maximize traditional cultural forms. The North-East of England provides a good example of how these contradictory dynamics of enterprise and heritage are developing. In this part of the country, it is over the symbolic body of Andy Capp that the two logics contest. 'Andy Capp is dead – Newcastle is alive'[32] – that is the message of enterprise. The region no longer has a place for Andy or for other cloth-capped local heroes like the late Tyneside comedian, Bobby Thompson. 'The real Northerner is no relation to Bobby or Andy', local celebrity Brendan Foster tells us.[33] The 'Great North' promotional campaign puts great emphasis on 'enterprise'and 'opportunity' and tries to play down the heritage of the region's old industrial, and later de-industrialized, past.[34] Newcastle City Council has recently employed J. Walter Thompson to change the city's image and to get rid of the old cloth-cap image once and for all. In order to position itself in the new global context, the region must re-image and, ultimately, re-imagine itself. The increasing Japanese presence in the region (around forty companies at the present moment) has become a key factor in this strategic identity switch. Japan is the very symbol of enterprise culture. In a little paean to Japanization, Labour MP Giles Radice invokes the buzz-words of 'quality', 'flexibility' and 'teamwork' to convey the benign influence of these foreign investors.[35] Japan is the key to constructing the new model Geordie. The region's history is now being reassessed to emphasize the special relationship between Japan and the North-East. 'The North-East aided Japan's progress towards modernisation in the late nineteenth and early twentieth centuries,' we are told, whilst, 'today Japanese investment is contributing to the revitalisation of a region that followed a very different course in the post-war period.' We must, it is stressed, 'adapt to changing times'.[36]

If the spirit of enterprise wants to kill off Andy Capp, there is, however, a counter-spirit that keeps him alive. The region's industrial past is its burden, but it is also its inheritance. It is clear that history can be made to pay. Beamish, The Land of the Prince Bishops, Roman Northumberland and Catherine Cookson Country are all heritage assets that can be exploited to attract tourists and investors alike. But if heritage is to be marketed, it becomes difficult to avoid the reality that the North-East was once a region of heavy engineering, shipbuilding and coal mining. And around these industries there developed a rich working-class culture. For many in the region, the conservation of local culture and traditions is extremely important. The photographic work of Newcastle's Side Gallery, and also the productions of film workshops like Amber and Trade, have paid great attention to working-class heritage. The work of writers like Jack Common, Sid Chaplin or Tom Hadaway has also contributed to the creation of a distinctive identify for the region.[37] Footballer Jackie Milburn is another powerful symbol of working-class heritage. And so too is the 'little waster,' Bobby Thompson, whom Brendan Foster sees as so much the embodiment of the Andy Capp myth. As Leslie Gofton argues, 'it is this image, and the perverse attachment of the people to the life which goes with it, which is being attacked by entrepreneurs such as John Hall'.[38] Yet it is a strangely irrepressible image, now resurfacing in Harry Enfield's character Buggerallmoney and in the comic magazine *Viz*. And it is also in many ways an affirmative image. The gritty and anarchic humour of Andy and Bobby distinguishes the region, gives it a positive sense of difference.

My objective is not to enter into a detailed account of enterprise and heritage cultures in the North-East, but rather to emphasize how the region's new global orientation is pulling its cultural identity in quite contradictory directions: it involves at once the devaluation and the valorization of tradition and heritage. There is an extreme ambivalence about the past. Working-class traditions are seen, just like 'pseudo-aristocratic values' as symptoms of the 'British disease', and as inimical to a 'post-industrial' enterprise ethos. But tradition and heritage are also things that entrepreneurs can exploit: they are 'products'. And they also have human meaning and significance that cannot easily be erased. At the heart of contemporary British culture is the problem of articulating national past and global future.

The burden of identity

I want, finally, to return to the question of what postmodern geographies might imply for the question of empire. Post-modernism, as Todd Gitlin argues, should be understood as 'a general orientation, as a way of apprehending or experiencing the world and our place, or placelessness, in it'.[39] Globalization is profoundly transforming our apprehension of the world: it is provoking a new experience of orientation and disorientation, new senses of placed and placeless identity. The global–local nexus is associated with new relations between space and place, fixity and mobility, centre and periphery, 'real' and 'virtual' space, 'inside' and 'outside', frontier and territory. This, inevitably, has implications for both individual and collective identities and for the meaning and coherence of community. Peter Emberley describes a momentous shift from a world of stable and continuous reference points to one where 'the notions of space as enclosure and time as duration are unsettled and redesigned as a field of infinitely experimental configurations of space-time'. In this new 'hyperreality', he suggests, 'the old order of prescriptive and exclusive places and meaning-endowed durations is dissolving' and we are consequently faced with the challenge of elaborating 'a new self-interpretation'.[40]

It is in this context that both enterprise and heritage cultures assume their significance. Older certainties and hierarchies of British identity have been called into question in a world of dissolving boundaries and disrupted continuities. In a country that is now a container of African and Asian cultures, the sense of what it is to be British can never again have the old confidence and surety. Other sources of identity are no less fragile. What does it mean to be European in a continent coloured not only by the cultures of its former colonies, but also by American and now Japanese cultures? Is not the very category of identity itself problematical? Is it at all possible, in global times, to regain a coherent and integral sense of identity? Continuity and historicity of identity are challenged by the immediacy and intensity of global cultural confrontations. The comforts of Tradition are fundamentally based upon the responsibilities of cultural Translation.

Neither enterprise nor heritage culture really confronts these responsibilities. Both represent protective strategies of response to global forces, centred around the conservation, rather than reinterpretation, of identities. The driving imperative is to salvage centred, bounded and coherent identities – placed identities for placeless times. This may take the form of the resuscitated patriotism and jingoism that we are now seeing in a resurgent Little Englandism. Alternatively, as I have already suggested, it may take a more progressive form in the cultivation of local and regional identities or in the project to construct a continental European identity. In each case, however, it is about the maintenance of protective illusion, about the struggle for wholeness and coherence through continuity. At the heart of this romantic aspiration is what Richard Sennett, in another context, calls the search for purity and purified identity. 'The effect of this defensive pattern', he argues, 'is to create in people a desire for a purification of the terms in which they see themselves in relation to others. The enterprise involved is an attempt to build an image or identity that coheres, is unified, and filters out threats in social experience.'[41] Purified identities are constructed through the purification of space, through the maintenance of territorial boundaries and frontiers. We can talk of 'a geography of rejection which appears to correspond to the purity of antagonistic communities'.[42]

Purified identities are also at the heart of empire. Purification aims to secure both protection from, and positional superiority over, the external other. Anxiety and power feed off each other. To question empire, then, is to call into question the very logic of identity itself. In this context, it is not difficult to understand the anxious and defensive efforts now being devoted to reinforce and buttress 'traditional' cultural identities.

Is it, then, possible to break this logic of identity? How do we begin to confront the challenge of postmodern geographies and the urgent question of cultural Translation? British enterprise and heritage cultures are inscribed in what Ian Baruma has called the 'antipolitical world of *Heimat*-seeking'.[43] Against this ideal of *Heimat*, however, another powerful motif of the contemporary world should be counterposed. It is in the experience of *diaspora* that we may begin to understand the way beyond empire. In the experience of migration, difference is confronted: boundaries are crossed; cultures are mingled; identities become blurred. The diaspora experience, Stuart Hall argues, is about 'unsettling, recombination, hybridization and "cut-and-mix"' and carries with it a trans-formed relation to Tradition, one in which 'there can be no simple "return" or "recovery" of the ancestral past which is not re-experienced through the categories of the present.'[44] The experience of diaspora, and also of exile, as Edward Said has powerfully argued, allows us to understand relations between cultures in new ways. The crossing of boundaries brings about a complexity of vision and also a sense of the permeability and contingency of cultures. It allows us 'to see others not as ontologically given but as historically constituted' and can, thereby, 'erode the exclusivist biases we so often ascribe to cultures, our own not least'.[45]

The experience of diaspora and exile is extreme, but, in the context of increasing cultural globalization, it is prefigurative, whilst the quest for *Heimat* is now regressive and restrictive. The notion of distinct, separate, and 'authentic' cultures is increasingly problematical. A culture, as Eric Wolf argues is 'better seen as a series of processes that construct, reconstruct, and dismantle cultural materials'; 'the concept of a fixed, unitary, and bounded culture must give way to a sense of the fluidity and permeability of cultural sets'.[46] Out of this context are emerging new forms of global culture. There is, to take one example, a new cosmopolitanism in the field of literature. Writers like Isabel Allende, Salman Rushdie or Mario Vargas Llosa are recording 'the global juxtapositions that have begun to force their way even into private experience', 'capturing a new world reality that has a definite social basis in immigration and international com-munications'.[47] For Rushdie, these literary exiles, migrants or expatriates are 'at one and the same time insiders and outsiders', offering 'stereoscopic vision' in place of 'whole sight'.[48]

The point is not at all to idealize this new cosmopolitanism (the Rushdie affair is eloquent testimony to its limits and to the real and profound difficulties of cultural Translation). It is, rather, to emphasize the profound insularity of enterprise and heritage cultures and to question the relevance of their different strategies to re-enchant the nation. As Dick Hebdige emphasizes, everybody is now 'more or less cosmopolitan; "mundane" cosmopolitanism is part of "ordin-ary" experience'.[49] If it is possible, then it is no longer meaningful, to hold on to older senses of identity and continuity. In these rapidly changing times, Hanif Kureishi writes, the British have to change: 'It is the British, the white British, who have to learn that being British isn't what it was. Now it is a more complex thing, involving new elements.'[50]

The argument of this chapter has been that the emergence of enterprise and heritage cultures has not been a matter of the purely endogenous evolution of British culture, but rather a response to the forces of globalization. If, however, over the past decade or so, both of these cultural developments have been provoked and shaped by those forces, neither has been open to them. The question for the 1990s is whether we will continue to insulate ourselves with protective and narcissistic illusions, or whether, in the new global arena, we can really find 'a new way of being British'.

Notes

1 R. Samuel, 'Introduction: exciting to be English', in R. Samuel (ed.), *Patriotism: The Making and Unmaking of British National Identity*, vol. 1, London, Routledge, 1989, p. xxxii.

2 P. Wright, 'Re-enchanting the nation: Prince Charles and architecture', *Modern Painters*, 1989, vol. 2, no. 3, p. 27. On Prince Charles's ideas about business in the community, see, for example, his article, 'Future of business in Britain', *Financial Times*, 20 November 1989.

3 E. W. Said, 'Representing the colonised: anthropology's interlocutors', *Critical Inquiry*, 1989, vol. 15, no. 2, p. 216.

4 ibid., p. 218.

5 H. Bhabha, 'Beyond fundamentalism and liberalism', *New Statesman and Society*, 3 March 1989, p. 35. On the Rushdie affair, see L. Appignanesi and S. Maitland (eds), *The Rushdie File*, London, Fourth Estate, 1989.

6 S. Shamoon, 'Mickey the Euro mouse', *Observer*, 17 September 1989.

7 Quoted in N. Fraser, 'Keeping the world covered', *Observer*, 12 November 1989. See also C. Schneider and B. Wallis (eds), *Global Television*, New York, Wedge Press, 1988.

8 J. Hughes, A. Mierzwa and C. Morgan, *Strategic Partnerships as a Way Forward in European Broadcasting*, London, Booz Allen & Hamilton, 1989; 'The entertainment industry', *Economist*, 23 December 1989. On global advertising corporations, see A. Rawsthorn, 'Internationalism, the watchword of the new cadre of global marketing groups', *Financial Times*, 7 December 1989; M. Kavanagh and S. Brierley, 'Media on the move', *Marketing*, 2 November 1989, pp. 28–32.

9 Quoted in C. Brown, 'Holston exports', *Broadcast*, 13 October 1989, p. 44.

10 Levitt, op. cit., pp. 30–1.

11 For discussion of *Magiciens de la Terre*, see the special issue of *Third Text*, 1989, no. 6; E. Heartney, 'The whole earth show: part II', *Art in America*, 1989, vol. 77, no. 7, pp. 91–6.

12 C. Fusco, 'About locating ourselves and our representations', *Framework*, 1989, no. 36, p. 13. On the appropriation of ethnic art, see A. M. Willis and T. Fry, 'Art as ethnocide: the case of Australia', *Third Text*, 1988–9, no. 5, pp. 3–20. See also K. Owusu, 'Voyages of rediscovery', *Marxism Today*, July 1989, pp. 34–5.

13 N. Ascherso, 'Europe 2000', *Marxism Today*, January 1990, p. 17.

14 On locality and localism, see A. Jonas, 'A new regional geography of localities?', *Area*, 1988, vol. 20, no. 2, pp. 101–10; on regionalism, A. Gilbert, 'The new regional

geography in English and French-speaking countries', *Progress in Human Geography*, 1988, vol. 12, no. 2, pp. 208–28.

15 On the 'flexible specialization' thesis, M. Piore and C. Sabel, *The Second Industrial Divide: Possibilities for Prosperity*, New York, Basic Books, 1984; P. Hirst and J. Zeitlin (eds), *Reversing Industrial Decline? Industrial Structure and Policy in Britain and her Competitors*, Oxford, Berg, 1989; P. Hirst and J. Zeitlin, 'Flexible special-isation and the competitive failure of UK manufacturing', *Political Quarterly*, 1989, vol. 60, no. 2, pp. 164–78; the special issue on 'Local industrial strategies' of *Economy and Society*, 1989, vol. 18, no. 4. For discussion and critique, see A. Amin and K. Robins, 'The re-emergence of regional economies? The mythical geography of flexible accumulation', *Environment and Planning D: Society and Space*, 1990, vol. 8, no. 1, pp. 7–34; A. Amin and K. Robins, 'Jeux sans frontières: verso un Europa delle regioni?', *Il Ponte*, 1990, vol. 46.

16 See A. Cochrane (ed.), *Developing Local Economic Strategies*, Milton Keynes, Open University Press, 1987; A. Cochrane, 'In and against the market? The development of socialist economic strategies in Britain, 1981–1986', *Policy and Politics*, 1988, vol. 16, no. 3, pp. 159–68.

17 Wright, op. cit.; Prince of Wales, *A Vision of Britain: A Personal View of Architecture*, London, Doubleday, 1989; C. Jencks, *The Prince, the Architects, and New Wave Monarchy*, London, Academy Editions, 1988.

18 D. Ley, 'Modernism, post-modernism, and the struggle for place', in Agnew and Duncan, op. cit., p. 60.

19 A. Pred, 'The locally spoken word and local struggles', *Environment and Planning D: Society and Space*, 1989, vol. 7, no. 2, pp. 211–33.

20 C. F. Alger, 'Perceiving, analysing and coping with the local-global nexus', *International Social Science Journal*, 1988, no. 117, pp. 321–40.

21 Quoted in W. Scobie, 'Carlo, suitor to La Grande Dame', *Observer*, 14 February 1988.

22 Wagstyl, op. cit.

23 Brown, op. cit., p. 44.

24 Home Office, *Broadcasting in the '90s: Competition, Choice and Quality*, Cm. 517, London, HMSO, 1988, p. 42.

25 Department of Trade and Industry, *DTI – the Department for Enterprise*, Cm. 278, London, HMSO, 1988 p. 3. See also M. J. Wiener, *English Culture and the Decline of the Industrial Spirit, 1850–1980*, Cambridge, Cambridge University Press, 1981.

26 For an extended discussion of the 'British disease' and of the counter-offensive being waged by the protagonists of enterprise culture, see K. Robins and F. Webster, *The Technical Fix: Education, Computers and Industry*, London, Macmillan, 1989, ch. 5.

27 S. Lash and J. Urry, *The End of Organised Capitalism*, Cambridge, Polity Press, 1987, pp. 86, 101–2.

28 H. Häussermann and W. Siebel, *Neue Urbanität*, Frankfurt, Suhrkamp, 1987. See also *An Urban Renaissance: The Role of the Arts in Inner City Regeneration and the Case for Increased Public/Private Sector Cooperation*, London, Arts Council of Great Britain, 1987.

29 M. Mayer, 'Local politics: from administration to management', paper presented to the Conference on Regulation, Innovation and Spatial Development, Cardiff, 13–15 September 1989, pp. 12–13. See also J. Esser and J. Hirsch, 'The crisis of fordism and the dimensions of a "postfordist" regional and urban structure', *International Journal of Urban and Regional Research*, 1989, vol. 13, no. 3, pp. 417–37; D. Harvey, 'Flexible accumulation through urbanization: reflections on "post-modernism" in the American city', *Antipode*, 1987, vol. 19, no. 3, pp. 260–86; D. Harvey, 'From managerialism to entrepreneurialism: the transformation of urban governance in late capitalism', *Geografiska Annaler*, 1989, vol. 71B, pp. 3–17; D. Harvey, *The Condition*

of Postmodernity, Oxford, Basil Blackwell, 1989, ch. 4; G. J. Ashworth and H. Voogd, 'Marketing the city', *Town Planning Review*, 1988, vol. 59, no. 1, pp. 65–79.

30 Even if it is the case that 'the serial reproduction of the same solution generates monotony in the name of diversity', D. Harvey, 'Down towns', *Marxism Today*, January 1989, p. 21.

31 I. Page, 'Tourism promotion in Bradford', *The Planner*, February 1986, p. 73.

32 J. Whelan, 'Destination Newcastle', *Intercity*, November 1989, p. 26.

33 B. Foster and D. Williams, 'Farewell to Andy Capp', *Observer*, 4 June 1989. See also, P. Young, 'Lets drive out Andy Capp!', *Evening Chronicle* (Newcastle), 22 November 1989; R. Baxter, 'Comment Thatcher a tué Andy Capp', *Politis*, 1989, no. 60, pp. 46–9.

34 See P. Hetherington and F. Robinson, 'Tyneside life', in F. Robinson (ed.), *Post-Industrial Tyneside: An Economic and Social Survey of Tyneside in the 1980s*, Newcastle upon Tyne, Newcastle upon Tyne City Libraries and Arts, 1989, pp. 189–210.

35 G. Radice, 'Fujitsu: just what we need in the NE', *The Journal* (Newcastle), 12 April 1989. For a more critical discussion of Japanese influence, see S. Crowther and P. Garrahan, 'Corporate power and the local economy: *Industrial Relations Journal*, 1988, vol. 19, pp. 51–9.

36 M. Conte-Helm, *Japan and the North East of England: From 1862 to the Present Day*, London, Athlone Press, 1989.

37 See M. Pickering and K. Robins, ' "A revolutionary materialist with a leg free": the autobiographical novels of Jack Common', in J. Hawthorn (ed.), *The British Working-Class Novel in the Twentieth Century*, London, Edward Arnold, 1984, pp. 77–92; M. Pickering and K. Robins, 'Between determinism and disruption: the working-class novels of Sid Chaplin', *College English*, 1989, vol. 51, no. 4, pp. 357–76.

38 L. Gofton, 'Back to the future', *Times Higher Education Supplement*, 20 January 1989.

39 T. Gitlin, 'Postmodernism: roots and politics', *Dissent*, Winter 1989, p. 101.

40 P. Emberley, 'Places and stories: the challenge of technology', *Social Research*, 1989, vol. 56, no. 3, pp. 755–6, 748.

41 R. Sennett, *The Uses of Disorder*, Harmondsworth, Penguin, 1971, p. 15.

42 D. Sibley, 'Purification of space', *Environment and Planning D: Society and Space*, 1988, vol. 6, no. 4, p. 410.

43 I. Baruma, 'From Hirohito to Heimat', *New York Review of Books*, 26 October 1989, p. 43.

44 S. Hall, 'New ethnicities', in *Black Film, British Cinema*, London Institute of Contemporary Arts, 1988, p. 30.

45 E. Said, 1989, op. cit., p. 225. See also, E. Said, 'Reflections on exile', *Granta*, 1984, no. 13, pp. 157–72.

46 E. Wolf, *Europe and the People Without History*, Berkeley, University of California Press, 1982, p. 387. See also T. Mitchell, 'Culture across borders', *Middle East Report*, July-August 1989, pp. 4–6.

47 T. Brennan, 'Cosmopolitans and celebrities', *Race and Class*, 1989, vol. 31, no. 1, pp. 4, 9.

48 S. Rushdie, 'Imaginary homelands', *London Review of Books*, 7–20 October 1982, p. 19.

49 D. Hebdige, 'Fax to the future', *Marxism Today*, January 1990, p. 20.

50 H. Kureishi, 'England, your England', *New Statesman and Society*, 21 July 1989, p. 29.

SECTION FIVE

IMAGINED PLACES

Editorial introduction

In the last section the idea of an imagined community was introduced to refer to the nation and to myths of original homelands. Here in Section Five we turn, in what seems initially to be a more straight-forward sense, to examine places that are literally imaginary, to images of places that have never existed like the imaginary Main Street and hometown created by the Disney corporation at Disneyland and Disney World, the pastiches of multiple exotic locations found side by side in theme parks and themed malls, and the constructed tourist environments all over the world where a vernacular polynesian atmosphere that does not exist is found, a sort of generalised exotic otherness, floating free from space and time, and catering for the transient pleasures of holiday makers who want to 'get away from it all'. Here are the classic examples of what might be termed the aestheticization of landscape to parallel the aestheticization of everyday life, identified by Featherstone (1991) and others as a key part of the shift from the modern to the postmodern period.

At the end of the century, when consumption has replaced production as the activity in the construction of identity and meaning, the shopping mall – that temple of escapist paradise – has replaced the factory or the office as the iconic building of postmodernity. In Chapter 19, **John Goss** discusses the production and consumption of the 'magic of the mall'. He works at the University of Hawaii, a location which may seem the quintessence of exotic otherness, at least as viewed from Manchester or Minneapolis. But images of exoticness, indeed even of Hawaii itself, are, as Goss shows, a common feature of shopping malls where developers construct magical representations of other places to persuade us to consume. Huge malls originated in the USA. Minneapolis was for many years the site of a mega-mall that could boast of being the largest in the world, before it was overtaken by the West Edmonton Mall in Canada (Shields, 1989 and see the special issue of the *Canadian Geographer*, volume 35, 1991 on this mall) and no doubt since then by a mall somewhere else. One of the most recent examples in Europe is CentrO, a mega-mall on a 200-acre site between Dussel-dorf and Dortmund in Germany which includes an adventure island, talking trees, a cinema complex, theme bars and hotels as well as stores and shops: a tourist attraction as much as a mall. Writing of reactions to this development among Germans who have never seen a real mall before, Newnham says 'not that "real" is a word which immediately springs to mind here, where reality itself is only skin-deep.

In CentrO, every surface is a veneer. Marketing and marquetry join forces to conjure an American dream out of thin air – an American dream made in Sheffield' (1996, p. 40). In fact the centre was built by a partnership between the Hull-based Stadium group and P&O on land once occupied by a Thyssen steel mill. In some senses it is easy to understand the despair felt by many commentators as industrial jobs are replaced by low-quality employment in the retail sector and as the dominance of Coca-Cola, McDonalds, and other world retailers produces a similar environment throughout the world. These malls have been criticised for producing a numbingly similar environment, cocooned from the 'real' world, but critics yearning for a more gritty reality might remember or try to imagine what it used to be like shopping in the rain, pushing a baby in a buggy and with a toddler, to numerous shops on a busy high street with uneven pavements and their postmodern romantic nostalgia for local difference might pale.

As the marketing of image and spectacle to encourage footloose capital to invest and the new middle class to consume comes to dominate the production of the urban landscape, images and imitations of other places, themselves real or imaginary, tend to produce environments that are increasingly interchangeable. In the old industrial areas of north-east USA and of Great Britain, coal-mines are turned into museums. Every port is turning its disused docklands into office or residential quarters, its tea wharfs and beached clippers into tourist attractions, and areas where in-migrants settle become simulacra of previous homescapes. Each and every global city seems to have a similar Chinatown or Little Tokyo where municipalities construct telephone boxes as pagodas as much to attract tourists as to serve as a reminder of home for the residents. In a witty analysis of several examples of these simulated landscapes in the United States, Umberto Eco (1986) recalls his journeys through constructed or imagined places on the west coast of the USA. Drawing on Baudrillard, he called his collection *Travels in Hyper Reality*. There he describes some of the wilder reaches of imagination that inspired such monumental follies as Hearst Castle, built by the newspaper magnate William Randolph Hearst in the 1930s where bits of European abbeys, palaces and convents have been reassembled to contain an eclectic mixture of fake and original classical statues, Italian ceilings, and Flemish tapestries. In an astonishing image, Eco describes the overwhelming impact of the place as 'like making love in a confessional with a prostitute dressed in a prelate's liturgical robes reciting Baudelaire while ten electronic organs reproduce the Well-Tempered Clavier played by Scriabin' (pp. 23–4). But Hearst Castle is not unusual: rather it 'fits into the California tourist landscape with perfect coherence, among the waxwork Last Suppers and Disneyland' (p. 24). Just a dozen miles down the coast Eco visited the Madonna Inn where overnight visitors might choose between the Prehistoric Room (a cavern with stalactites), the Safari Room with a zebra skin on the bed, the Old Mill or other themes.

While the California tourist landscape might represent the extreme versions of placeless imaginary landscapes, in the analyses of postmodern architecture and of imagined postmodern spaces and places as a whole, the west coast, California, and Los Angeles in particular, tends to feature prominently as the exemplar of the future. Indeed Los Angeles has become a metaphor for a postmodern sense of place that is placeless or nowhere: with no centre and no edge as it eats into the high desert around it: an amorphous, hetereogeneous non-place. In Jameson's (1991) classic essay 'Postmodernism or the cultural logic of late capitalism' a hotel in the centre of that city – the Westin Bonaventure Hotel, built by architect and developer John Portman – is analysed as the symbolic image of postmodern difference: a place without logic, unmappable and unknowable, designed to confuse, perplex and tease, with 'curious unmarked entries' (Jameson, 1991, p. 39). Ed Soja, perhaps the best-known geographer writing about Los Angeles, is also drawn to this hotel as metaphor and symbol of a postmodern urbanity in both his *Postmodern geographies* (1989) and his more recent *ThirdSpace* (1996). And Los Angeles as a whole is often seen as the quintessential postmodern city in its lack of centre, its own rambling layout, its architecture which includes buildings by some of the most well-known postmodern designers – Charles Moore's eclectic mixture of styles, and Frank Gehry's buildings which are a mixture of pastiche and visual jokes, for example. For Christopher Jenks, Los Angeles is a 'heteropolis': a city which is defined by heterogeneity and difference. In LA 'there are only minorities' (Jenks, 1993, p. 7). He believes that the new style of architecture and urban design emerging in LA may become part of a positive and creative response to heterogeneity:

> it suggests a way of using otherness, hybridization and informality as creative responses to what is now an impasse: the conflict of dominant cultures with their subordinate minorities. Obviously it does not hold answers to the larger political questions but it does suggest methods for confronting oppositions by creative displacement and creative eclecticism. (p. 9)

Other analysts of postmodern architecture and urban planning are less optimistic. For Sharon Zukin, these postmodern landscapes, whether in reconstructed parts of old capitals and other cities – in London, Paris and New York for example or in the new cities of the USA like Los Angeles, Miami and Houston – function primarily as mechanisms of social control. Similarly, Mike Davis (1990) argued in his forceful, pessimistic picture of contemporary Los Angeles, that the glitter of the 'City of Quartz' is only a surface distraction from the hard and repressive mechanisms of control that are necessary to deal with the consequences of increasing inequalities. For Davis, the Bonaventure is a playground for the rich and its fortress-like architecture acts not to playfully confuse but to exclude the growing population of the poor and the homeless who live in the surrounding inner city. Similarly Disneyland, that playground of pleasure for middle America, located in Orange County, and Disney World in Florida also contain and control difference in their reconstruction of a vision of a 'decent' country and simple values. As **Sharon Zukin** argues in her analysis of Disney's tourist parks, included here as Chapter 20, Disney recreated the vernacular landscape of small-town America where women and other multiple 'others' knew their place.

If, however, we are to take the arguments of postmodern and deconstructivist theorists seriously, it is necessary to challenge the binary distinction between the 'real' and the 'imagined'. In important ways our perception of, awareness of and reactions to landscapes and to places, the connections between them and the sets of meanings associated with them are always imagined. Places and landscapes have no intrinsic meaning. Instead they are socially constructed, embedded within the sets of social relations and value systems of a period. The sublime landscapes of mountain areas that we now see as romantic wilderness, the Lake District in the UK for example, in an earlier era were seen as bleak and desolate (Urry, 1995). The religious and symbolic meanings of the distribution of interior and exterior spaces of medieval cities have disappeared in a secular world which has altered the careful spatial balance by, for example, opening up spaces and piazzas in front of medieval cathedrals designed not to be seen from a vista but instead to spring unexpectedly from the muddle of medieval streets to startle the onlooker with the power of God.

The recognition that places are interpreted differently over time, and by different viewers or participants, has led to the application of literary methods of deconstructive analysis to geographical locations, providing multiple readings of places as if they were texts (Cosgrove and Daniels, 1988; Duncan and Duncan, 1988). This textual move has also led to a growing interest by geographers in fictional representations of place, places created in text and images, rather than the imagined but indubitably real spaces and places of Disney and the mall. The regional fiction of, for example, Alan Sillitoe (Daniels and Rycroft, 1993) and John Harvey (Howell, 1997), both of whom write about Nottingham, has been analysed by geographers. The latter author is one of that growing number of crime writers for whom a specific city is an essential element in the construction of place and meaning as well as the backdrop to the plot. In his book *Into the Badlands*, John Williams (1991) has analysed the role of places as various as Miami, Chicago, New Orleans, New York and Los Angeles in contemporary crime fiction. Los Angeles has, of course, been a key location in the development of the 'hard-boiled' genre of crime fiction. Raymond Chandler's Phillip Marlowe is perhaps the private dick who has inspired the greatest number of imitators. As James Donald (1992) has suggested the single male detective is a key figure of modernity, bringing order to chaos in his commitment to restore a particular version of the moral order. While he may have changed gender lately, the female equivalents are equally modern figures, isolated heroines forever caught in the light of the open, empty fridge door, resolving uncertainty for readers who have become ever more distrustful of a final resolution. That Los Angeles, now the quintessential postmodern city, has been so significant in this genre, blurs the distinction between periods and forces us again to reassess the oversimple binary associations between period and place. 'Postmodern' Los Angeles also figures as a dystopian virtual reality in another genre: that of film. Ridley Scott's *Bladerunner* has become the epitome of the urban nightmare and dream of the end of the twentieth century, in a similar way to crime fiction in the middle years of the century.

In the final extract in this section, we pursue the notion of imaginary spaces, in this case neither novels nor films, but 'real' virtual reality. In the continuing compression of space by time that has been characteristic of the later decades of the twentieth century, the advent of virtual or cyberspace in which individuals and groups are 'free' to take on whatever identities they choose, fictional or not, to roam or surf the worldwide web, it seems that the end of geography, or more prosaically, the end of the friction of space is at hand. The internet has deepened and vastly increased the scale and pace of change that began more than a half century earlier with the introduction of television. The world now enters the sitting room and the office – and the necessity of travel to access information, to earn money or for leisure – is correspondingly reduced. Anthropologist **Christopher Pinney** reimagines the world we have lost, or may very well have done, when he comes to retire in 2029. In this new virtual world, fixed and solid places – universities among them – may disappear or melt into air, as a new form of academic exchange replaces old place-based discourses. In the academy of the future, it may be as common to link into a lecture given by, say, David Harvey in Baltimore, Alan Scott in Los Angeles or Doreen Massey in Milton Keynes, perhaps years ago or at the moment in which you access it, as it is to stumble into the first-year lecture room on a Monday morning. As Pinney concludes, the detemporalisation of space and the abolition of Cartesian conceptions may be the moment when 'I' (as subject and traveller) ceases to exist. In his paper, Pinney traces the development of Enlightenment views of space and travel and their transformation at the end of the twentieth century and beyond. So we end Section Five almost where we began in Section One, assessing the impact of electronic media on conceptions of space, place and community.

References and further reading

Anderson, K. and Gale, F. 1992: *Inventing places: studies in cultural geography.* London: Longman.

Barnes, T. and Duncan, J. (eds) 1992: *Writing worlds: discourse, text and metaphor in the representation of landscape.* London: Routledge.

Chaney, D. 1991: Subtopia in Gateshead: the MetroCentre as a cultural form. *Theory, Culture and Society* 7.

Cosgrove, D. and Daniels, S. 1988: *The iconography of landscape: essays on the symbolic representation, design and use of past landscapes.* Cambridge: Cambridge University Press.

Crang, M. 1994: On the heritage trail: maps of and journeys to olde Englande. *Environment and Planning D: Society and Space* 12.

Daniels, S. and Rycroft, S. 1993: Mapping the city: Alan Sillitoe's Nottingham novels. *Transactions, Institute of British Georgraphers* 18, 460–80.

Davis, M. 1990: *City of quartz.* London: Verso.

Donald, J. 1992: Metropolis: the city as text. In R. Bocock and K. Thompson (eds), *Social and cultural forms of modernity.* Cambridge: Polity, 417–61.

Duncan, J. and Duncan, N. 1988: (Re)reading the landscape. *Environment and Planning D: Society and Space* 6, 117–26.

Duncan, J. and Ley, D. 1992: *Place/culture/representation.* London: Routledge.

Eco, U. 1986: *Travels in hyper reality*. San Diego, California: Harcourt, Brace, Jovanovich.

Featherstone, M. 1991: *Global culture*. London: Sage.

Fiske, J. 1989: *Reading the popular*. London: Unwin Hyman.

Game, A. 1991: *Undoing the social: towards a deconstructive sociology*. Milton Keynes: Open University Press.

Howell, P. 1997: Crime and the city solution: crime fiction, urban knowledge and radical geography. *Antipode* (forthcoming).

Jameson, F. 1991: *Postmodernism or the cultural logic of late capitalism*. London: Verso.

Jenks, C. 1993: *Heteropolis: Los Angeles: the riots and the strange beauty of heterarchitecture*. London: Academy Editions, Ernst and Sohn.

Kearns, G. and Philo, C. (eds) 1993: *Selling places: the city as cultural capital, past and present*. London: Pergamon Press.

Newnham, D. 1996: Vorsprung durch shopping. *The Guardian Weekend*, 23 November, 38–44.

Robins, K. 1991: Prisoners of the city: whatever could a postmodern landscape be. *New Formations* 5, 1–22.

Shields, R. 1989: Social spatialisation and the built environment: the West Edmonton Mall. *Environment and Planning D: Society and Space* 7, 147–64.

Short, J. 1992: *Imagined country: society, culture and environment*. London: Routledge.

Soja, E. 1989: *Postmodern geographies*. London: Verso.

Soja, E. 1996: *ThirdSpace: journeys to Los Angeles and other real-and-imagined places*. Oxford: Blackwell.

Sorkin, M. 1992: *Variations on a theme park: the new American city and the end of public space*. New York: Noonday Press.

Urry, J. 1995: *Consuming places*. London: Routledge.

Williams, J. 1991: Los Angeles: looking for the big nowhere. In *Into the badlands: a journey through the American dream*. London: Paladin, Collins, 84–110.

19 Jon Goss

The 'Magic of the Mall': An Analysis of Form, Function, and Meaning in the Contemporary Retail Built Environment

Excerpts from: *Annals of the Association of American Geographers* **83**, 18–47 (1993)

And the truth-sayers of the shopping mall, as the death of the social, are all those lonely people, caught like whirling flotsam in a force field which they don't understand, but which fascinates with the coldness of its brilliance (Kroker et al. 1989, 210).

Shopping is the second most important leisure activity in North America, and although watching television is indisputably the first, much of its programming actually promotes shopping, both through advertising and the depiction of model consumer lifestyles. The existential significance of shopping is proclaimed in popular slogans such as: 'Born to Shop', 'Shop 'Til You Drop', and 'I Shop Therefore I am'. An advertisement for Tyson's Corner, Virginia, asks: 'The joy of cooking? The joy of sex? What's left?' and the answer provided is, of course, 'The joy of shopping'! As Tyson's obviously knows, recent market research shows that many Americans prefer shopping to sex (Levine 1990, 187).

Despite increases in catalog sales, shopping remains essentially a spatial activity – we still 'go' shopping – and the shopping center is its chosen place. The time spent in shopping centers by North Americans follows only that spent at home and at work/school. Centers have already become tourist destinations, complete with tour guides and souvenirs, and some include hotels so that vacationers and conferees need not leave the premises during their stay. Downtown retail complexes often include condominia, and residential development above the suburban mall is predicted to be an inevitable new trend ('The PUD Market Guarantee' 1991, 32). Their residents can literally shop without leaving home (or be at home without leaving the shops?). Moreover, planned retail space is colonizing other privately owned public spaces such as hotels, railway stations, airports, office buildings and hospitals, as shopping has become the dominant mode of contemporary public life.

Nevertheless, there persists a high-cultural disdain for conspicuous mass-consumption resulting from the legacy of a puritanical fear of the moral corruption inherent in commercialism and materialism, and sustained by a modern intellectual contempt for consumer society. This latter critique condemns the system of correspondences between material possessions and social worth (Veblen 1953; Boorstin 1973), the homogenization of culture and alienation of the individual (Adorno and Horkheimer 1969; Marcuse 1964) and the distortion of human needs through the manipulation of desire (Haug 1986). The contemporary shopper, while taking pleasure in consumption, cannot but be aware of this authoritative censure, and is therefore, like the tourist (Frow 1991, 127), driven by a simultaneous desire and self-contempt, constantly alternating between assertion and denial of identity. This ambivalence is, I think, precisely expressed in the play of the slogans cited above, which cock a snook at the dominant order of values, but in so doing also acknowledge its inevitable authority.

This paper argues that developers have sought to assuage this collective guilt over conspicuous consumption by designing into the retail built environment the means for a fantasized dissociation from the act of shopping. That is, in recognition of the culturally perceived emptiness of the activity for which they provide the main social space, designers manufacture the illusion that something else other than mere shopping in going on, while also mediating the materialist relations of mass consumption and disguising the identity and rootedness of the shopping center in the contemporary capitalist social order. The product is effectively a *pseudoplace* which works through spatial strategies of dissemblance and duplicity.[1]

The analysis proceeds in several parts. First, I briefly consider the contemporary cultural context and the connection between the techniques of environmental design and image making in (post)modern society. Second, I examine the retail built environment as an object of value; that is, a private, instrumental space designed for the efficient circulation of commodities which is itself a commodity produced for profit. Third, I discuss the means by which developers have obscured this logic by constructing shopping centers as idealized representations of past or distant public spaces. Finally, I consider the retail built environment as a system of signification that gives symbolic expression to the cultural values of consumer capitalism, refers to other times and places, and attaches preferred meanings to commodity displays.

This account is necessarily limited to the workings of the design, that is the assumptions made about the retail built environment and its users, and the intent of the developers as inferred from a reading of their professional literature and of the landscape itself. This requires some care lest we fall into the same trap that compromises the modernist critique of consumption, a critique which holds much intellectual force but little political potential. This is to conceive of the consumer as cultural dupe and helpless object of technical control, exactly as the (mostly) male middle-class designers imagine them. Consumers are constructed as passive, sensual, and vulnerable victims of the 'force field which they don't understand', just as the designers' discourse is both manifestly elitist and gendered – from 'market penetration analysis' to the persistent tropes of seduction, stimulation, and physical manipulation.

Given the gender division of labor and the exploitation of women's social insecurities by the commodity aesthetic, the stereotypical shopper is female – in fact, 67 percent of shopping center users are female (Stores 1989, 43) – and my choice of pronoun form reflects this reality while also recognizing that the label 'shopper' applies increasingly to males.[2] The key point is that the shopper is not merely the object of a technical and patriarchal discourse and design, but is also a subject who may interpret the design aberrantly or intentionally appropriate meaning for her/his own purposes. The manner in which the shopping center is read by consumers, both as individuals and social subjects, is a complex and politically vital question in dire need of research, and I agree with Meaghan

Morris (1988, 206) that the object of our analysis should be precisely the intersection of the instrumental discourse of design, and its reception and active use by the consumer.

The commodification of reality

All human societies invest physical objects with sociopsychological meaning, and consumer goods have long marked 'invidious distinctions' (Veblen 1953), as well as provided for the satisfaction of socially defined needs. It is only under contemporary capitalism, however, that material and symbolic production occupy the same site – productive activity is organized to produce simultaneously the objects of consumption and the social subjects to consume them (Sahlins 1976, 216). Thus 'you are what you buy' *us much us* 'you are what you do'. All human societies also recognize a specialized class that mediates between the material and symbolic worlds, but again it is only recently that this class can control both sides of this relation, and that they are able to persuade us that our 'self-concept' as well as social status is defined by the commodity. Contemporary commodities simultaneously express the social organization of production, communicate social distinctions, sublimate contradictions of the psychological self, and constitute identity (McCracken 1988, 118–19; Csikszentmihalyi and Rochberg-Halton 1981, 22–38).

The 'captains of consciousness' (Ewen 1976) apply highly sophisticated technologies to achieve effects directly analogous to, although infinitely more intensive and far-reaching than, the magic systems of preliterate societies (Diggins 1977, 368; Williams 1980, 185). This modern magic involves a collective superstition that it is the object itself – much like the 'primitive' fetish – that confers upon the owner a power over nature and others; whereas such power, in fact, lies in the social relations that ascribe the power of possession. Both religious and secular traditions harbor moral tales about the danger of wrongful possession, from the shameful exposure of the pretentious who acquire objects above their station to the wrath of gods visited upon the unrighteous holder of sacred icons, but with the necessary personal and social qualifications – cultivated taste and cash or credit – the consumer can invoke the magic power of the commodity (Goss 1992).

Advertising does not have to directly instruct its audience, but need only highlight latent correspondences (Sahlins 1976, 217) or homologies (Bourdieu 1984, 137) between the commodity and common cultural symbols, for contemporary consumers are expected to have accumulated a considerable store of cultural knowledge and acquired the skills necessary to interpret complex texts and subtle rhetorical devices used to elicit cultural meaning (Bourdieu 1984, 66). And if the audience is predisposed to believe, the real magic of advertising is to mask the materiality of the commodity – fetishism in the marxist sense – that is, to sever it from the social and spatial relations that structure its productions and the human labor it embodies. This is especially so for mass-produced commodities, which threaten to invalidate the conditions required for rightful and righteous possession, so advertising necessarily divorces the commodity from the labor process that produced it. Whether it is high fashion sewn in immigrant sweatshops or electronic gadgets assembled in Third World factories, few consumers, therefore, know or can afford to give thought to what the commodity is composed of, or where, how and by whom it was made (Jhally 1987, 49).

Critical to the processes by which the commodity is simultaneously severed from its origins and associated with desirable sociocultural attributes is its context – the real or imagined landscape in which it is presented (Sack 1988, 643–44). Advertisers draw upon knowledge of places, and upon the structuration of social space, to create an imaginary setting that elicits from us an appropriate social disposition or action. With the collapse of time-space produced by global electronic media and tourism (Meyrowitz 1985), the stock of place imagery in the consumer's *musée imaginaire* (Jencks 1987, 95), has expanded dramatically, and we are able to read with facility a vast array of clichéd signs of real and fictitious elsewheres.

In the contemporary shopping center, there is a close connection between the means of the 'consciousness industry' (Enzenberger 1974) and environmental design: they are both media of mass communication, employing rhetorical devices to effect hidden persuasions; both may be experienced passively; they both belong unobtrusively to everyday life; and they are both motivated by profit (Eco 1986, 77). Developers readily employ the glitz and showcraft of entertainment – literally 'learning from Las Vegas' (Venturi et al. 1972); the iconography of advertising (Frampton 1983, 19) – 'learning from Madison Avenue'; and the 'imagineering' of North American theme parks (Relph 1987) – 'learning from Disney.' Sophisticated techniques of illusion and allusion enable them to create an appropriate and convincing context where the relationship of the individual to mass consumption and of the commodity to its context is mystified. This technical capacity, the predispositions of contemporary consumers (increasingly well understood due to market research), and the economic and political capacity of speculative capital combine to manufacture a total retail built environment and a total cultural experience.

The making of the mall

The developer's profit accrues from the construction and sale of shopping centers, lease rent, and deductions from retail revenues. Unlike other forms of real estate, where markets have been rapidly saturated and are dependent upon urban and regional economic fortunes, shopping center construction has been a relatively secure investment, whether in the suburbs, always provided a big name department store could be enticed to sign an agreement (Frieden and Sagalyn 1989, 79), or downtown, provided subsidies could be negotiated from cooperative municipal governments.

It is important at the outset to realize the scale and detail of the conception. Shopping centers are typically produced by huge corporations or ad hoc coalitions of finance, construction, and commercial capital (typically pension funds, developers, and department stores), and are meticulously planned. They usually involve state agencies and teams of market researchers, geo-demographers, accountants, asset managers, lawyers, engineers, architects, landscape artists, interior designers, traffic analysts, security consultants, and leasing agents. Development, therefore, involves the coordination of a complex of concerns, although always overdetermined by the goals of retail profit.

While individual retailers may pursue their own strategies for profit within limited bounds, the center operates as a whole to maximize 'foot traffic' by attracting the target consumers and keeping them on the premises for as long as possible.

The task begins with the manufacture and marketing of an appropriate sense of place (Richards 1990, 24), an attractive place image that will entice people from their suburban homes and downtown offices, keep them contentedly on the premises, and encourage them to return.

Imag(in)ing the mall

In constructing an attractive place image for the shopping center, developers have, with remarkable persistence, exploited a modernist nostalgia for authentic community, perceived to exist only in past and distant places, and have promoted the conceit of the shopping center as an alternative focus for modern community life. Shopping districts of the early years of this century, for example, were based on traditional market towns and villages, and a strong sense of place was evoked using stylized historical architecture and landscaping (typically evoking the village green). They were built on a modest scale, functionally and spatially integrated into local communities, in order to provide an idyllic context for consumption by the new gentry (Rowe 1991, 141). With the contemporary postmodernist penchant for the vernacular, this original form is undergoing a renaissance in the specialty center, a collection of high-end outlets that pursue a particular retail and architectural theme. Typically these are also idealizations of villages and small towns, chock-full of historical and regional details to convince the consumer of their authenticity (Goss 1992, 172).

In contrast, the modern regional shopping center was built on a large scale with regular, unified architecture. Its harsh exterior modernism and automobile-focused landscaping refused any compromise with the rustic aesthetic. As Relph (1987, 215) notes, however, 'modernism ... never wholly succeeded in the landscape of retailing', and the interior contained pedestrian walkways, courts, fountains and statuary that referred reassuringly to the traditional urbanism of southern Europe (Gruen 1973; Rowe 1991, 126), Victorian Britain or New England. According to Victor Gruen, the acknowledged pioneer of the modern mall, his 'shopping towns' would be not only pleasant places to shop, but also centers of cultural enrichment, education, and relaxation, a suburban alternative to the decaying downtown (Gruen and Smith 1960).

Gruen's shopping centers proved phenomenally successful, and he later argued that by applying the lessons of environmental design learned in the suburbs to downtown, 'we can restore the lost sense of commitment and belonging; we can counteract the phenomenon of alienation, isolation and loneliness and achieve a sense of identity' (Gruen 1973, 11).

The new downtown retail built environment has taken two essential forms, which in practice may be mixed. First is the commercial gentrification of decaying historical business and waterfront districts, pioneered by James Rouse with Quincy Market in Boston. Its opening in 1979 supposedly marked 'the day the urban renaissance began' (Rouse, cited in Teaford 1990, 253) and subsequently no self-respecting city seems complete without its own festival marketplace, replicating more or less the original formula.[3] Historical landmarks and 'water exposure' (Scott 1989, 185) are critical features, as this retail environment is consciously reminiscent of the commercial world city, with its quaysides and urban produce markets replete with open stalls, colorful awnings, costermonger barrows, and nautical paraphernalia liberally scattered around.

A second form is the galleria, the historic referent of which is the Victorian shopping arcade and especially the famous Galleria Vittorio Emanuele II in Milan. After Cesar Pelli pioneered the galleried arcade in the early 1970s (at The Commons in Columbus, Ohio and the Winter Gardens in Niagara Falls, New York), glazed gallery and atria became standard feature in downtown mixed-use developments, their huge vaulted spaces suggesting a sacred-liturgical or secular-civic function. They have since been retrofitted to suburban malls and natural daylight has enabled support of softscapes – interiorized palms, trees, and shrubs – reminiscent of the street in the model garden city, the courts of Babylon, and most especially, the tropical vacation setting (Fig. 19.1a and b). Enclosed streetscapes refer to the idealized, historic middle-American Main Street or to exotic streets of faraway cities, including Parisian boulevards, Mexican paseos, and Arabic souks or casbahs, if only because the contemporary North American street invokes fear and loathing in the middle classes. They reclaim, for the middle-class imagination, 'The Street' – an idealized social space free, by virtue of private property, planning, and strict control, from the inconvenience of the weather and the danger and pollution of the automobile, but most important from the terror of crime associated with today's urban environment.

In creating these spaces, developers and public officials articulate an ideology of nostalgia, a reactionary modernism that expresses the 'dis-ease' of the present (see Stewart 1984, 23), a lament on the perceived loss of the moral conviction, authenticity, spontaneity, and community of the past; a profound disillusionment with contemporary society and fear of the future. More specifically, we collectively miss a public space organized on a pedestrian scale, that is, a setting for free personal expression and association, for collective cultural expression and transgression, and for unencumbered human interaction and material transaction. Such spaces no longer exist in the city, where open spaces are windswept tunnels between towering buildings, abandoned in fear to marginal populations; nor were they found after all in the suburb, which is subdivided and segregated, dominated by the automobile, and repressively predictable and safe. Such spaces only exist intact in our *musées imaginaire*, but their forms can now be expertly reproduced for us in the retail built environment. Below, I discuss the form and the contradictions inherent in the reproduction of such spaces as conceived in their idealized civic and liminal forms.

The shopping center as civic space

By virtue of their scale, design, and function, shopping centers appear to be public spaces, more or less open to anyone and relatively sanitary and safe. This

(A)

(B)

Figure 19.1 Interior of Miller Hill in Duluth (A) as constructed in 1973 (B) after renovation and 'interior softscaping' in 1988. Source: Retail Reporting Corporation.

appearance is important to their success for they aim to offer to middle-Americans a third place beyond home and work/school, a venue where people, old and young, can congregate, commune, and 'see and be seen' (Oldenburg 1989, 17). Several strategies enhance the appearance of vital public space, and foremost is the metaphor of the urban street sustained by streetsigns, streetlamps, benches, shrubbery, and statuary – all well-kept and protected from vandalism. Also like the ideal, benign civic government, shopping centers are extremely sensitive to the needs of the shopper, providing a range of 'inconspicuous artifacts of consideration' (Tuan 1988, 316), such as rest areas and special facilities for the handicapped, elderly, and shoppers with young children (recently including diaper changing stations). For a fee they may provide other conveniences such as gift wrapping and shipping, coat checking, valet parking, strollers, electric shopping carts, lockers, customer service centers, and videotext information kiosks. They may house post offices, satellite municipal halls, automated government services, and public libraries; space is sometimes provided for public meetings or religious services. They stage events not only to directly promote consumption (fashion and car shows), but also for public edification (educational exhibits and musical recitals). Many open their doors early to provide a safe sheltered space for morning constitutionals – mall-walking – and some have public exercise stations with health and fitness programs sponsored by the American Heart Association and YMCAs (Jacobs 1988, 12). Some even offer adult literacy classes and university courses.

Such services obviously address the needs of the public and attest to the responsiveness of management. Many facilities, however, are not so much civic gestures as political maneuvers to persuade local government to permit construction on the desired scale. This is particularly the case with day care facilities now featured in many shopping centers (Reynolds 1990c, 30).[4] It is also clear from the professional literature that many concessions are made in order to enhance the atmosphere of public concern precisely because it significantly increases retail traffic (McCloud 1991, 25). Public services not consistent with the context of consumption are omitted or only reluctantly provided, often inadequate to actual needs and relegated to the periphery. This includes, for example: drinking fountains, which would reduce soft drink sales; restrooms, which are costly to maintain and which attract activities such as drug dealing and sex that are offensive to the legitimate patrons of the mall (Hazel 1992, 28); and public telephones, which may be monopolized by teenagers or drug dealers. As a result, telephones in some malls only allow outgoing calls (Hazel 1992, 29).

The idealized public street is a relatively democractic space with all citizens enjoying access, with participatory entertainment and opportunities for social mixing, and the shopping center re-presents a similarly liberal vision of consumption, in which credit-card citizenship allows all to buy an identity and vicariously experience preferred lifestyles, without principles of exclusion based on accumulated wealth or cultural capital (Zukin 1990, 41). It is however, a strongly bounded or purified social space (Sibley 1988, 409) that excludes a significant minority of the population and so protects patrons from the moral confusion that a confrontation with social difference might provoke (see also Lewis 1990). Suburban malls, in particular, are essentially spaces for *white* middle classes.[5]

There have been several court cases claiming that shopping centers actively discriminate against potential minority tenants, employees, and mall users. Copley Place in Boston, for example, has been charged with excluding minority tenants ('Race Is not the Issue' 1990, 32); a Columbia, South Carolina mall was accused of discriminatory hiring practices ('NAACP in Hiring Pact . . .' 1991, A20), and security personnel have been widely suspected of harassing minority teenagers.

'Street people' are harassed because their appearance, panhandling and inappropriate use of bathrooms (Pawlak et al, 1985) offend the sensibility of shoppers, their presence subverting the normality of conspicuous consumption and perverting the pleasure of consumption by challenging our righteous possession of commodities. Even the Salvation Army may be excluded from making its traditional Christmas collections, perhaps because they remind the consumer of the existence of less-privileged populations and so diminish the joy of buying.

Finally, the politics of exclusion involves the exclusion of politics, and there is an ongoing struggle by political and civil liberties organizations to require shopping centers to permit handbilling, picketing, and demonstrations on their premises, on the grounds that they cannot pretend to be public spaces without assuming the responsibility of such, including recognition of freedom of expression and assembly. Courts have generally found in favor of free speech in shopping centers by virtue of their scale and similarity to public places, provided that the activities do not seriously impair their commercial function (Peterson 1985). The Supreme Court, however, has ruled that it is up to individual states to decide (Kowinski 1985, 357), and in a recent case, an anti-war group was successfully banned from leafletting in New Jersey malls ('Judge Bars Group . . .' 1991, 31).

The shopping center as liminal space

The market, standing between the sacred and secular, the mundane and exotic, and the local and global, has always been a place of liminality; that is, according to Turner (1982), a state between social stations, a transitional moment in which established rules and norms are temporarily suspended (see also Zukin 1991 and Shields 1989). The marketplace is a liminoid zone, a place where potentiality and transgression is engendered by the exciting diversity of humanity, the mystique of exotic objects, the intoxicating energy of the crowd channeled within the confined public space, the prospects of fortunes to be made and lost in trade, the possibility of unplanned meetings and spontaneous adventures, and the continuous assertion of collective rights and freedoms or *communitas* (Bahktin 1984, 8–9). The market thrives on the possibility of 'letting yourself go,' 'treating yourself', and of 'trying it on' without risk of moral censure, and free from institutional surveillance.

Places traditionally associated with liminoid experiences are liberally quoted in the contemporary retail built environment, including most notably seaports and exotic tropical tourist destinations, and Greek agora, Italian piazzas, and other traditional marketplaces. Colorful banners, balloons and flags, clowns and street theater, games and fun rides, are evocative of a permanent carnival or festival.

Lavish expenditure on state-of-the-art entertainment and historic reconstruction, and the explosion of apparent liminality is perfectly consistent with the logic of the shopping center, for it is designed explicitly to attract shoppers and keep them on the premises for as long as possible.

This strategy reaches its contemporary apotheosis in the monster malls that contrive to combine with retailing the experiences of carnival, festival, and tourism in a single, total environment. This includes, most famous, the West Edmonton Mall (WEM), Canada, which has already become a special concern of contemporary culture studies ('Special Issue on the WEM' 1991; Hopkins 1990; Shields 1989; Wiebe and Wiebe 1989; Blomeyer 1988), and others inspired by its extravagant excess: Franklin Hills in Philadelphia, River Falls in Clarksville, Tennessee; the controversial new Mall of the Americas in Bloomington, Minnesota; Meadowhall in Sheffield and Metrocentre in Gateshead in England; and Lotte World in Seoul, South Korea. The shopping center has become hedonopolis (Sommer 1975). Shopping centers have become tourist resorts in their own right, recreating the archetypical modern liminal zone by providing the multiple attractions, accommodations, guided tours, and souvenirs essential to the mass touristic experience, *all* under a single roof.[6]

There are necessarily strict limits to any experience of liminality in these environments. Developers are well aware of the 'more unsavory trappings of carnival life' (McCloud 1989b, 35), and order must be preserved. The contrived retail carnival denies the potentiality for disorder and collective social transgression of the liminal zone at the same time that it celebrates its form.

The shopping center and signification

Elements of the built environment are signifiers which refer, through culturally determined systems of association, to abstract concepts, social relations, or ideologies. In combination, they constitute texts which communicate social meaning to acculturated readers. The built environment first denotes its function, informing the user of its practical purpose. Thus, for example, the shopping center announces itself through its location and its conventional form as a p(a)lace of consumption. A wide range of styles is practicable, however, in realizing this basic function, and even the most technologically constrained architectural solutions give symbolic expression (Winner 1980, 127). The built environmental is also always, therefore, connotative of meaning, consistent with, but extending beyond its immediate function. As Barthes (1979, 6) expresses it: 'architecture is always dream and function, [an] expression of a utopia and instrument of a convenience.' I have suggested, for example, that shopping centers present an image of civic and liminal spaces, forms consistent with, but not identical to, the function of selling commodities. In addition to the thematic imaging of space, however, carefully selected and highlighted elements of design communicate specific meaning, which, through the operation of an environmental rhetoric, can float free and attach to the act of consumption or to the commodities on display.

The plants and water features of the shopping center ask to be contrasted with the degraded nature of the suburbs and the decaying second nature of the city.

They also apparently soothe tired shoppers, enhance the sense of a natural outdoor setting, create exotic contexts for the commodity, imply freshness and cleanliness, and promote a sense of establishment (Maitland 1985). More important, the presence of nature, albeit tamed in a garden setting, naturalizes consumption, and mitigates the alienation inherent in commodity production and consumption. Hence the recent proliferation of natural products stores and the extraordinary lengths developers may go to in order to capture and display commodified nature for this premium.[6]

Water seems to be particularly important. Fountains signify civilized urban space, while on a larger scale, the importance of the waterfront to retail environments is due to their association with sport and recreation, historic trade, and the potential for a new life of adventure (being cast away, press-ganged or ticketed on a departing ship). Perhaps also the value of water is due to the fact that the ocean is the only remaining natural environment and is the habitat of the only uncolonized minds, or noble Other (whales and dolphins). Hence shopping centers with large-scale aquaria, including Scottsdale Galleria, the Mall of the America (with its walk-through adventure, 'Underwater World'), Newport City (with the Cousteau Ocean Center), and, of course, WEM (with its definitive Deep Sea Adventure).

Similarly pervasive is the signification of the past in the retail built environment, as the 'heritage industry' (Hewison 1989) exploits our collective nostalgia for real places and historic roots. This is best illustrated in the festival marketplaces, which reproduce historical landscapes in the city with restored architectural details, antique material artifacts strewn almost casually on the landscape, and professional actors in period costume portraying historical characters (Rouse Co.'s 'Art in the Marketplace' program, for example, now has a performing arts programs to create 'Living History'). Even the older suburban centers are now retrofitted with Victorian, Colonial, or Art Deco detailing. Needless to say this historical vernacular effects an idealization of the past and mystifies its relationship with the present. Although extreme attention is paid to minor details, the reconstruction is fitted with modern facilities, and no reference is made to exploitative social relations that may have actually structured life at the time. No attempt is made to critically interrogate the present through creative juxtaposition – it offers pastiche, not parody (Jameson 1984, 64–65). Ironically too, even while idealizing noncommodified social relations, this historicism normalizes the commodity aesthetic by projecting it backwards into the past and rewrites history as a sequence of style.

If the sense of history is violated in the shopping center, so is time itself. A symptomatic and almost universal new feature of the postmodern retail environment is the clock (Goss 1992, 174). Previously banished because of its reminder of the precious value of time and the power of its regime over the modern individual, it is now often set prominently in a plaza or court, where it quotes public places of the past, or is mounted on towers and bracketed to facades, quoting the respectable historical institutions of the church and main street business. It is, therefore, almost invariably an antique analog clock, visually punning history and the way things were – referring literally to *times* past. The

Figure 19.2   A clock as a focal point, reminiscent of a Victorian railway station. The Galleria, San Francisco. Source: author.

passage of time is recorded, but the time of the antique clock is not a threat to the idyll of consumption for it always stands at the threshold of the present, or just before the dreaded future began. For the postmodern consumer, temporality has collapsed, time is an extension of the moment, and, punning again, past time signifies the *pastime* of shopping. Finally, the combination of the prominent clock and atria or gallery bears a resemblance to the nineteenth-century railway station (see Fig. 19.2), a place that marks liminality, with its prospects for romance and mystery extolled in countless popular novels and movies. Shopping centers occupy restored railway stations (such as St. Louis Station), and miniature railways – as 'people movers' – are increasingly common. The railway rhetoric may go further in the analogy between window shopping and the gaze upon exotic landscapes passing by the carriage window.

In such retail playgrounds, the 'pleasure of innocence is meant to leak outside its sphere' (Chaney 1990, 62) for, as I have noted, the magic of the commodity depends upon an innocence about the relations of production and the social construction of consumption. The sense of innocent fun mitigates the guilt of conspicuous consumption and a residual innocence may similarly attach to the commodities for sale.

Art, on the other hand, has always had a place in the retail built environment because it symbolizes a noncommercialized aesthetic, and because it is a form of object display sanctioned by high culture (Harris 1990). Its auratic content is also meant to spill over into the commodities on sale and to sanctify shopping by association with the legitimate activity of aesthetic appreciation. Hence shopping centers host symphonies (Southland hosted the Minneapolis Symphony Orchestra), operas (the Bel Canto competition is held in shopping centers across the country) and Shakespearean plays (staged, for example, at Lakeforest Mall in Gaithersburg, Maryland) (see Goss 1992, 174). Considerable sums are invested in fixed art displays. Art is like the commodities on display abstracted from its origins and so fetishized. This art rarely demands to be interpreted, so one suspects that its purpose is merely to be recognized as a sign of what it is – that is Art, a mystified quality of high culture. At the same time, of course, it does allow those with cultivated tastes to exercise and display their cultural capital and so mark their distinction from the mass consumer (see Bourdieu 1984). Art exhibits also act as focal features drawing customers along the mall, filling empty spaces, and enhancing the sense of public space ('City's Love of Art ...' 1991, 103). Like other corporate art, shopping center displays signify a commitment to public edification expected of a benevolent authority and are a means to express and legitimate the power of the owner.

More recent too is continuous reference to the television and the emulation of televisual experience within the retail built environment. The shopper strolls through experiences as he or she might scan through TV channels (see Kroker et al. 1989, 109; Davis 1991, 5; Kowinkski 1985, 71–73) and is bombarded by simultaneous images of multiple places and times; spatial narratives dissolve and individual pedestrian trajectories or narratives are constantly broken by contrived obstacles. Developers recognize that their customers expect drama, excitement, and constant visual stimulation thanks to the effect of television, and they seek to provide a surrogate televisual experience (O'Connor 1989, 290).

The connection with TV goes further. As noted earlier, both the retail built environment and the TV function primarily to display and sell commodities and the lifestyles associated with them; both are escapes from suburban everyday life, a means of transport from reality; and both are highly controlled media that play to the cultural bottom-line, presuming a passive, psychologically manipulatable public. Commercial video walls – banks of TV screens – provide a new point-of-purchase promotion so that the shopper can watch TV in the mall, and in fact the TV is impossible to escape. There has been recently a phenomenal televisualization of the retailing concept, that is, the direct lifting of retail concepts from television shows in stores such as Cartoon junction, the Disney Store, Hanna Barbera Shop, NFL for Kids, Circle Gallery of Animation and Cartoon Art, Sanrio Co., and the Sesame Street General Store. Thus it can truly be said that 'shopping malls are liquid TVs for the end of the twentieth century' (Kroker et al. 1989, 208).

Also analogous to the TV, in its capacity to allow viewers to be simultaneously in multiple times and places even while sitting at home, the shopping center creates a diverse range of temporal and spatial experiences within a comfortable landscape for consumption. Hopkins (1990), for example, has described the metonymical strategies by which shopping centers exploit 'myths of elsewhere' to elicit specific behavioral responses. First, they employ semantic metonyms or place names. Typically, early centers favored names redolent of Arcadia or pastoral scenes (Country Club, Highland Park, Farmers Market etc.), while modern suburban malls employ Utopian, placeless names (Northland, Southland etc.), and contemporary centers may imply tourist and other liminal destinations (Harborfront, Seaport Village, Forest Fair, etc.). More effective perhaps is the use of iconic metonyms, or objects which function as signs of other places and times, to evoke stereotypical associations. Standing synecdochally for other places, such icons are also metaphors for the spatial experience of other places, in the manner by which, say, the Eiffel Tower, which first stands for Paris, then evokes *haute cuisine*, cosmopolitan sophistication, and relaxed elegance. Generic icons such as fountains, benches, statuary, and clocks signify traditional urban public space, and evoke notions of community and civic pride. More elaborate reconstructions of other places, whether generic – such as fairgrounds, Greek agora, Italian piazzas, Parisian sidewalk cafes, and Mediterranean villages – or specific – such as Bourbon Street and Miami Beach – quote well-worn clichés of place from our collective *musée imaginaire*. These simulated places exude an 'aura of familiarity' (Davis 1991, 2) and provoke predictable associations and dispositions facilitating consumption. Drinking capuccino coffee in a sidewalk cafe on Europa Boulevard, WEM, for example, is likely to elicit fantasies conducive to the purchase of luxury items in the nearby fashion stores. These simulations exploit another contemporary dis-ease, that is *otherwheritis*, the spatial equivalent of nostalgia, a social condition in which a distant place is preferable to here and now.

Moving on the mall: reclaiming the shopping center

The shopping center appears to be everything that it is not. It contrives to be a public, civic place even though it is private and run for profit; it offers a place to commune and recreate, while it seeks retail dollars; and it borrows signs of other places and times to obscure its rootedness in contemporary capitalism. The shopping center sells paradoxical experiences to its customers, who can safely experience danger, confront the Other as a familiar, be tourists without going on vacation, go to the beach in the depths of winter, and be outside when in. It is quite literally a fantastic place, and I suspect the disappointment that some experience at the mall may result from the impossibility of these paradoxes (psychoanalysis tells us, of course, that the inability to attain the goal of the desire is precisely its necessary condition). It is a representation of space masquerading as a representational space (Lefebvre 1991, 38–39); that is, a space conceptualized, planned scientifically and realized through strict technical control, pretending to be a space imaginatively created by its inhabitants.

The question then is how to retrieve these spaces from such calculated control, and there are a number of possible tactics. First is the exposure of the fetishism of the commodity and the re-problematization of the relations of consumption. Consumer activists, for example, have exposed the materialism of the commodity by organizing information campaigns and consumer boycotts.

Figure 19.3 Constructing a narrative and directing the shopper through the mall:
Pier 39 in San Francisco (courtesy of the author).

Second is the attempt to resist the economic and spatial logic of the shopping
center. The struggle of community groups against large-scale retail development
in their neighborhoods has had some limited success in some parts of the U.S.,
particularly the north-west, through delaying construction and negotiating environmental concessions from developers ('Building Despite the Obstacles ...'
1990). In most places, however, these strategies are more limited due to what is
generally perceived by localities and states to be the need for capital investment
in their communities. Most urban communities, in particular, are only too happy,
if financially capable, to provide developers with incentives, and in this context,
development offers an undoubted improvement in environmental quality. Moreover, environmental and community-based resistance is only effective against new
development, while renovation of existing centers is fast becoming the dominant
trend.

Third is the struggle to open the shopping center to all activities consistent
with public space, even those that may affront the sensibilities of the consumer or
disrupt the smooth process of consumption.

Fourth is the tactical occupation of spaces, particularly by actions that would
be excluded by the signs and security guards of private property. No architectural
form is entirely effective, and all spaces must open up some possibilities as they
shut others down. As Eco (1986, 77) notes, architecture fluctuates between being

coercive, forcing one to live and behave in a particular way, and indifferent, allowing one freedom to move, express oneself, and dream. Users of the shopping center may pursue such freedoms and exploit the opportunities that shopping centers present. It is only the overwhelming normalcy of everyone and everything in the shopping center that allows the will of the plan to remain unquestioned.

Finally there is the attempt to subvert the systems of signification operating in the retail built environment. This involves recognizing the intention behind the sign – which I have attempted – and a far more creative appropriation, or reassignation of meaning. The built environment is always complexly and multiply coded, and the assignation of specific meaning depends upon the predisposition of the reader. There is always the potential for consciously perverted interpretation, a challenge to the meaning of sign and to the class structuration of the signification system.

More significant then will be attempts by the users themselves to subvert meaning through strategies of social parody and 'detournement of pre-existing aesthetic elements' (Knabb 1981, 45; see also Bonnett 1989, 135). This is to accept the limitations placed upon resistance: that there is no possibility of a critique from outside the dominant representational discourses, for there is no position that is not implicated in the object code; and that there is also no possible alternative to the totalizing logic of social relations within this society of the spectacle. One cannot imagine, say, the rational planning, construction, and successful operation of a genuinely alternative shopping center, nor practically conceive of nonalienating forms of consumption. Instead, effective tactics can only employ the means of the strategy against itself, by taking its word to extremes. Users already do this to a limited extent, treating the mall as the social space it pretends to be, 25 percent freely enjoying its facilities without making a purchase ('Who Shops in Shopping Malls?' 1989, 43). What I have in mind, however, is the construction of situations; that is, the collective staging of games and farcical events, by artists, activists, and the shopping center patrons themselves, that can temporarily bring carnival into the shopping center, upsetting the conventional play of signification, subverting the cultural codes that are strategically deployed.

Ultimately, however, we must realize that the nostalgia we experience for authenticity, commerce, and carnival lies precisely in the loss of our ability to collectively create meaning by occupying and using social spaces for ourselves. While developers may design the retail built environment in order to satisfy this nostalgia, our real desire, as Frow (1991, 129) notes, is for community and social space free from instrumental calculus of design.

Notes

1 The term is borrowed from Wood (1985, 81), who uses it to describe 'places made over to be something they never were'.

2 Although the industry wishes to assure us that 'Real Men Do Like Shopping' (International Council of Shopping Centers 1990, 16), it is significant that there are at least two games designed for girls based on the mall experience. The object of the

board game 'Meet Me at the Mall' is to fill shopping bags with merchandise and 'shop 'til you drop'; an electronic game, 'Mall Madness', uses pretend credit cards and automated voice product descriptions ('The Short Run' 1992, 5).

3 Rouse's other schemes include Harborplace in Baltimore, South Street Seaport in New York, Santa Monica Place in Santa Monica, California; the Tivoli Brewery in Denver, the Grand Avenue in Milwaukee, St Louis Union Station in St Louis, Portside in Toledo, and Waterside in Norfolk. Other festival marketplaces based on this model include, for example, Harbor Island in Tampa, Trapper's Alley in Detroit, Rainbow center in Niagara Falls, New York, Marina Marketplace in Buffalo, Pier 39 in San Francisco, and Charleston Place in Charleston, South Carolina. Less publicized failures include 6th St Marketplace in Richmond and Water St Pavillion in Flint (Sawicki 1989, 348).

4 Although required by some local governments, day care is not proving very successful because of the difficulty and cost of obtaining liability insurance and of parental distrust of strangers (Reynolds 1990b, 29). More in keeping with the commodified setting is a novel enterprise, part day care, part entertainment for children, pioneered in Evergreen Plaza in Rolling Meadows, Illinois. Children may be deposited at 'Kids Only Cartoon Theaters,' where they are 'barcoded' with identification tags and constantly monitored by video while they watch continuous cartoons. The parent is given a pager in case problems arise or they fail to return within the prescribed time. Electronic doors prevent children leaving, and the same adult must collect them with the barcode on the pager also matched with that on the child.

5 Note, however, that downtown developers have recently discovered the 'positive demographics' of minorities and have designed centers and tenant mix explicitly to capture these underserved markets. This applies especially to Hispanics: centers directed explicitly at Hispanic markets include, for example, the Mercado in Phoenix, Fiesta Marketplace in Santa Ana, California, Galleria of the Americas in New York, and Palm Plaza in Hialeah, Florida.

6 The developer personally chose the 29 thirty-foot palms that grace Tyson's Corner in Fairfax County, Virginia. Trees were dug up in Florida and kept for 18 months in shade houses to gradually acclimatize them to indoors before being taken in temperature-controlled trucks to their new home.

References

Adorno, T., and Horkheimer, M. 1969. *The dialectic of enlightenment.* New York: Continuum.

Bakhtin, M. M. 1984. *Rabelais and his world.* Bloomington: Indiana University Press.

Barthes, R. 1979. *The Eiffel Tower and other mythologies.* New York: Hill and Wang.

Blomeyer, G. 1988. Myths of malls and men. *The Architect's Journal* May: 38–45.

Bonnett, A. 1989. Situationism, geography, and poststructuralism. *Environment and Planning D: Society and Space* 7: 131–46.

Boorstin, D. J. 1973. *The Americans: The democratic experience.* New York: Vintage Books.

—— 1961. *The image: A guide to pseudo-events in America.* New York: Harper Colophon.

Bourdieu, P. 1984. *Distinction: A social critique of the judgement of taste.* Cambridge: Harvard University Press.

Chaney, D. 1990. Subtopia in Gateshead: The MetroCentre as a cultural form. *Theory, Culture and Society* 7(4): 49–68.

Csikszentmihalyi, M., and Rochberg-Halton, E. 1981. *The meaning of things: Domestic symbols of the self.* New York: Cambridge University Press.

Diggins, J. P. 1977. Reification and the cultural hegemony of capitalism: The perspectives of Marx and Veblen. *Social Research* 44(2): 354–83.

Eco, U. 1986. Function and sign: The semiotics of architecture. In *The city and the sign*, ed. M. Gottdiener and A. Ph. Lagopoulos, pp. 55–86. New York: Columbia University Press.

Enzenberger, H-M. 1974. *The consciousness industry.* New York: Seabury.

Frampton, K. 1983. Towards a critical regionalism: Six points for an architecture of resistance. In *Postmodern culture*, ed. H. Foster, pp. 16–56. London: Pluto Press.

Frieden, B. J., and Sagalyn, L. B. 1989. *Downtown, Inc.: How America rebuilds cities.* Cambridge: MIT Press.

Frow, J. 1991. Tourism and the semiotics of nostalgia. *October* 57: 123–51.

Goss, J. D. 1992. Modernity and postmodernity in the retail built environment. In *Ways of seeing the world*, ed. F. Gayle and K. Anderson. London: Unwin Hyman.

—— 1988. The built environment and social theory: Towards an architectural geography. *The Professional Geographer* 40: 392–403.

Gruen, V. 1973. *Centers for the urban environment: Survival of the cities.* New York: Van Nostrand Reinhold.

—— 1978. The sad story of shopping centers. *Town and Country Planning* 46(7/8): 350–52.

—— and Smith, L. 1960. *Shopping towns USA: The planning of shopping centers.* New York: Van Nostrand Reinhold.

Harris, N. 1990. *Cultural excursions: Marketing appetites and cultural tastes in modern America.* Chicago: University of Chicago Press.

Haug, W. F. 1986. *Critique of commodity aesthetics.* Minneapolis: University of Minnesota Press.

Hazel, D. 1992. Crime in the malls: A new and growing concern. *Chain Store Age Executive*, February 27–29.

—— 1989. After the fall: Lessons from L. J. Hooker, *Shopping Center Age*, September: 27–30.

Hewison, S. 1989. *The heritage industry.* London: Methuen.

Hopkins, J. S. P. 1990. West Edmonton Mall: Landscape of myths and elsewhereness. *Canadian Geographer* 34(1): 2–17.

—— 1991. West Edmonton Mall as a centre for social interaction. *Canadian Geographer* 35(3): 268–79.

Jacobs, J. 1988. *The mall: An attempted escape from everyday life.* Prospect Heights, IL: Waveland Press.

Jencks, C. 1987. *The language of post-modern architecture.* Harmondsworth, U.K.: Penguin.

—— 1986. *Modern movements in architecture.* London: Academy Editions.

Jhally, S. 1987. *The codes of advertising.* London: Frances Pinter.

Judge bars group from leafletting in malls. 1991. *New York Times*, July 28, Sec. 1, p. 31.

Knabb, K. 1981. *Situationist international anthology.* Berkeley, CA: Bureau of Public Secrets.

Kowinski, W. S. 1985. *The malling of America: An inside look at the great consumer paradise.* New York: William Morrow.

Kroker, A.; Kroker, M.; and Cooke, D. 1989. *Panic encyclopedia: The definitive guide to the postmodern scene.* New York: St. Martin's.

Levine, J. 1990. Lessons from Tysons Corner. *Forbes* April 30: 186–87.

Lewis, G. H. 1990. Community through exclusion and illusion: The creation of social worlds in an American shopping center. *Journal of Popular Culture* 24: 121–36.

McCloud, J. 1989a. Hotels check in to stay. *Shopping Center World*, April: 22–32.

—— 1989b. Fun and games is serious business. *Shopping Center World*, July: 28–35.

—— 1991. Today's high-tech amenities can increase owners' profits. *Shopping Center World*, July: 25–28.

McCracken, G. 1988. *Culture and consumption: New approaches to the symbolic character of consumer goods and activities.* Bloomington: Indiana University Press.

Maitland, B. 1985. *Shopping malls: Planning and design.* New York: Nichols.

—— 1990. The new architecture of the retail mall. New York: Van Nostrand Reinhold.

Marcuse, H. 1964. *One-dimensional man.* Boston: Beacon Press.

Meyrowitz, J. 1985. *No sense of place: The impact of the electronic media on social behavior.* New York. Oxford University Press.

Morris, M. 1988. Things to do in shopping centres. In *Grafts: Feminist cultural criticism*, ed. S. Sheridan, pp. 193–225. New York: Verso.

Oldenburg, R. 1989. *The great good life.* New York: Paragon House.

Pawlak, E. J., et al. 1985. A view of the mall. *Social Service Review* June: 305–17.

Peterson, E. C. 1985. Diverse special interest groups may have access to center property. *Shopping Center World*, May: 85.

The PUD market guarantee. 1991. *Chain Store Age Executive*, April: 31–32.

Relph, E. 1987. *The modern urban landscape.* Baltimore: Johns Hopkins University Press.

Richards, G. 1990. Atmosphere key to mall design. *Shopping Center World*, August: 23–29.

Rowe, P. G. 1991. *Making a middle landscape.* Cambridge: MIT Press.

Sack, R. 1988. The consumer's world: Place as context. *Annals of the Association of American Geographers* 78: 642–64.

Sahlins, M. 1976. *Culture and practical reason.* Chicago: University of Chicago Press.

Scott, N. K. 1989. *Shopping Center Design.* London: Van Nostrand Reinhold.

Shields, R. 1989. Social spatialization and the built environment: West Edmonton Mall. *Environment and Planning D: Society and Space* 7: 147–64.

Sibley, D. 1988. Survey 13: The purification of space. *Environment and Planning D: Society and Space* 6: 409–21.

Sommer, J. W. 1975. Fat city and hedonopolis: The American urban future. In *Human geography in a shrinking world*, ed. R. Abler et al., pp. 132–48. North Scituate, MA: Duxbury Press.

Special issue on the West Edmonton Mall. 1991. *Canadian Geographer* 35(3).

Stewart, S. 1984. *On longing: Narratives of the miniature, the gigantic, the souvenir, the collection.* Baltimore: Johns Hopkins University Press.

Teaford, J. C. 1990. *The rough road to renaissance: Urban revitalization in America, 1940–1985.* Baltimore, Johns Hopkins University Press.

Tuan, Y-F. 1988. The city as a moral universe. *Geographical Review* 78(3): 316–24.

Turner, V. 1982. *From ritual to theater.* New York: Performing Arts Publications.

Veblen, T. 1953. *The theory of the leisure class.* New York: Mentor Books.

Venturi, R.; Scott-Brown, D.; and Izenour, S. 1972. *Learning from Las Vegas.* Cambridge: MIT Press.

Wiebe, R., and Wiebe, C. 1989. Mall. *Alberta* 2(1): 81–90.

Williams, R. 1980. *Problems in materialism and culture: Selected essays.* London: New Left Books.

Winner, L. 1980. Do artifacts have politics? *Daedalus* 109: 121–35.

Wood, J. S. 1985. Nothing should stand for something that never existed. *Places* 2(2): 81–87.

Zukin, S. 1991. *Landscapes of power: From Detroit to Disney World.* Berkeley: University of California Press.

—— 1990. Socio-spatial prototypes of a new organization of consumption: The role of real cultural capital. *Sociology* 24(1): 37–56.

20   Sharon Zukin

Disney World: The Power of Facade/The Facade of Power

Excerpts from: *Landscapes of power: from Detroit to Disney World,* pp. 217–50. Berkeley and Los Angeles: University of California Press (1991)

The modern city's image of centrality is turned inside out by the landscape of power at the bicoastal extremities of the Sunbelt. Without a traditional center or downtown, cities like Los Angeles and Miami can only be seen in fragments. Intensive real estate development does not produce the usual vertical skyline. Seashore, freeways, and canals erupt, instead, in dense clusters of suburban-style housing like the ranch house, beach house, villa, and bungalow, isolated office buildings, and low-lying industrial parks. There is little difference between 'city' and 'country' in this clustered diffusion.

Under pressure from investment in the built environment, space both expands and contracts. Collective spatial forms expand to include relatively unstructured cities like Los Angeles and Miami as well as increasingly structured exurban areas (Orange County in southern California and Fort Lauderdale – Palm Beach in Florida). But the same forms are also individualized and collapsed into specific journeys: no one would claim to appropriate the entire experience of L.A. or Miami. The automobility that flows through these cities makes some people excited or uneasy, for there is little connection – according to modernist expectations – between built form and urban identity. In Miami, in the architect Michael Sorkin's view, 'the pattern of settlement [is] obviously in total thrall to the landscape. Clearly, this [is] a zone to be grasped not through the familiar repertoire of urban categories but through a far broader sense of territory'.[1]

Beginning in the 1920s, and with greater force since the recognition of a 'power shift' to the Sunbelt in the 1970s, the material landscape of contemporary cities has indeed demanded a broader set of categories than modernism offers. The dynamic interplay between nature and artifice in the built environment of Los Angeles or Miami has forced us to visualize urban development as a set of multiple, decentered processes, and to acknowledge the strong force of derivation over originality in architecture. Furthermore, the forging of a metropolis out of many private jurisdictions has challenged the primacy of public space as an organizing principle of social life. From the outset, moreover, Los Angeles and Miami have grown from service rather than manufacturing economies, wrenching the landscape of power from its nineteenth-century roots. This must be our image of postmodern urbanity. Yet it has been a long time building. Palm trees, fast cars on freeways, hot flamingo pinks and dazzling white villas: the symbols of post-World War II choice and alienation stretch continuously from *Play It as It Lays* to *Less than Zero*. This is the world of Raymond Chandler colonized by Walt Disney's world, where the fantasies of the powerless are magically projected onto landscape developed by the powerful.

The postmodern synergy between landscape and vernacular is structured by the apparent falsity that Henry James long ago derided as 'hotel-civilization'. When he saw the expensive resort architecture of Palm Beach in 1907, James called it 'a Nile without the least little implication of a Sphinx', a set of the 'costlicst reproductions . . . as if to put to shame those remembered villas of the Lake of Como, of the Borromean Islands'. 'Instant Portofino,' Michael Sorkin likewise snapped about Miami's affluent villas eighty years later. In Los Angeles, a more benign Reyner Banham found 'an instant architecture and an instant townscape' in the 1960s. 'Most of its buildings are the first and only structures on their particular parcels of land', he wrote, 'they are couched in a dozen different styles, most of them imported, exploited, and ruined in living memory'.[2] It is impossible to view this landscape from the perspective of a modern city like New York, Chicago, or San Francisco.

In the built environment today, the stage-prop quality of a landscape of consumption takes more ingenious root in the social imagination. This landscape is built up by the electronic image that faithfully transmits, yet also renders more abstract, the architectural facade that both mirrors and recedes, the Disneyland that re-creates a built environment for mass leisure consumption. This is a landscape that asks more imagination of its viewers than it offers.[3]

The entire landscape of cities like Miami and Los Angeles visually projects the liminality between *market* and *place*. The usual forms of social control – by police, employers, corporate elites – are embedded in an amusing architecture and individualized means of consumption like automobiles. Although L.A. and Miami are real cities, they are built on the power of dreamscape, collective fantasy, and facade. This landscape is explicitly produced for visual consumption. Moreover, it is *self-consciously* produced.

Miami was, after all, invented three times in this century. Developed out of swampland as a socially, ethnically, and racially exclusive southern resort in the 1920s, Miami differed little in intention from its more exclusive neighbor, Palm Beach. After World War II, a second Miami, monopolized by Miami Beach, developed as a cheaper vacation land for middle-class northerners, mainly ethnic and eventually Jewish; Miami offered a more exotic and more individualized consumption of leisure than social camps in the Catskills. From 1960, following Castro's revolution, a new wave of Cuban exiles and other Central Americans swelled the racially and ethnically segregated population in the inner city. By 1980, bypassing the native black population in business and political organization, they made the Miami area a Latin metropolis, an adjunct to Caracas and Rio and an alternative to Havana. Ethnic recruits and Latin capital created in the third Miami a microcosm of global exchange. While the city always had a service economy, hotels became less important than foreign investment in banks, weapons traffic, and the illegal drug trade.[4]

Los Angeles has also been invented three times in this century. Visualized by celluloid fantasy, Los Angeles first figured as backdrop and back lot in countless silent Westerns. The prewar dream capital was created as production space for the manufacture of films, which was, like tourism, a mass leisure industry. But Los Angeles was also created by making desert and hills habitable for immigrant labor, both skilled and unskilled, outside the film industry, in oil and gas refineries, car making, and light manufacturing. The first Los Angeles was an amalgam of East European Jews, Asians, Chicanos, African-Americans, Okies, and most of all, midwesterners. Invention of the second Los Angeles, which dates from 1940, reflected the new labor force required for military and industrial activities in the port and airport and large, unionized plants. This is the Los Angeles of middle-class affluence and social mobility, an American dream. As in Miami, however, much of the population that was drawn to the city in this period has been passed over in the growth of the third Los Angeles.

Both decentralized and recentralized, the new Los Angeles has a coherent landscape downtown, built by new Asian immigrants and mainly foreign investment in banking and financial services. But it also has inner suburbs where manufacturing branch plants of U.S. firms have rapidly shut down, displacing and cheapening union labor, and other industrial suburbs where electronics and garment plants have continued to grow even during economic recessions. These suburbs are integrated by clusters of high-tech industries, yet they are segregated at work by race and gender and segregated in residential communities by race and class.[5]

The constant reinvention of landscape furnishes a narrative for 'footloose' capital. It provides a social geography for the shifting landscapes of the global economy. Yet while the real history of Los Angeles and Miami illustrates the processes of structural change in the U.S. economy, these cities are more interesting as spatial metaphors. We are fascinated by Los Angeles and Miami because we think they show us the future. Their freeways, their 'decentering,' their 'Mondo Condo' pursuit of private leisure: this is the way the future looks. These cities stun because of their unique ability to abstract an image of desire from the landscape and reflect it back through the vernacular. Just as they show the power of facade to lure the imagination, so they also represent the facade of global corporate power.

No single image symbolizes a postmodern city like Los Angeles or Miami the way the steel mill symbolizes the company town or downtown symbolizes the modern city. 'Ain't no skyline,' an ironic songwriter says about L.A. We see these landscapes, instead, by modes of visual consumption that play with image and reality. Visual consumption of landscape rests on an interplay of nature and artifice that goes back many years. Toward the end of the eighteenth century, the 'Eidophusikon' – a construction of pasteboard cut-outs, lights, and sound – created scenes of London, the countryside, and the seashore. In the nineteenth century, dioramas and panoramas manipulated light to make translucent images –

often images of city streets – on a screen. Their twentieth-century descendants, Hollywood films, widened spectators' power over darkness while strengthening the hold of the image over its viewers. After the reign of still photography during the Depression and World War II, electronic media completed the transformation of image into power. Broadcast television broke all existing barriers between public and private, local and global, the living room and the world, until the viewers rather than the image became the product.[6]

The domestication of fantasy in visual consumption is inseparable from centralized structures of economic power. Just as the earlier power of the state illuminated public space – the streets – by artificial lamplight, so the economic power of CBS, Sony, and the Disney Company illuminates private space at home by electronic images. With the means of production so concentrated and the means of consumption so diffused, communication of these images becomes a way of controlling both knowledge and imagination, a form of corporate social control over technology and symbolic expressions of power.

Fantasy as a landscape of power

While Walt Disney won fame as a founder of Hollywood's animation industry, his real genius was to transform an old form of collective entertainment – the amusement park – into a landscape of power. All his life Disney wanted to create his own amusement park. But to construct this playground, he wanted no mere thrill rides or country fair: he wanted to project the vernacular of the American small town as an image of social harmony. 'The idea of Disneyland is a simple one. It will be a place for people to find happiness and knowledge', Disney said. But 'in fact', a recent essay on Disney points out, 'it was the appearance of Disneyland, not the idea, that was simple'.[7]

Appearance nonetheless caused the great animator some concern when he began to plan Disneyland, the archetypal theme park, which was finally built in Orange County, California, in 1955. Disney hired two architects whose plans didn't quite capture what he had in mind, and giving up on them, used an animator from his studio to draw up architectural designs to his own specifications. Disney's peculiar vision was based on a highly selective consumption of the American landscape. Anchored by a castle and a railroad station, Disneyland evoked the fantasies of domesticity and illicit mobility that were found in the vernacular architecture of southern California. The castle and station were joined on an axis by 'Main Street USA', an ensemble of archaic commercial facades. This mock-up in fact idealized the vernacular architecture Disney remembered from his childhood in Marceline, Missouri, before World War I.

Disney's fantasy both restored and invented collective memory. 'This is what the real Main Street should have been like', one of Disneyland's planners or 'imagineers' says. 'What we create', according to another, 'is a "Disney realism", sort of Utopian in nature, where we carefully program out all the negative, unwanted elements and program in the positive elements'.[8] And Disneyland succeeded on the basis of this totalitarian image-making, projecting the collective desires of the powerless into a corporate landscape of power. In this way it paralleled the creation of a mass consumption society. Disney's designs also included an element of play that was pure Hollywood construction, for Disneyland featured five different stage-set amusement parks, organized around separate themes: Adventureland, Lilliputian Land, Fantasyland, Frontier Land, and Holiday Land. The unique combination borrowed motifs from carnivals, children's literature, and U.S. history.

That Disneyland significantly departed from the dominant fantasy landscape of the time was dramatized when Disney failed to arouse enthusiasm in a convention of amusement park owners that previewed plans for the park in 1953. They criticized the small number of rides, the large amount of open space that wouldn't generate revenue, and the need for constant, expensive maintenance in the theme parks.[9] In their view, Disneyland would never succeed as a business venture. They objected – had they used the now-fashionable word *concept* in those days – that Disneyland was too self-conscious and unrealistic a concept to be an amusement park. But from our point of view, they failed to understand that Disneyland was an ideal object for visual consumption, a landscape of social power. Despite their criticism, Disneyland was commercially successful from the day it opened.

Disneyland offered a multidimensional collage of the American landscape. The playgrounds organized around a theme provided consumers with their first opportunity to view several different landscapes – some imaginative historical recreations and others purely imaginary – simultaneously. With this variety for individual visitors to choose from, Disneyland differed from historical dioramas and reconstructions like Greenfield Village and Colonial Williamsburg, where costumed actors re-created the routines of daily life in an earlier age. Disneyland, moreover, had no educational veneer. It merely told a story, offering the selective consumption of space and time as entertainment. This was the 'wienie', as Walt Disney and amusement park owners alike called the lure that attracted customers to a paying event.

Visitors to Disneyland paid for a variety of entertainment experiences linked by the narrative of the different themes. These in turn provided a narrative for different program segments on the Disney Studio's weekly television series. Combining narrative with serial expectations, each visual product of the Disney Company fed into the others. Although commercial spin-offs were not a new creation, this commercialization was the most extensive to take place under a single corporate sponsor. Disney's business growth also related to important processes of change in the larger society: notably, the demographic growth of the baby boom, the spread of television, and the increase in domestic consumption. Moreover, it contributed to the development of both Orange County and the tourist industry. Disney's success coincided with the expansion of the suburbs and a population shift to the Southwest, the growth of the service sector, and a boom in leisure-time activities, including sales of recreational land and travel. Just as the real landscape reflected the intensive, unplanned development of the country by subdivision and mass construction, so the imaginary landscape of Disneyland reflected the growth of mass communications built on visual consumption.

While this kind of entertainment invited escape from the modern world, it also relied upon the centralization of economic power typical of modern society.

Consumption at Disneyland was part of a service-sector complex relating automobiles and airplanes, highways, standardized hotels, movies, and television. Furthermore, the social production of Disneyland related a major corporate presence – the Disney Company – to entertainment 'creation', real estate development and construction, and product franchising. In all these senses, Disneyland suggested the social and economic potential of liminality in the modern American marketplace.

Disney's second amusement park, Disney World, near Orlando, Florida, was conceived on a somewhat different model. Here the centerpiece was a 'wienie' that would lure people to a new utopian vernacular, an imaginary landscape of the future. 'EPCOT [the Experimental Prototype Community of Tomorrow] will be like the city of tomorrow ought to be', Walt Disney said. This vision in part reflected Disney's childhood experiences, as the son of a man with several unsuccessful careers and business failures. It also responded to the lasting significance of social upheaval during the Great Depression. A conservative utopia, EPCOT from the outset joined entertainment values to motifs of social control.

Yet EPCOT was not built exactly to Disney's original specifications. For one thing, the company decided in 1966 that building a residential community involved too much legal responsibility. It decided instead to make its ideal community a temporary haven – that is, a resort colony. Another problem was the lack of available technology to build EPCOT as envisioned by Disney and his imagineers. Although the biggest U.S. corporations signed on as exhibit sponsors to enhance their corporate images, the automated technology to build the attractions in Disney's own pavilions was too expensive for the Disney Company. Disney's Hall of Presidents, for example, was designed in the late 1950s as an automated historical diorama in which the presidents are 'played' by robots, but it wasn't feasible for the company to build it until twenty years later. EPCOT itself didn't open as part of Florida's Disney World until 1982.[10]

The fantasy theme parks in Orlando (Captain Nemo, Pirates of the Carribean, Space Mountain) are joined by facades of touristic architecture (the Eiffel Tower for France, Italy's St. Mark's Square) where food and souvenirs are sold. Thus Paul Bocuse's restaurant at Disney World expands on the international cultural villages of the 1893 World's Columbian Exposition.

People-movers at Disney World use sophisticated technology in a strategy of social control over consumption. At Disney World, however, the solar-powered 'traveling theater cars' that propel blocks of sixty visitors through the sights, sounds, and smells of the exhibits literally hold the public captive to the image. The sensation of consuming has become more intense, but also more ambiguous, so that visitors to Disney World feel that while they can control technology, they are also being controlled. Mobility in Disney World is a movement from one enveloping theatrical environment into another, where information blends into an implicit call to consume: Feel! Marvel! Buy!

The stage sets of the theme parks, however, are dwarfed by the entertainment architecture of Disney World's hotels. The Magic Kingdom at Disney World includes ten resort hotels with 5,700 rooms, 1,190 camp sites, 580 vacation villas sold on time shares, and three convention centers. Lake Buena Vista and Walt

Disney World Village provide an additional seven hotels, with 3,500 rooms. Although the Orlando complex was completed in the 1960s, new hotels are added continuously. From 1988 to 1990, three super-hotels were built within the grounds: the Grand Floridian, a 'Victorian' theme hotel, with 900 rooms; and the Dolphin and Swan hotels, which together hold 2,300 rooms and a 200,000 square foot convention center. The Dolphin and Swan form a temporary residential community for 10,000 people, making the Disney Company the largest hotel and convention center developer in the southeastern United States.

Disney World's newest hotels are theme parks in themselves. Designed by the postmodern architect Michael Graves, the Dolphin and Swan look like the rest of Graves's work – a strong terra cotta facade, pastel colors, and playful decoration – and also resonate with the production values of Disney's animated cartoons. 'There's no other commercial project you can compare with Graves's design', says John Tishman, the developer of the hotel complex (in partnership with Metropolitan Life Insurance and the Aoki Corporation, a Japanese construction and real estate development firm). 'It is a Disney building, fun and frivolous like a stage set'. This stage set is organized around distinctive animal motifs. One hotel is surmounted on the roof by a five-story-tall dolphin; the other, by a giant swan. Although Graves chose these animals, he says, as classical symbols of nature (i.e., Florida's waterfront vernacular), they also symbolize the trademark landscape of the Disney Studios' artifice (e.g., Mickey and Minnie, Pluto, Donald Duck). 'They have the kind of warmth that the whole Disney experience gives', Graves says. They show, in other words, the friendly face of power.[11]

The water motif also claims to appropriate the south Florida vernacular. So the Dolphin Hotel features a cascading waterfall that runs down the facade into a series of huge clamshells and ends in a shell-shaped pool supported by four large, sculpted dolphins. The hotels' exterior walls are covered with a pattern of waves and banana leaves, colors are 'tropical' blue-green and coral (which are also Graves's signature colors). 'We want to create a sense of place that is unique,' says Michael Eisner, the chairman of the board of Walt Disney Company.

Hotel civilization is significant to structural transformation in the market economy because it furnishes an imaginary landscape, a prototype design for a collective life based on domestic consumption. It camouflages behind a friendly facade the global, yet centralized, economic power that makes this consumption possible. The replicas of Borromean villas that James derided are now the Dolphin and the Swan; our common culture consists of Mickey Mouse and pink flamingos. 'Obviously, there are political implications', E. L. Doctorow observes about the social control implicit in this derivative market culture.

> What Disneyland proposes is a technique of abbreviated shorthand culture for the masses, a mindless thrill, like an electric shock, that insists at the same time on the recipient's rich psychic relation to his country's history and language and literature. In a forthcoming time of highly governed masses in an overpopulated world, this technique may be extremely useful both as a substitute for education and, eventually, as a substitute for experience. Disney symbols, in other words, determine the limits of consumers' imagination.[12]

In the present, however, Disney's imaginary landscape uses the symbols of a common culture to significant economic effect. The opening of EPCOT in 1982 put the Disney Company back in the spotlight on Wall Street, reversing the declining stock prices of previous years. With the company's diversification, moreover, into new entertainment areas (e.g., the production of adult movies and a cooperative venture with MGM at Disney World that includes both movie production facilities and a Hollywood theme park), Disney received glowing reviews in the business press. Licensing arrangements proliferated, for the nature of Mickey and Minnie ensure that 'Disney is expandable to anything in the consumer realm.'[13]

Yet it is primarily as a creator of landscape that Disney has been praised. Walt's original ability to abstract the desires of the powerless from the vernacular of Main Street and the Midway, and project them as a landscape for mass visual consumption, mapped a new vernacular image of a postmodern society. This has influenced both the real landscape and the perspective from which we view it. On the one hand, Disneyland and Disney World present an imaginary landscape based on a manipulated collective memory and consumption that we also find in downtown shopping malls like Faneuil Hall, Inner Harbor, and South Street Seaport. As early as 1963, James Rouse, the developer of these malls, praised Disney as an urban planner at a conference at Harvard University. On the other hand, Disneyland and Disney World have stimulated a whole regional complex of service-sector activities around tourism and real estate development. No wonder Walt Disney was named by *Los Angeles* magazine in April 1986 as one of '25 people who changed Los Angeles'.

Reyner Banham relates the underlying structure of Disney's landscape to the movie studio lots of Los Angeles. Fantasy presented as public architecture, movie studios were eventually 'elevated to the status of cultural monuments, which now form the basis for tourist excursions.' Banham suggests that the economic robustness of visual consumption also influenced companies to offer guided tours of their plants, and build theme parks for visitors around their main consumer products. The circle is completed by the Disney-MGM movie studio lot at Disney World, presenting 'a Hollywood that never was, and will always be'.[14]

The stage-set landscape is a liminal space between nature and artifice, and market and place. It mediates between producer and consumer, a cultural object with real economic effect. The Disney landscape has in fact become a model for establishing both the economic value of cultural goods and the cultural value of consumer products. Just as the 'World of Coca Cola' museum at the corporate headquarters in Atlanta places an established consumer product in a narrative framework that renews its cultural legitimacy, so art museums have replaced their encyclopedic manner of display, which induces 'museum fatigue', with story-telling strategies. The Disney World narrative is ubiquitous. Ironically, Disney's market-oriented landscape evokes a strong sense of place.[15]

Feeding the synergy between consumer markets, a self-conscious sense of place is also highly desired. Ski resorts combine the perennial alpine resort architecture with entertainment facilities borrowed from Disney fantasy: 'Vail (the town, not the mountain) always reminded me of a cross between a fancy hotel and Disneyland', someone writes in a letter to the editor of the *New York*

Times Magazine, 'no cars, people in funny costumes, entertainment based on ostentation and make-believe.' Upscale housing developments, moreover, borrow Disney's abstraction of the Main Street vernacular: 'The newest idea in planning is the nineteenth-century town', says the Miami architect Andres Duany, who has become successful as a town planner of historically derived reconstructions. 'That's what is really selling'.[16]

With conflict designed out, and comfort designed in, the image of a service-sector society is a utopian dream. It is self-consciously produced not to be disturbed by such problems as homelessness, low wages in service-sector jobs (although there have been strikes by Disney World employees), and racial competition. This projection of desire taunts the image of reality. 'If you come down here [to Miami after the riots in black neighborhoods that preceded the Super Bowl in 1989] and think this is Disneyland, that's not real', a prominent football player says. But for many consumers, the self-conscious production of the imaginary landscape is what they perceive as real. And it is becoming real, for even though Walt Disney never built the residential community of his dreams at EPCOT, the Disney Development Company, which started up in the mid 1980s, plans to build real residential communities near Disney World. Where does the theme park end?[17]

It has taken forty years to perfect this imaginary landscape. From the two world's fairs of 1893 and 1939, fantasy architecture for mass entertainment moved steadily toward a utopian visual consumption, with strong motifs of mobility, populism, and social control. But it was Disney World that abstracted the desire for security from the vernacular and projected it into a coherent landscape of corporate power. Not surprisingly, most visitors to Disney World are from the upper middle class, bosses rather than workers, and white. Consumers from this social base are most likely to believe they can control history and technology if only they submit to the control of the guiding corporate hand.[18]

Fredric Jameson is wrong about the postmodern landscape of visual consumption. Disney World suggests that architecture is important, not because it is a symbol of capitalism, but because it is the capital of symbolism.

Notes

1 Michael Sorkin, 'Travel: Miami Virtues: Sun, Sea and Dazzling Urban Design', *Vogue*, January 1986, p. 140; lack of correlation between urban form and social form from Reyner Banham, *Los Angeles: The Architecture of Four Ecologies* (London: Allen Lane, Penguin Press, 1971), p. 237. Also see E. W. Soja, 'Taking Los Angeles Apart: Some Fragments of a Critical Human Geography,' *Environment and Planning D: Society and Space* 4 (1986): 255–72 and James J. Flink, *The Automobile Age* (Cambridge, Mass.: MIT Press, 1988), pp. 140–48.

2 Henry James, *The American Scene* [1907] (Bloomington: Indiana University Press, 1968), pp. 447, 462; Sorkin, 'Travel', p. 142; Banham, *Los Angeles*, p. 21.

3 James, *American Scene*, p. 442; cf. David Rieff, *Going to Miami: Exiles, Tourists, and Refugees in the New America* (Boston: Little, Brown, 1987), pp. 11, 24–27. James on Lake Como from *Italian Hours* (Boston: Houghton Mifflin, 1909), pp. 131–32.

4 See Raymond A. Mohl, 'Miami: The Ethnic Cauldron,' in *Sunbelt Cities: Politics and Growth since World War II*, ed. Richard M. Bernard and Bradley R. Rice (Austin: University of Texas Press, 1983), pp. 58–99; Penny Lernoux, 'The Miami Connection', *Nation*, February 18, 1984, pp. 186–98; Joan Didion, *Miami* (New York: Simon & Schuster, 1987); Rieff, *Going to Miami*.

5 See Edward W. Soja, 'Economic Restructuring and the Internationalization of the Los Angeles Region', in *The Capitalist City*, ed. Michael Peter Smith and Joe R. Feagin (New York: Blackwell, 1987), pp. 178–98; Allen J. Scott, *Metropolis: From the Division of Labor to Urban Form* (Berkeley and Los Angeles: University of California Press, 1988), pp. 91–202; Mike Davis, *City of Quartz: Excavating the Future in Los Angeles* (London: Verso, 1990).

6 Eidophusikon, dioramas, and panoramas described in Wolfgang Schievelbusch, *Disenchanted Night: The Industrialization of Light in the Nineteenth Century*, trans. Angela Davies (Berkeley and Los Angeles: University of California Press, 1988); also see Francis D. Klingender, *Art and the Industrial Revolution* (London: Paladin, 1972), pp. 86–87; Joshua Meyerowitz, *No Sense of Place: The Impact of Electronic Media on Social Behavior* (New York: Oxford University Press, 1985), p. 73.

7 Martin Pawley, 'Tourism: The Last Resort', *Blueprint*, October 1988, p. 38.

8 Cleaning the past of its contradictions is developed in Mike Wallace, 'Mickey Mouse History: Portraying the Past at Disney World', *Radical History Review* 32 (1985): 33–57; imagineers quoted on pp. 35–36.

9 Pawley, 'Tourism: The Last Resort', p. 39.

10 Wallace, 'Mickey Mouse History', pp. 38, 42.

11 Pawley, 'Tourism: The Last Resort'; quotations from Joseph Giovannini, 'At Disney, Playful Architecture Is Very Serious Busines', *New York Times*, January 28, 1988. These motifs are congruent with Walt Disney's intentions from the very beginning. The circular plan, 'deliberate sensory overload', and lack of differentiation between real and surreal surroundings were meant to make people forget their everyday lives. See Patricia Leigh Brown, 'In Fairy Dust, Disney Finds New Realism', *New York Times*, July 20, 1989.

12 E. L. Doctorow, *The Book of Daniel* (New York: Random House, 1971), p. 289. Product tie-ins have convinced some critics that the colonization of fantasy foreseen by Doctorow has already arrived. Janet Maslin, 'Like the Toy? See the Movie', *New York Times*, December 17, 1989.

13 Entertainment industry analyst quoted in Douglas C. McGill, 'Mickey Sells; Is He Now Oversold?' *New York Times*, May 20, 1989. On connections between new products, licensing, profits, and stock values, also see *New York Times*, August 29 and November 6, 1989.

14 Banham, *Los Angeles*, pp. 127ff.; quotation on Hollywood theme park from Michael Eisner's introduction to 'Magical World of Disney' television program celebrating opening of the Disney–MGM Studios, April 29, 1989.

15 Trish Hall, 'Making Memories Go Better: Coke Plans Museum in Atlanta', *New York Times*, November 23, 1988; Allon Schoener, 'Can Museums Learn from Mickey and Friends?' *New York Times*, October 30, 1988.

16 Quotation on Vail from *New York Times Magazine*, December 25, 1988; Duany quoted in Philip Langdon, 'A Good Place to Live,' *Atlantic*, March 1988, p. 45. Duany's team come into a development site for a day, consult with local architects and the appropriate developers, feed on-site data into their computer model, and generate a town plan based on elements of vernacular architecture and local desires (*Progressive Architecture*, May 1989, pp. 84–88).

17 Football player quoted in George Vecsey, 'Amid Uneasiness, Super Bowl Visitors Stream In,' *New York Times*, January 19, 1989; see also 'Report: Disney in Florida,' *Progressive Architecture*, March 1990, p. 80.
18 In the mid 1980s, after EPCOT opened, more than 90 percent of visitors were white, their median income was $35,700, 48 percent held professional and technical jobs, and 26 percent were managers and administrators (Wallace, 'Mickey Mouse History', p. 53).

21   Christopher Pinney
Future Travel: Anthropology and Cultural Distance in an Age of Virtual Reality Or, A Past Seen From a Possible Future

Excerpts from: *Visual Anthropology Review* 8, pp. 38–55 (1992) (also reprinted in L. Taylor (ed.), *Visualizing Theory*, pp. 409–28. London: Routledge (1994))

[The Year is 2029 in instruction unit 47,
London subsection C of the Citizenship Foundation.]

I am honored to be able to address you on the occasion of my retirement. It gives me even greater pleasure to record that next week I shall be leaving for Reforestation Zone 27 of the Himalayan foothills where I will be able to spend my last years in quiet solitude and contemplation. When the time comes my ashes will be taken to Kashi where they will be cast into the holy River Ganges. You will all be aware that had I had the misfortune to have retired ten years ago, in 2019, I would have, on this occasion, presented a brief Cryogenic Suspension De-inauguration Lecture after which I would have been frozen pending further scientific breakthroughs. This reversal of fortune is remarkable and it is this change with which I will be concerned here. It is this reinvention of the world we lost, and our reattainment of our nonsubjectivities which I want to examine in this talk.

Until 2020 Virtuality consumers were largely Western and metropolitan and – as we will all be painfully aware – the world's wealth was increasingly consumed within a parallel world of virtuality. The merging of the Nintendo and Fujitsu Corporations with what was left of IBM in 1998 lead to a proliferation throughout the capitalist world of millions of 'journey program units' and 'teledildonic units'. These immobile devices with head mounted visual displays and full body data-gloves (later known as 'senso-sheaths') allowed cyberspace foreign holidays in real and imaginary locations[1] and sexual encounters with imaginary or real persons. By the following year three million head-mounted displays and senso-gloves were privately owned in the USA alone.

Tourism, a major source of Third World income, collapsed. Travellers were reluctant to contract Delhi-belly when they could experience the subcontinent pain-free through a virtuality journey program. Disease-free 'teledildonics' ('interactive tactile telepresence[s]' – Rheingold 1991: 348) made sex an increasingly sterile pursuit. This finally limited the spread and mortality from AIDS, but the care of a residual and aging population became an increasing economic burden.

Given over to the simulatory pleasures of immediate and unlimited gratification, this world came to care even less for the agony of the non-metropolitan world.[2] Global food sources were secured, but the punitive measures required to maintain these and their cheap labor supplies became (even more) invisible, lost in a world of simulation which no longer required a veneer of humanity in which to ensconce its exotic holiday locations. The process of periphero-cide apparent throughout the fifteenth to twentieth centuries was obscenely intensified, taken to the point of destruction. By 2004 the Fujitsu syndicate had sold 18 million Personal Virtual Worlds – the new lightweight data suits that produced perfect resolution personal worlds. Two years after this the Global Cyberspace Foundation was established and in the next 15 years a further 450 million Personal Virtual Worlds were sold. Throughout this period we dissidents struggled uselessly for the rights of the non-Virtual world. But what did the Virtual world care when it could experience anything at the touch of a button?

Between 2015 and 2025 this situation began to change however. Across all of Global Zone A there were popular movements in support of debt cancellations and tariff agreement changes to benefit the Southern Global Zone. Citizen migration controls were abolished, huge transfers of wealth arranged and massive restitution and unilateral movements of cultural and historical collections were undertaken.

Younger members of my audience may be amazed to reflect on the fact that only forty years ago when I delivered a paper on the subject of cyberspace[3] I had to explain to the audience in very basic terms what this technology involved. Indeed, younger members of this audience will find it hard to perceive the circumstances of that seminar. It had taken me two hours to travel a mere sixty miles to work that morning and during that journey I saw over 10,000 private motor cars. The sky above London was full of aeroplanes carrying non-Governmental personnel. The audience's only experience with Virtual Reality was a crude device in a local amusement arcade and the academics had still to accept Walter Benjamin's attempt of fifty years previous in his Arcades Project to 'bridge the gap between everyday experience and traditional academic concerns, to achieve [a] phenomenological hermeneutics of the profane world' (Buck-Morss 1989: 3). When I gave that seminar it was [still] only eight years since the writer William Gibson had introduced the term cyberspace in his execrable novel *Neuromancer*. There he had described it as:

Cyberspace. A consensual hallucination experienced daily by billions of legitimate operators, in every nation . . . A graphic representation of data abstracted from the bank of every computer in the human system. Unthinkable complexity. Lines of light ranged in the nonspace of the mind, clusters and constellations of data (Gibson 1986: 67)

In the 1990s the world was posed on the brink of 'Virtuality,' a technological world of simulation made possible by 'Virtual Reality' (VR).[4] This technological apotheosis of man in his role as *Homo depictor* allowed the cybernetic creation of complete simulatory environments communicated to the body via three-dimensional glasses/headsets and body suits.[5] The Japanese Ministy of International Trade and Industry had already organized a 'virtual reality study commission' at the beginning of the 1990s which met monthly (Rheingold 1991: 310), and by 1992 some British schoolchildren were learning languages through immersion in a foreign 'virtual city'.[6] At that time critical responses tended to emphasize its role as future entertainment (or 'edutainment'), as a 'teledildonic'[7] or glorified video game, and there had been almost no consideration of the implications of this new form of travel and cultural experience for either anthropology or even mass tourism. In any event, those few analyses which had appeared by that stage showed no prescience of coming changes.

It was already believed in the 1990s that the increasing military use of VR might also have practical benefit.[8] As Rheingold observed, 'war games in cyberspace . . . use less petroleum and create fewer conditions for fatal accidents' (Rheingold 1991: 361). But it had still not been realized that VR would come to displace some 'real' wars into the bargain. At that time people were still incensed by Baudrillard's claim that THE GULF WAR [OF 1990] [HAD] NOT TAKE[N] PLACE. Many people's unwillingness to subsume the charred bodies of 150,000 Iraqi conscripts under 'simulation'[9] blinded them to the correctness of Baudrillard's claim that it was television (whose interests ran parallel to a recession hit military machine) which made it necessary.

Virtual Reality however – as we now know – took this full circle and provided a cyberspace where such suturing can be produced without recourse to death and destruction. When cyberspace became (perceptually and sensually) more 'real' than reality it was this – peopled merely by electronic shadows – which became the arena for the resolutions and terminations of politicians and armaments manufacturers.

Heidegger and Cartesian perspectivalism

The sorry story which I will relate is that of what a German philosopher of the last century – Martin Heidegger – termed 'the age of the world picture.' It is the history of this and its destruction by Virtual Reality about which I shall speak. It has been argued by some that Virtual Reality was first created in the 'anamorphic' rock paintings of the Upper Paleolithic (Rheingold 1991: 380–381). It is claimed that these anamorphic representations were 'painted in a precisely distorted manner on natural protuberances and depressions in the limestone in order to give the rendering a three dimensional appearance when viewed from the proper light and angle' (ibid). How interesting that the cavemen of Lascaux should have prefigured the contemporary activities of our contemporary global Cyberspace Foundations.

And how interesting that, in the intervening thirty thousand years, this Virtual Reality could be so effectively suppressed. It is the history of this suppression about which I shall also tell.

Heidegger traced a homology between the bounded visuality of Renaissance and post-Renaissance representations and the Cartesian objectification of the world on which its exploitation by science depended. It is from this stage that 'human capability . . . is given over to measuring and executing, for the purpose of gaining mastery over that which is as a whole' (Heidegger 1977: 132). Science, rooted in mathematical knowing in advance, is concerned with the projection of objects within a spatiotemporal field of certainty.

Heidegger argued that although Plato's stress on *eidos* (1977: 131) prefigured later developments, it is Enlightenment rationalism – manifested through what Martin Jay was later to call Cartesian perspectivalism (1988: 4) – that truly begins to view the world as picture – 'the fact that the world becomes a picture at all is what distinguishes the modern age' (1977: 130). Whereas for Parmenides 'man is the one who is looked upon by that which is' (1977: 131), in the modern age 'that which is . . . come[s] into being . . . through the fact that man first looks upon it, in the sense of a representing that has the character of subjective perception' (*ibid*). In the modern age man sets himself *before* and also *against* nature and the world is 'placed in the realm of man's knowing and of his having disposal' (1977: 130). There begins, Heidegger argues,

> that way of being human which mans the realm of human capability as a domain given over to measuring and executing, for the purpose of gaining mastery over that which is as a whole. (1977: 132)

It is an easy matter to trace parallel developments in the practice and theory of Western travel in which, as the world entered the modern period, travel became increasingly systematized and rule-bound. One important function of this methodization occurring in the sixteenth century was to produce the object of travel as a pictorial surface and the outward and return journey as parts of the picture frame.

The magnitude of the transformation of the spatial imagery of travel can be conceived through the opposition of the earlier pilgrim's view of the world centered on the Holy Land or Rome with a later stress on a grid of equivalent sites each of which should be surveyed from the highest vantage point so as to immediately perceive the terrain as a whole – what Kenneth Littlejohn termed the 'hodological' approach. We may recall the opposition he drew between Temne conceptions of the landscape – a poetics of space – and Western surveying grounded in a massive 'dequalification' of this poetic character. European determinations of space, he argued:

> are responsible for an initial dequalification of any landscape they take over. Surveying rests on this initial dequalification, space being reduced to 'points' which are all alike. The best method of surveying is by beacons – 'points of light' – on hilltops at night, when the qualities of intervening landscape are totally blotted out. (Littlejohn 1967: 335)

In the earlier view the Holy Land is the *axis mundi*, a cosmological and cosmogonic center in which, as Mary W. Helms remarked, to travel through space is also to travel through time back to *illud tempus* (1988: 34ff).

In the sixteenth century – although, as Justin Stagl argued, the travellers' world remained 'qualitatively heterogeneous' (1990: 321) – major transformations are apparent.[10] Chief among these is the development of a 'formal doctrine of travel' by Zwinger, Turler, Pyrckmair and Blotius, all working under the influence of Ramist methodology.[11] Joan-Pau Rubies Mirabet has described this process as 'teaching the eye to see' (n.d.). As Stagal noted the terms they used to described their doctrine – *ars apodemica* or *prudentia peregrinandi* – remainded in use until the eighteenth century (the 'art of travel', *Reisekunst* – 1990: 316), and survived until the end of the twentieth century in the form of anthropological fieldwork, 'the most archaic scientific method [then] in existence' (1990: 328).

Stagl studied 300 treatises – manuals for proper and correct travel – published during the sixteenth century and it is clear that their advice mirrors the more general process described by Heidegger in which:

> to represent [*vor-stellen*] means to bring what is present at hand [*das Vorhandene*] before oneself as something standing over against, to relate it to oneself, to the one representing it, and to force it back into this relationship to oneself as the normative realm. (1977: 131)

Thus the new 'true' travelling (*peregrinari*) was privileged over the older 'aimless and useless rambling (*vagari*).' Movement through space was now justified in terms of education and 'useful knowledge' although Stagl suggested that the methodizers were also arguing against those who associated travel with pilgrimage (which had been discouraged during the Reformation) and rising states eager to restrict the freedoms of their subjects (1990: 317).

The *ars apodemica* provided instructions on what and how to observe, on the posing of suitable questions and the recording and evaluation of information (1990: 320). They urged the traveller to 'ascertain interesting facts from everywhere and everyone':

> he should communicate with people of all estates and trades . . . and be as persistent as possible without becoming irksome or suspicious to them. There is . . . advice on how to enter into conversation, to pose questions and to elicit information from people, without them noticing it. (1990: 320)

In addition to systematizing the method of encounter with other societies, the manuals also gave advice – some of it based on earlier traditions (1990: 318), concerning behavior at the end of one's travels:

> After returning the traveller should resume his native dress and customs, not show off with foreign expressions or despise old friends, but should remain in epistolary contact with his new won friends abroad. (1990: 318)

What was clear here is *the importance of the journey* – of a movement through space – regulated by the appropriate rules in which the full, embodied traveller could practice his 'art.' Although this is emphasized even more after 1800, even in this early empiricism Stagl noted that great importance was placed on the relative invisibility of the traveller to those among whom he travelled. Appropriate native dress and disguise were recommended which did not arouse the suspicions of local peoples. In the later period of rigidly methodized travel one can perhaps detect the intensification of the perspectival voyage and the subjectivity (predicated precisely on the requirements of *objectivity*) it engendered. As Heidegger wrote:

> Science as research is an absolutely necessary form of this establishing of self in the world; it is one of the pathways upon which the modern age rages to fulfillment of its essence with a velocity unknown to its participants. (1977: 135)

World as picture, and simulation

Writing in 1887, Sir John Lubbock (author of *The Origin of Civilization and the Primitive Condition of Man* etc.) claimed in an address on the subject of 'The Pleasures of Travel' that:

> We may have read the most vivid and accurate description, we may have pored over maps and plans and pictures, and yet the reality will burst on us like a revelation . . . like everyone else, I had read descriptions of the Pyramids. Their form is simplicity itself. I do not know that I could put into words any characteristic of the original for which I was not prepared. It was not that they were larger; it was not that they differed in form, in color, or situation. And yet, the moment I saw them, I felt that my previous impression had been but a faint shadow of the reality. The actual sight seemed to give life to the idea. (1887: 117–8)

This claim is interesting only because it runs so directly counter to the consensus of most pre-Virtuality travelers that the places to which they traveled could only really be judged in terms of their simulatory reference to the expectations created by 'maps and plans and pictures'.[12] Lubbock's stance is however only provisional and temporary, for having praised 'reality' over its faint representational shadows, he then also acknowledged that representations of other places could 'help us to see much more than we should perhaps perceive for ourselves' (1887: 119):

> It may even be doubted whether some persons do not derive a more correct impression from a good drawing or description, which brings out the salient points, than they would from actual but unaided, inspection. The idea may gain in accuracy, in character, and even in detail, more than it misses in vividness. (1887: 119)

The manner in which European travelers to the Middle East mediated the real by means of simulation has been brilliantly discussed by Timothy Mitchell. Marx noted that man 'raises his structure in imagination before he erects it in reality' (cited in Mitchell 1988: 21) and Mitchell – as also Linda Nochlin has done – unearths ample evidence of this in travelers' responses to the Orient – what was, ostensibly, an external reality – the 'great signified.' As products of a representational world they were unable to grasp the real except when conceived as an exhibition, a painting, a photograph – 'a picture world set apart from its observer' (1988: 23).[13]

One means to this was the adoption of what was termed the 'point of view,' 'a position set apart and outside'. Lane for instance sketched Cairo and its suburbs in 1825 from the summit of a large hill used for military telegraphing, echoing the advice given in 1642 by James Howell that upon encountering a strange city the true traveler should ascend the highest summit in order to 'gain a general impression of the place and to make a quick drawing' (Stagl 1990: 320).[14]

This clearly exemplifies that Cartesian need to see things as 'a whole', as a picture, but this adoption of the bird's eye view also suggests a striving by the imperial cartographer for a divine omniscience, for a 'point of view' which, prior to the invention of the aeroplane could only have been available to divine vision.[15]

Within twentieth-century non-Virtual anthropology what were almost literally birds'-eye views were surprisingly common. Evans-Pritchard credits the Royal Air Force for an air view of the Nuong village in his classic monograph *The Nuer* (1940: plate XVI). Edmund Leach used a similar view to illustrate 'the surroundings of Pul Eliya' in his 1961 monograph *Pul Eliya* (1961: plate 1). Marcel Griaule developed a well-known penchant for aerial photography which he himself ascribed to his experiences during his first job as an air force navigator. 'Perhaps,' he wrote, 'it is a quirk acquired in military aircraft, but I always resent having to explore unknown terrain on foot. Seen from high in the air, a district holds few secrets' (cited by Clifford 1988: 68).

The rise of the frame

A once fashionable but now little read French philosopher named Jacques Derrida once posed this question:

> imagine the damage caused by a theft which robbed you only of your frames, or rather of their joints, and of any possibility of reframing your valuables or your art-objects. (Derrida 1987: 18)

From the perspective of 2029 it is quite clear that Virtual Reality performed such a theft of our own frames – of our 'World Picture – but the thirty thousand years between Lascaux and the Global Cyberspace Foundation of the twenty-first century has to be accounted for. We need to account, in other words, for the rise of the 'frame' which transformed earlier virtual realities into mere 'pictures' for so long.

To raise the possibility of the 'picture' is to presuppose a frame. The nature of this frame is necessarily problematic, poised as it is between the inside and the outside. John Berger (1972), an art historian-turned-professional French peasant of the last century, once made a convincing argument concerning the historical role of the frame and the migration of value in European painting. What Berger also stressed – and subsequent events demonstrated the veracity of this – is that the values defined within painting came also to determine the wider philosophical and political values of society.

What Berger also noted was the manner in which the oil painting permitted the 'imitation' of nature in a manner that convinced and pleased the Cartesian mind. Within this system of Cartesian perspectivalism, 'nature was predestined for man's use and was the ideal object of his observation' (1972: 214). For Berger, as for Heidegger and also Lévi-Strauss,[16] European techniques of representation as exemplified and moulded by oil-painting, 'refer to the experience of taking possession': Berger stressed the importance of the ability of oil paint to 'render the tangibility, the texture, the luster, the solidity of what it depicts' (1972: 88), and his argument concerning the depiction of mobile wealth in Holbein's *The Ambassadors* (1533) is still remembered. Within this representational appropriation, however, value was also articulated with reference to the frame. Berger noted how the positioning of the owner-spectator changed. Whereas in earlier paintings the commissioner of a crucifixion or nativity might have been depicted *within* 'standing at the foot of the cross or kneeling around the crib' (1972: 216), this tactic was later made redundant 'because physical ownership of the painting guaranteed their immanent presence within it.'

Within anthropology over the last several centuries there have clearly been similar migrations of value and – until recently – a parallel reliance on the frame. 'Strangeness' and 'difference' – anthropology's signifiers of value – have been increasingly pushed to the margins – where, although they continued to play the same indispensable and constitutive role, their strategic function had changed. Within twentieth-century ethnography 'remoteness' (inexplicability/strangeness) was not permitted *within* the frame. The anthropologist tabooed strangeness (the value created by 'difference') within the frame – for it was also a part of his professional competence to make the 'strange' familiar and coherent. But, located now along the edges of his account – conceptualized in terms of the 'remoteness' of the place or society described through the journeyings to and fro – the value of strangeness was permitted, indeed required.[17]

A Poussin and Claude in every back-yard

Rheingold observed that the first teledildonic experiences would be *communications* with other senso-sheath wearers (over the telephone) rather than experiences with artificial bodies. The frequency of such communications – as we now know – altered previous estimates of value and: 'the physical commingling of genital sensations [came] to be regarded as a less intimate act than the sharing of the data structures of [our] innermost self-representations' (Rheingold 1991: 352).

Further fundamental shifts and transformations of notions of value rooted in an earlier scarcity also occurred. Lévi-Strauss once argued that both Poussin and the Impressionists projected the 'same amount of beauty' (Lévi-Strauss 1969: 136) onto their subjects.[18] The difference between them was that the access to the subjects depicted had changed:

> The ennoblement and aesthetic promotion of suburban landscapes can be explained perhaps by the fact that they, too, are beautiful, although this had previously not been recognized, but it is mainly a consequence of the fact that the great landscapes which had inspired Poussin were no longer available. (Lévi-Strauss 1969: 136)

Cyberspace threw this process – the progressive beautification of the mundane – into reverse and triggered the decay of the once elegant. The democratization of the previously scarce instituted ever-receding foci of value as the landscapes of Claude and Poussin became common-place scrublands, despised because of their omnipresence.

Chorology in the brain

The incarnated viewer of the world as picture fused easily with a longer Western tradition of chorology through which a mythic cultural self-presence was instituted. From classical times onward *the travel account* has emerged from the geographic vanishing point to incarnate society and its individuals. Michel de Certeau identified the standard form of such accounts in Montaigne's essay 'On Cannibals':

> First comes the outbound journey: the search for the strange, which is presumed to be different [by] the place assigned it in the beginning by the discourse of culture. This *a priori* of difference, the postulate of the voyage, results in a rhetoric of distance in travel accounts. It is illustrated by a series of surprises and intervals (Monsters, storms, lapses of time, etc.) which at the same time substantiate the alterity of the savage, and empower the text to speak from elsewhere and command belief. (de Certeau 1986: 69)

From the perspective of 2029 it is easy to see to see the Journey Programs of cyberspace as the culmination of two millennia of human traveling. From its origins as a sacred journey, travel has come – via various epochs that have moulded it differently – finally in this age of cyberspace to reside in the brain – a neurological chorology. As Paul Virilio wrote many years ago, 'from now on everything will happen without our even moving, without us even having to set out' (Virilio 1989: 112). We are by now used to this proposition, we have spent the last two decades never having even to set out. As Virilio noted we have replaced the ascendancy granted in the nineteenth and twentieth centuries to distance/time with the ascendence of the '*distance/speed* of the electronic picture factories' (Virilio 1989: 112).[19]

But this *intensification* and *culmination* is in no sense grounded in a continuity with these earlier modes. Rather this intensification and increasing speed of the journey has involved a transformation. Exceeding the speed of light, for instance, cannot be comprehended simply in terms of its supercession of speeds less than that of light. There is *a going beyond, a rupture* which must be comprehended. The bullock cart, the steam railway engine and the head-mounted display of the visual component of cyberspace from one perspective suggest obvious affinities, but beyond these lie more significant chasms. Such was understood by Virilio even in the late 1980s:

> If automobile vehicles, that is, all air, land and sea vehicles are today also less 'riding animals' than *frames* in the optician's sense, then it is because the self-propelled vehicle is becoming less and less a vector of change in physical location than a means of representation, the channel for an increasingly rapid optical effect of the surrounding space. (Virilio 1989: 114)

Space has been detemporalized.[20] In other words, travels in cyberspace have permitted 'cannibals' without the 'monsters, storms, [and] lapses of time.'

The frame of the Other

De Certeau suggested that the Other always figured as a form of vanishing point whose converging lines were the 'accounts of the outward journey and the return' (de Certeau 1986: 69). The journey – the setting off and returning – also constituted the historical 'frame' of the 'picture of a new body' (the Other) which assured the 'strangeness of the picture'. The parallels with Heidegger's disparaged World as Picture are striking, for here also distance marks the separation of the world from those who should be living *within* it. Travel accounts support this view of the *journey* as the 'frame' of the other. This is made most explicitly clear in Augustus Klein's *Among the Gods* (1895) in which he refers to India and Ceylon as 'the picture' and the Indian Ocean as 'the frame.'

It is now forty years since UNC researchers commenced their work on the possibility of ultrasound images accessed via a head-mounted display which would allow them to 'see directly into the living tissue' (Rheingold 1991: 26). Today we are all familiar with the huge range of such diagnostic technology, but there is a more important metaphorical sense in which inside and outside came to lose their fixity. In the short term Virtual Reality dramatically intensified the metropolitan world's oppression of, and indifference to, the 'Third World' but it contained within it the seeds of its own destruction. Virtuality was uniquely different from other forms of representation inasmuch as it is 'unframed'. Unlike paintings, photographs, books, fruit machines, and the cinema, it had no apparent frame, or window. Now, at least, the experiencer of Virtuality does not spectate, but is immersed within, and part of, the visual and experiential field. There had of course been Virtuality journey programs which permitted the traveler to remain an invisible presence, but these were destroyed by the Pandora virus which by a series of complex and brilliant mutations transformed invisible travelers into the visible.[21]

With no window, with no 'framing,' what were the consequences for cyberanthropology to be? These possibilities were anticipated in Flaubert's letter from Cairo of 1850:

> What can I say about it? What can I write you? As yet I am scarcely over the initial bedazzlement . . . each detail reaches out to grip you; it pinches you; and the more you concentrate on it the less you grasp the whole. Then gradually all this becomes harmonious and the pieces fall into place of themselves, in accordance with the laws of perspective. But the first days, by God it was such a bewildering chaos of colors. (cited by Mitchell 1988: 21)

Flaubert's initial disorientation is manifested as 'an absence of pictorial order'. During the period of bedazzlement, 'each detail reaches out to grip you; it pinches you . . .' (1988: 21). As Timothy Mitchell noted, there is insufficient distance between the viewer and the view and hence the eyes are 'reduced' to organs of touch – 'without a separation of the self from a picture moreover it becomes impossible to grasp "the whole"' (1988: 22).

Cultural distance without distance: or arriving without setting out

Mark Poster had also presciently suggested some of the possible consequences of this in 1990 when he observed that electronic writing[22] would 'destabilize the figure of the subject ... the Cartesian subject who stands outside the world of objects in a position that enables certain knowledges of an opposing world of objects ...' (1990: 99). Similarly Howard Rheingold had hinted – via a brief New Age excursion into anthropology – at such a possibility and argued that Virtuality might well become the avenue to a recasting of the familiar buccaneering cogito. The thirty-thousand-year-old 'anamorphic' paintings in the Lascaux cave may have strived for Virtuality effects through the use of 'special theatrical rituals and light-and-sound spectacles intended to alter ... consciousness' (Rheingold 1991: 382).

Until representation through cyberspace became the dominant mode however, modern Western selves – as we have already suggested – consolidated themselves through an accumulating externality, an ineluctable accretion of possessions – presences – which effaced (through a displacement) the absence of the very self they purported to reflect.[23] Gayatri Spivak suggested that the self was 'always a production rather than [a] ground' (1987: 212), a conclusion that subsequent events amply demonstrated. To recapitulate, among the most important of these possessions for the nineteenth- and twentieth-century bourgeoisie (the most secured selves in history) were representations such as oil paintings (Berger 1972) and the immanent historicity produced by photographic portraiture (Trachtenberg 1985). Critical to these representations' complicity in producing Western selves was their 'framing,' their ability to set themselves up as bounded objects against which the self produces its illusion of presence. The frame secures the converging lines which institute the vanishing point which is 'the anchor of a system which *incarnates* the viewer.'[24]

As the visual and haptic qualities of the medium were refined virtuality quickly came to exceed the reality-effects offered by the non-virtual world. The simulation of, for instance, the 1990s was always largely metaphoric, a world hallucinated by French professors slumped in front of their American TV sets with two six-packs. For most people it was a world whose perceptual seamlessness was always liable to be ruptured by the shaking hand that holds the photograph, the cough from the back row of the cinema auditorium. The imaginary worlds of Virtual Reality by contrast are only 'framed' through physical insertion into the 'senso-sheath' and the ability to finance it. In the 1990s, many people were arguing that cyberspace would always be 'framed' socioeconomically, technologically, and culturally.[25] This was in many ways true, but phenomenologically the Virtual and non-Virtual were blurred. The technology of VR must of course always be consciously entered but once within – especially for the addict – *there is no turning back, no turning away.* There is none of the distance engineered by the frame.

The frame that Virtual Reality lost, however, was not only visual. It is clear that cyberspace worlds could no longer be viewed as pictures. But in addition a *temporal* frame was also shed. The frame imposed by the voyage, the journey, was also lost. In Paul Virilio's words the traveler arrived *without having to set out* and one result of this was that cultural distance came to shed its temporal aspects. Retrospectively we can see that the *temporal* framing provided by the journey was crucial to the framing of certain Western subjectivities.

Subjective identities themselves were grounded in narrational journeys as the sagacious neurologist Oliver Sacks observed. Writing in the 1980s about a patient who criss-crossed hundreds of different personae in his daily existence, Sacks mused thus:

> Each of us is a singular narrative, which is constructed, continually, unconsciously, by, through, and in us ... We must recollect ourselves, recollect the inner drama, the narrative, of ourselves. A man *needs* such a narrative, a continuous inner narrative, to maintain his identity, his self. (1986: 105–106)

Since 1839 ever increasing numbers of the world's population have come to construct these narratives and recollect their histories through photographic reproductions, isolated pictures organized in frames that were usually called 'albums'. Sack's patient who lacked such 'a quiet, continuous, inner narrative' was driven to a 'narrational frenzy.' Our experience of Virtual Reality has shown that this causality can operate in reverse – the phantom-worlds and pseudo-worlds of cyberspace could also destroy (have also destroyed) the quiet inner narrative on which the nineteenth- and twentieth-century self was built.

In only forty years Virtual reality made 'distance-travel' a reality while 'time-travel' has remained the domain of science fiction. Some of today's distance-loop paradoxes are as remarkable as any time-loop paradox dreamt up by Ray Bradbury or the makers of *Terminator*. The collapse of space and distance has been almost as cognitively disorienting as the possibilities offered by the time-loop paradox. The removal of the outward journey and the return – the removal of the frame and the incarnated viewer – have indeed threatened the 'entire gridding of the system [which presupposes] that there is a *place* for every *figure*' (de Certeau 1986: 70). Whereas time travel blurred parent and child, Virtual Reality's distance-travel has blurred self and other.

Conclusion

How surprised should we be by the remarkable changes of recent years? Is it a consequence of certain formal features of this new means of representation (and specifically its detemporalization of space) or is it the result of some as yet completely unknowable factor such as the diffusion of the Pandora computer virus?

The contemplative thinker could perhaps have known what would happen without any direct experience of Virtual Reality. Heidegger's litany of complaints about Cartesian perspectivalism concluded with a prescient glimpse of future change. He noted that science as research effloresced in what he termed 'giganticism':

> everywhere and in the most varied forms and disguises the gigantic is making its appearance. In so doing, it evidences itself simultaneously in the tendency toward the increasingly small. We have only to think of numbers in atomic physics. The gigantic

presses forward in a form that actually seems to make it disappear – in the annihilation of great distances by the airplane, in the setting before us of foreign and remote worlds in their everydayness, which is produced at random through radio by the flick of the hand. (1977: 135)

Today the flick of a hand produces worlds not merely conjured by the crackle and static of international radio stations, but seamless three-dimensional worlds peopled by casts of millions and all characterized by an absorbing and hypnotic intensity.

Yet for Heidegger this giganticism – at first apparently the triumph of the world as picture – also came to exist as the 'invisible shadow that is cast around all things everywhere when man has been transformed into *subjectum* and the world as picture' (*ibid*). This 'shadow,' Heidegger suggested, 'extends itself into a space withdrawn from representation' [i.e., from the world as picture] and – he concluded somewhat mysteriously – it may lead to 'genuine reflection' (1977: 136). Heidegger's obscurity however is dispensable for there are other and much clearer prefigurations of what has happened over the last forty years.

The first of these dates from 1764, when Oliver Goldsmith published his poem *The Traveler; Or, A Prospect of Society* in which the vantage point, or the 'point of view,' meets its nemesis. In the poem a traveler sits among the Alps and looks down upon the various countries whose boundaries meet underneath his gaze. As he looks down 'where an hundred realms appear' (line 34) he reaches the relativistic conclusion presaged in his dedicatory foreword that 'there may be equal happiness in states that are differently governed from our own; that every state has a particular principle of happiness' (1906: 4). Denied a vantage point over a singular terrain the wanderer is faced with a plural multiplicity of worlds whose equal merits he is forced to recognize.[26]

It was precisely this fecundity of worlds and perspectives which would later engage theorists considering the role of narrative in cinematic film. Anticipating the displacements created by Virtual Reality, film theorists – such as Stephen Heath – highlighted the role of narrative in assuring 'that the subject positioning brought about by perspective will not be undone by the movement of the film' (Carroll 1988: 162). It was 'narrative' and its ability to restore equilibrium which subdued and restored the disruptions created by character and camera movement and editing. It sutured the potentially destabilizing effects of the contradictory lines of force unleashed by devices such as shot/reverse shot sequences.[27] Narrative restores the frame:

. . . the spectator is situated by a constant renewal of perspective. Thus organized [by narrative], the frame imposes a coherence and continuity, forestalling the risk of textual and subjective chaos. Indeed 'the narrative is the very triumph of framing'. (Lapsey and Westlake 1988: 139, referring to Stephen Heath)

However, in much the same way that a vantage point which permits more than one simultaneous 'prospect' negates the subject's power and control, it has also become evident that the absence of narrative within Virtual Reality has also undercut the position of the subject. Located within cyberspace this subject is not only a participant – rather than simply a spectator poised at the apex of converging lines of vision – but has also lost the temporal frame. In previous epochs this was provided by the narrative of the journey referred to by de Certeau. Today it is gone. We always arrive without setting out, and we are catapulated – disorganized, unpositioned, and *split*.

The second prefiguration of the changes which have occurred recently has an even earlier point of reference. Forty years ago when I gave the talk on this subject in London[28] any one of the audience could have strolled down to the newly constructed Sainsbury wing of what was then the National Gallery and viewed Hans Holbein the Younger's *The Ambassadors*. Now of course they could not because it's in the Musuem of New World Art at Delhi. Had they done so, they would have seen the foretelling of everything I have just said. As Homi Bhabha wrote at the time – partly paraphrasing Jacques Lacan:

> The two still figures stand at the center of their world, surrounded by the accoutrements of *vanitas* – a globe, a lute, books and compasses, unfolding wealth [and, he might have added, the signs of travel]. They also stand at that moment of temporal instantaneity where the Cartesian subject emerges – in the same figural time – as the subjectifying relation of geometrical perspective . . . the *depth* of the image of identity. But off-center, in the foreground [. . .] there is a flat spherical object, obliquely angled . . . the disc is a skull; the reminder (and remainder) of death, that makes visible, nothing more than the alienation of the subject, the *anamorphic ghost*. (1987: 7; *emphasis added*)

Perhaps then in addition to Rheingold's list of the cave paintings at Lascaux, and the Pueblo *kivas* – the pre-technological manifestations of Virtual Reality – we need to add Holbein's anamorphic skull. This signifies the death of the subject, and also the death of the traveler. What we have learned is that if I can arrive without ever having to set out, that self-same 'I' ceases to exist.

Notes

1 A trend which had been forecast as early as the 1980s in the film *Total Recall* (Dir. James Cameron, based on the Philip K. Dick novel).

2 VR's capability as a sedative or hallucinogen was much touted in the early 1990s. Rheingold suggested that 'in a world of tens of billions of people, perhaps cyberspace is a better place to keep most of the population relatively happy, most of the time.' A global system which was not prepared to provide food for half its population proved unwilling, however, to provide (greatly more expensive) cyberspace.

3 This was given on 17th February 1992 in the Department of Anthropology and Sociology, School of Oriental and African Studies, University of London. I would like to thank all surviving seminar participants, particularly Kit Davis, Tim Screech, David Parkin, Claire Harris & Lisa Croll for their kind suggestions. I am also grateful to Chris Wright, Lucien Taylor and an anonymous reviewer for *VAR* for their help.

4 The subject of Howard Rheingold's fascinating book, *Virtual Reality*, Summit Books, $22.05; Secker and Warburg, £16.99.

5 'Embedded in the inner surface of the suit, using a technology that does not exist, is an array of intelligent sensor-effectors – a mesh of tiny tactile vibrators of varying degrees of hardness, hundreds of them per square inch, that can receive and transmit a realistic sense of tactile presence' (Rheingold 1991).

6 This was a project funded by the British Department of Employment in West Denton School, Newcastle upon Tyne but at that stage involved merely the use of desktop color monitors rather than VPL Eyephones and Datagloves (*The Guardian*, 16.1.92.: 35).

7 A term first coined in 1974 by Theodor Nelson to describe How Wachspes's patent for the conversion of sound into tactile sensation (Rheingold 1991: 345).

8 Although just as many analyses stressed its origins as a form of military technology. Had a talk such as this been written in the 1990s without the benefit of hindsight it would very probably have been attacked for its effacement of its own 'logistics of perception' to use Virilio's term. Other critics – unable to see what has actually occurred – might have derided its mock-apocalyptic tones as the kind of futurology which might well have revealed itself to be McLuhanism part-two – a critical frenzy which demonstrated the critic's distance from the quotidian inertias which always scupper the logics of 'innovation.' However, had they said any of these things, time would have proved them wrong.

9 Baudrillard argued that it was a 'televisual subterfuge', a simulacral contest in which 'what was at stake was not political or territorial domination, but the very status of war itself, its meaning and its future' (Gilbert Adair, 'Fading of the Looking Glass War', *The Guardian* 11.7.91).

10 I do not wish to overstate this however. Stagl notes that although the *ars apodemica* represented the 'secularization of pilgrimage,' they retained certain aspects of an earlier genre – directions for pilgrims (1990: 317). Readers are recommended to refer to Stagl's original and important research.

11 According to the 'practically oriented universal method for all arts and sciences' (Stagl 1990: 312) developed by Petrus Ramus (1515–1572).

12 Perhaps some sense of the process of textual recuperation and generation centered on travel is given by Boswell's remark that he made journeys only 'in order to keep a journal' (Carter 1988: 47). There are clearly also profound and formative links between the concepts of travel, writing and authorship which (to coin a phrase) *space* does not permit me to *pursue*.

13 Mitchell also notes that Lane's *Account of the Manners and Customs of the Modern Egyptians* (1935) an early piece of written para-ethnography, had originated as part of a larger visual record of Egypt made with a *camera lucida*. This was abandoned after Lane was unable to find a publisher who could reproduce 'the minute and mechanical accuracy of the drawings' (1988: 23).

14 Mitchell cites Herman Melville's complaint during a visit to Constantinople: 'Perfect labyrinth. Narrow. Close, shut in. If one could but get *up* aloft' (1988: 32).

15 However, it should be noted that in the sixteenth century Zwinger had described travel by air, referring to the 'examples of Daedalus and of the Angels' (Stagl 1990: 316).

16 Berger wrote that 'significantly enough' it was an anthropologist (Lévi-Strauss) who first came close to recognizing the 'analogy between *possessing* and the way of seeing which is incorporated in oil painting' (1972: 83).

17 See Ardener (1987) for the view that his 'remoteness' was not an exclusive function of geography.

18 Sasha Markovic has pointed out the fallaciousness of the argument here, given that Poussin's landscapes were 'imaginary.' Lévi-Strauss would have done better to cite an artist such as Constable.

19 Similarly, Mark Poster when anticipating the decentering effects of 'computer writing' in 1990, stressed its antecedents in the forms of earlier conventional transport systems which 'increase the speed with which bodies move in space.' (1990: 128). This

however suggests a false continuity, a continuum which the ruptures of recent years have invalidated. See below.

20 See Soja (1989).

21 Pandora's roots can be traced as far back as the Michelangelo virus which corrupted vast amounts of corporate software on 6th March 1992.

22 He was referring here to word-processor inscribed text.

23 See C. B. MacPherson (1962) and James Clifford's interesting use of this (1988: 217–18).

24 Norman Bryson (1983: 106) cited by Rotman (1987: 14). Bryson makes an evolutionary distinction between the 'glance' and the 'gaze,' the 'vanishing point' and the 'punctum,' and a corporeal and disembodied spectator (see Rotman 1987: 32–33). Among the various representational modalities discussed by Rotman, it is perspectival painting coded by the vanishing point ('founded on a fiction of a framed portion of nature, a detached fragment of some prior visual "reality," being represented with truth and accuracy' 1987: 40) which most clearly parallels the realist strategies of travel and anthropological accounts. The distinctions drawn by commentators such as Bryson, Alpers and Jay between various forms of subject positioning and perspective contingent on specific historical moments and movements in Western artistic and other representation do not, I believe, overturn the general trajectory and hegemony of Cartesian perspectivalism described in this talk.

25 One of these people was Lucien Taylor, the then editor of *V.A.R.* (personal communication).

26 Travelling without ever reaching his goal – his vanishing point:

> My fortune leads to traverse realms alone,
> And find no spot of all the world my own.

– the wanderer desolately ascends the mountains only to discover the reasons for his own rootlessness and lack of placed-ness – the misrule of his own country and the migration of villagers:

> Forced from their homes, a melancholy train.
> To traverse climes beyond the western main.

27 A pair of shots, each of which represents 'from a more or less oblique angle, one endpoint of an imaginary 180-degree line running through the scenographic space' (Jean Pierre Oudart, cited by Carroll 1988: 183).

28 See note 4.

References

Ardener, Edwin. 1987: 'Remote Areas: Some Theoretical Considerations.' In *Anthropology at Home*, ed. A. Jackson, ASA 25. London: Tavistock.

—— 1972: *Ways of Seeing*. Harmondsworth: Penguin Books and BBC Publishing.

Bhabha, Homi. 1987: 'Interrogating Identity.' In *Identity*, ed. Homi Bhabha, London: Institute of Contemporary Arts.

Brown, Paul. 1991: 'Metamedia and Cyberspace: Advanced Computers and the Future of Art.' In *Culture and Technology in the Late Twentieth Century*, ed. Philip Hayward. London: John Libbey.

Bryson, Norman. 1983: *Vision and Painting: The Logic of the Gaze*. London: Macmillan.

Buck-Morss, Susan. 1989: *The Dialectics of Seeing: Walter Benjamin and the Arcades Project*. Cambridge, Mass.: MIT Press.

Carroll, Noel. 1988: *Mystifying Movies: Fads and Fallacies in Contemporary Film Theory*. New York: Columbia University Press.

Carter, Paul. 1988: 'Invisible Journey: Exploration and Photography in Australia, 1839–1889.' In *Island in the Stream: Myths of Place in Australian Culture*, ed. Paul Foss. Leichhardt (NSW): Pluto Press.

Clifford, James. 1988: 'Tell Me About Your Trip: Michel Leiris.' In his *The Predicament of Culture: Twentieth-Century Ethnography, Literature and Art*. Cambridge: Harvard University Press.

Cornwell, Regina. 1991: 'Where is the Window? Virtual Reality Technologies Now.' *Artscribe* (January): 52–54.

Cuignon, Charles. 1983: *Heidegger and the Problem of Knowledge*. Indianapolis: Hackett Publishing Co.

D'Amato, Brian. 1992: 'The Last Medium: the Virtues of Virtual Reality.' *Flash Art* XXV, 162 (Jan–Feb): 96–98.

De Certeau, Michel. 1986: 'Montaigne's "Of Cannibals": The savage "I".' In his *Heterologies: Discourse on the Other*. Manchester: Manchester University Press.

Derrida, Jacques. 1987: *The Truth in Painting*. Chicago: Chicago University Press.

Evans-Pritchard, Edward. 1940: *The Nuer*. Oxford: Clarendon Press.

Gibson, William. 1986: *Neuromancer*. London: Grafton (first published 1984).

Goldsmith, Oliver. 1906: *The Traveler and The Deserted Village*. Ed. W. Murison. Cambridge: Cambridge University Press.

Hamilton, Norah Rowan. 1915: *Through Wonderful India and Beyond*. London: Holden and Hardingham.

Heidegger, Martin. 1977: 'The Age of the World Picture.' In his *The Question Concerning Technology and Other Essays*. New York: Harper Torchbooks.

Jay, Martin. 1988. 'The Scopic Regimes of Modernity.' In *Vision and Visuality*, ed. Hal Foster. Seattle: Dia Press.

Klein, Augustus. 1895. *Among the Gods: Scenes of India – With Legends Along the Way*. Edinburgh: William Blackwood.

Lapsey, R. & M. Westlake. 1988. *Film Theory: An Introduction*. Manchester: Manchester University Press.

Leach, Edmund R. 1961. *Pul Eliya: A Village in Ceylon*. Cambridge: Cambridge University Press.

Lear, Edward. 1953. *Indian Journal: Watercolours and Extracts from the Diary of Edward Lear, 1873–1875*. Ed. R. Murphy. London: Jarrolds.

Lévi-Strauss, Claude & Charbonnier, Georges. 1969. *Conversations with Claude Lévi-Strauss*. London: Cape.

Littlejohn, Kenneth. 1967. 'The Temne House.' In *Myth and Cosmos*, ed. J. Middleton. New York: Natural History Press.

Lubbock, John. 1887. 'The Pleasures of Travel.' In his *The Pleasures of Life*. London: Macmillan.

Lucas, E. V. 1924. *Roving East and Roving West*. London: Methuen.

Macpherson, C. B. 1962. *The Political Theory of Possessive Individualism*. Oxford: Oxford University Press.

Malinowski, Bronislaw. 1922. *Argonauts of the Western Pacific*. London: Routledge.

Metcalf, Thomas. 1984. 'Architecture and the Representation of Empire: India 1860–1910.' *Representations* 6: 37–65.

Mitchell, Timothy. 1988. *Colonizing Egypt*. Cambridge: Cambridge University Press.

Muller, Max. 1919. *India: What Can it Teach Us?* London: Longmans, Green and Co.

Pasolini, Pier Paolo. 1984. *The Scent of India*. London: Olive Press.

Penley, Constance. 1989. 'Time Travel, Primal Scene, and the Critical Dystopia.' In her *The Future of an Illusion: Film, Feminism and Psychoanalysis*. London: Routledge.

Poignant, Roslyn. 1980. *Observers of Man*. London: Royal Anthropological Institute.

Poster, Mark. 1990. *The Mode of Information: Poststructuralism and Social Context*. Oxford: Polity.

Rheingold, Howard. 1991. *Virtual Reality*. New York: Summit Books.

Ronaldshay, Earl of. 1924. *India: A Bird's Eye View*. London: Constable.

Rubies i Mirabet, Joan-Pau. n.d. 'Teaching the Eye to See.' Unpublished ms.

Sacks, Oliver. 1986. *The Man Who Mistook His Wife for a Hat*. London: Picador.

Soja, Edward. 1989. *Postmodern Geographies: The Reassertion of Space in Critical Social Theory*. London: Verso.

Somerville, Henry B. T. 1928. 'Surveying in the South Seas.' Unpublished ms. [Photographic Collection. Royal Anthropological Institute, London].

Spivak, Gayatri Chakravorty. 1987. 'Subaltern Studies: Deconstructing Historiography.' In her *In Other Worlds: Essays in Cultural Politics*. London: Methuen.

Stagl, Justin. 1990. 'The Methodising of Travel in the 16th Century: A tale of three cities.' *History and Anthropology* 4(2): 303–338.

Trachtenberg, Alan. 1985. 'Albums of War: On Reading Civil War Photographs' *Representations* 9: 1–32.

Williams, Monier. 1879. *Modern India and the Indians*. London: Trubner's Oriental Series.

Virilio, Paul. 1989. 'The Last Vehicle.' In his *Looking Back on the End of the World*, ed. Dietmar Kamper & Christoph Wulf. NY: Semiotext(e).

SECTION SIX

POSTSCRIPT: THE POSSIBILITY OF A POLITICS OF PLACE

Editorial introduction

In this final section I want to retreat from the future gazing of the end of Section Five, and instead address some of the more immediate implications of the socio-spatial changes that have been described and analysed in the previous sections. I want to turn here specifically to questions about power and control in a world marked by heterogeneity and difference, and by gross inequalities. Who is it who has power in the formation of places from spaces, given the multiple ways in which this occurs, the variety in the bases of place formation and their fluidity and flexibility? How are places contested and fought over? How are they protected against external threat? What is at stake in these contestations? How might we develop a different definition of place, one that is not based on fixed and excluding boundaries? Is a relational sense of place possible? A non-hierarchical sense of place without fixed exclusionary boundaries, as Iris Marion Young asks; even a progressive sense of place, as Doreen Massey has suggested? The particular paper in which Massey addresses this question has been widely reprinted. As it is also included in one of the companion readers in this series (Daniels, S. and Lee, R. 1996: *Exploring Human Geography: A Reader*), I have not included it here.

These questions are clearly extremely important; indeed even beginning to approach possible answers is difficult. In the pages left to me, I have chosen two articles that begin to sketch in the outline to some answers. The particular criticism often levelled at postmodern thought and at notions of an identity politics is that it leads to relativism, to an inability to judge between arguments and actions. If, it is argued, we abandon the notion of a singular and universal set of moral principles, then we evacuate the moral high ground which enables certain arguments and acts to be judged as better than others, even correct or morally right, opposed to something else which is wrong. It seems to me, however, that it is a great leap from pointing out that the notions of rationality, objectivity and knowledge that were established in western enlightenment thought might exclude the points of view of, say, women, people of colour or imperial subjects, to sliding headlong into a morass of relativism from wherein it is impossible to take a position. As Judith Squires (1992) argued, limited or contextual principled positions are possible (and see also my article McDowell, 1996). How we make judgements and choose to act often depends on the circumstances.

In the modern/postmodern controversies, counter positions are fiercely debated. On the one hand, demands to hold on to old certainties (perhaps David Harvey's unwavering belief in Marxist dialectics and class politics is the most obvious example in geographic discourse) are opposed to the wholesale adoption of a postmodern scepticism of progressive or utopian projects on the other. While the 'old left' despairs of the lack of 'authenticity' in contemporary consumer spaces, the 'new left' (New Labour perhaps in the UK or Clinton democrats in the USA) celebrates style and the construction of identity through consumption. When weighing up these counter claims: for or against material reality, or the assertion that 'there is nothing outside the text', I find myself adopting the sort of pragmatic position adopted by many of the contributors to Judith Squires' collection *Principled Positions*. This pragmatic position is eloquently outlined by Kate Soper in the introduction to her wonderful book *Troubled Pleasures* (Soper, 1990; see also Soper, 1993).

It is important to bring together insights from both sides of that divide [between Marxism and postmodernism] even at the cost of being exposed to accusations of equivocation. As far as Marxism is concerned, in fact, I would say that any serious Marxist-influenced analysis of our times has long recognised the need to qualify the classic nineteenth-century visions of the socialist future and the means of arriving thereat [and even more in 1996 than in 1990 when Soper was writing], and in this sense subscribes to postmodernist critiques of classical Marxism, while refusing for that reason to settle for nihilist or escapist modes of political response. The fact that we have been made wise to a certain glibness in the solutions offered in the traditional left wing canon to the problems of economic inequality, national and racial tension, sexuality, morality and hedonism is not for this more self-critical Marxist approach a cause to give up on the very form of socialist rationality or to abandon any remedial efforts whatsoever. Equally, it seems to me, there are rather few who espouse the postmodernist position and are able to practise the modes of political dissociation that are the logical implication of some of their critiques. There is too much bodily pain and mental suffering already, and too much of it being heaped up for future generations, for anyone of average sensibility to find a genuine escape in the pleasurable flux of the actual (which is to say in a mode of living entirely freed from all forms of troubled conscience). Conversely there is still too much promise held out in the resources now at our command, both spiritual and material, to make the nihilist response seem anything but rather spurious (an excuse for inaction or a voguish bravado rather than a thought out philosophy of indifference to all values). A more authentic reaction is that of the 'troubled pleasure' perspective – a reaction, in other words, which is sensitive to the difficulties which dog all efforts at improvement without giving up on the task itself, and without confusing the critique of various forms of rationality, objectivity and power, with the triumph of irrationalism, relativism and impotence. (pp. 12–13)

The two pieces I have chosen to conclude this reader are, I believe, good examples of the sort of politics that might arise from Soper's notion of 'troubled pleasure' or my belief in a pragmatic politics based on context-dependent principled positions. The first is by **Doreen Massey** in which she defends her insistence on local action against critics who are anxious about the limited nature of locally based single action issues and the reactionary nature of the celebration of the local, immediate, and the particular (Harvey, 1989). Her paper is a particularly appropriate conclusion to this reader as in it she defines the reasons why the 'locality' programme was initiated in the 1980s in the UK. It arose from the very socio-spatial and political changes that have been addressed here: economic restructuring from an economy based in the main on manufacturing to one based on services, the differential impacts on communities and localities, the changing nature of gender relations, new forms of identity politics based on consumption rather than production, and the supposed decline of the working class. Massey's conceptual and practical defence of the importance of thinking and acting through place is an extremely important argument for the continuing significance of geographical differentiation and an inspiration for retheorising the links between self, identity and place.

The final chapter is by **Iris Marion Young** from Squires' (1993) edited book *Principled Positions*. Here Young draws out the problems and biases which she sees as inherent in traditional class-based politics and argues instead for a group-based politics based on the recognition and acceptance of difference. Although her focus is not explicitly geographic like Massey's, she is a strong adherent of 'a conception of self as socially constituted and embedded in particular communities' (p. 132 in original) and many, although not all, of these communities are place-based. Young's argument here is primarily theoretical. I have not included the second and third parts of her chapter where she illustrates her theoretical arguments through two case studies based on conflict in Eastern Europe and the position of Maori people in New Zealand, as these are outside the boundaries of the Anglo-American perspective here. Her arguments apply equally, however to the aspatial politics of gender and to a range of locationally based issues: to the politics of 'race' for example, and to 'eco-' and green movements in the USA and the UK, and to locally based single issue political movements. Her challenge to both segregationist and assimilationist strategies on the part of the excluded as well as the powerful take us back to the issues discussed in Sections Two and Four and provide a spur for radical rethinking about a progressive sense of place.

I believe that the sort of politics of difference that has grown out of contemporary changes in the relationships between place and identity is an optimistic sign of progress rather than a retreat from the solidarity of working-class politics. As Soper suggests it is possible to build a new form of politics which may be 'viewed as all the more progressive and utopian for bringing some newer forms of scepticism and irony on the old ideas of progress and utopia' (p. 13).

References and further reading

Harvey, D. 1989: *The condition of postmodernity*. Oxford: Blackwell.

Massey, D. 1991: Is a progressive sense of place possible? In Bird, J. et al. *Mapping the futures*. London: Routledge (originally in *Marxism Today* and reprinted in Barnes, T. and Gregory, D. (eds), *Reading human geography*. London: Arnold).

McDowell, L. 1996: Understanding diversity: the problems of/for theory. In Johnston, R., Taylor, P. and Watts, M. (eds), *Geographies of global change*. Oxford: Blackwell (reprinted in Barnes, T. and Gregory, D. (eds), *Reading human geography*. London: Arnold).

Soper, K. 1990: *Troubled pleasures: writings on politics, gender and hedonism*. London: Verso.

Soper, K. 1993: Postmodernism, subjectivity and the question of value. In Squires, J. (ed.). *Principled positions*. London: Lawrence and Wishart, 17–30.

Strathern, M. 1992: *After nature*. Cambridge: Cambridge University Press.

Yeatman, A. 1994: *Postmodern revisionings of the political*. London: Routledge.

Young, I. M. 1991: *Justice and the politics of difference*. Princeton, New Jersey: Princeton University Press.

22 Doreen Massey
The Political Place of Locality Studies

Excerpts from: *Environment and Planning A* **23**, 267–81 (1991)

Introduction: space, politics, and locality research

At a number of points in the rich debate about locality studies in the United Kingdom, various authors have made various assumptions about the *reasons* for pursuing this kind of research in the first place. It is difficult adequately to assess research without understanding its aims in the first place, both in order to have something to evaluate it against and because the objectives may themselves be open to evaluation. For another thing, this issue of the aims of locality studies links into a wider debate about our role as academics or intellectuals and the relationship of our work to current political issues and debates (Walker, 1989). The purpose of the present paper is simply to reflect upon some of the reasons why the programme of locality studies called the Changing Urban and Regional System (CURS) was first proposed and developed.[1]

One of the most striking things about the assumptions most often made by commentators about the reasons for the programme is that, although most of them come from people who would define themselves as being 'on the left', they almost never refer to politics, and more particularly to the political situation in which the issue of locality studies was being raised. If politics does enter the question, then it usually does so in one of two ways (and sometimes both at once). On the one hand it is assumed by some that Marxist theory or the mechanics of accumulation at a grand scale are always and everywhere politically 'OK' things to work on. On the other hand it is argued, or asserted, that studying 'the local' or 'place' is necessarily politically problematic. [Among the few exceptions is the interesting discussion by Jonas (1988).]

The idea for the focus on locality studies arose in the United Kingdom of the early 1980s. From the end of the 1960s there had been clear intimations that the economy at least, and maybe society more widely, was entering a period of significant change. There was an accelerating shift away from manufacturing, a noticeable increase in registered unemployment, a continuing transformation of the occupational structure, and so on. The major social changes which appeared to be heralded by these economic shifts also provoked political reflection. It was a set of processes which were further heightened by the events of the early 1980s. The debates about flexibilisation and 'post-Fordism' and the continued presence in power of a right-wing Conservative government, and the failure of the Labour Party to exercise any hold over the imaginations of the majority, reinforced the feeling that an era was at an end.

Although these debates did not take place primarily within human geography as an academic discipline, they related to it in a number of ways. For one thing, among the significant changes under way in British society, some of the most important ones were geographical. There was a spatial restructuring as an integral part of the social and economic. The economies of the big manufacturing cities went into severe decline. The bases of the heavy-industry regions were undermined. There was decentralisation of both population and employment from big cities outwards to more rural areas and, in some parts of the period, from core regions to the old industrial periphery. The increase in paid employment for women, and the shifts in balance between male and female employment, happened differentially across the country. The so-called high-technology industries, and the hugely expanding banking, finance, and professional services sectors transformed the South East region. More recently, there has been a noticeable, if spatially restricted, transformation of parts of once declining inner cities. Waterfronts everywhere are being revitalised into expensive housing and trendy offices; the Docklands, in London's eastern area, became for a while so much a symbol of the transformative impact of Mrs Thatcher's government that she began her 1987 election campaign there. Indeed, much of what was going on seemed to be about 'places' and their reconstitution in some way or another.

Moreover, and more urgently significant in a political sense for some, the organisational base of the left was being affected by spatial changes as well as by changes in the national economy and society (Lane, 1982; Massey, 1983; Massey and Miles, 1984). Perhaps more than anything else, the very fact that the national structural changes themselves involved a geographical restructuring meant that people in different parts of the country were experiencing highly contrasting shifts, and that even the trajectories of change (for example, in class structure) could be quite different in one place from in another. And, especially because it is not simply final outcomes but processes of change which are significant to people's experience of their world, this meant that the political implications of these 'structural changes' were likely also to be highly contrasting between one place and another. Moreover, this spatial variation was reinforced by the fact that people in different parts of the country had distinct traditions and resources to draw on in their interpretation of, and their response to, these changes.

It was also the case that a great deal of immediate politics, both on the part of the government and in terms of oppositional political activity had a clearly and, more importantly, *explicitly* local base. Perhaps the most obvious example was the rise of what came to be called 'the new urban left'. In a number of major cities, of which London and the Greater London Council were only the most prominent example, a new radical left (both within and independent of the local state) became one of the main foci of opposition both to the government and to the labourist politics of the leadership of the main opposition party.

The complexity of this geography of restructuring, its reverberations, and the political responses to it, had a number of important implications. *First*, it meant that some of the debates being conducted solely at national level, and some of the conclusions being drawn from them, were quite simply unsubstantiatable at that level in any rigorous sense. Across the political spectrum, casual connections were being made between changes in employment and occupational structure and wider social, ideological, and political changes. We were facing the end of the working class, the end of class politics, a new ideology of individualism, a politics of consumption, the dominance of what were referred to as 'new social movements'. All this was being argued, most frequently, from national-level statistics. Yet, quite apart from the difficulty of establishing such causal connections in the first place and the dubiousness of the economistic form in which they were usually proposed, the issues of spatial scale and spatial variation were usually ignored. And yet presumptions of cause and effect made at national level were clearly untenable when each of the component causal processes, which were supposedly interacting, was taking place unevenly (and differently so) over the national space. Further, the relation between political, cultural, and economic changes may have an important local level of operation. In other words, some of the causal processes which were being appealed to in the debate could not be seen as operating at national level only.

Second, spatial variation meant that the potential, the problems, and even the style of political response and organisation would be different in distinct parts of the country. Conclusions drawn at national level about policy implications and changes in political strategy could not be assumed to be universally applicable, to resonate in the same way with the particular traditions and circumstances in different parts of the country. At perhaps the most trivial, but certainly the most easily documented, political level it was clear that the voting patterns of individual social groups were becoming increasingly geographically differentiated.

Third, recognising variation in no way implies abandoning wider movements or wider levels of organisation. But local contrasts did mean that it was not possible to construct them by simply proclaiming that each local change was underlain by capitalism – that is, by simply asserting 'the general'. It also required, for a solid foundation, a recognition and understanding of the reality and conditions of diversity, and of the actual processes which linked the local particularities (Massey, 1983).

The fact of spatial variation in national change, in other words, had immediate and obvious political importance.[2] It became important to know just how differently national and international changes were impacting on different parts of the country. Something that might be called 'restructuring' was clearly going on, but its implications both for everyday life and for the mode and potential of political organising were clearly highly differentiated and we needed to know how. It was in this context that the localities projects in the United Kingdom were first imagined and proposed. It was research with an immediate, even urgent, relevance beyond academe.

The local, the concrete, and the postmodern

This history has a number of implications. It contradicts a number of other retrospective interpretations.

It is not, for instance, the case that the study of locality is a necessary vehicle for, nor equivalent to, empirical research or the study of concrete phenomena. This question has generated confusion. The issues of specificity and empirical uniqueness were on the agenda in the same period, and again as part of wider movements, in philosophy, the social sciences beyond geography, and the humanities. Localities are certainly 'specific' in this context in the sense that one of the prime aims, given the social and political background outlined above, was precisely to understand their differences. [This does not, of course, mean that they are unrelated and one of the aims of such locality research has to be (and was in this case) to understand, not just the interdependencies between localities in the sense of direct links, but the ways in which, in part, the changes going on in them were products of a wider restructuring.] In this case, then, the counter-position is between general (meaning wider) and specific (meaning more local). Some commentators, however, have at this point fallen into the trap of eliding the fact of being specific in this sense with that of being 'concrete', the product of many determinations. They then reason that, because localities are in this sense concrete, *only* localities are concrete. Here, the elision is between the dimensions specific–general and concrete–abstract. The current world economy, for instance, is no less concrete than a local one. The world economy is *general* in the sense of being a geographically large-scale phenomenon to which can be counterposed internal variations. But it is also unequivocally concrete as opposed to abstract. It is no more than a local economy, the simple manifestation of the capitalist mode of production. It is, just as much, a specific product of many determinations. Those who conflate the local with the concrete, therefore, are confusing geo-graphical scale with processes of abstraction in thought.

Moreover, those who make this mistake then frequently rush headlong into another: they confuse the study of the local with description, which they oppose to theoretical work. Smith (1987), for instance, seems to be arguing that locality research is necessarily descriptive in these terms. There are a number of problems with this argument. First, in the form in which Smith puts it, it is an accusation which could only ever be made from a view of the world which equated empirical generalisability with explanation, a position which the theoretical basis of the CURS localities approach most clearly rejected. There is an assumption behind it that 'theory' is 'opposed to a concern with specificity or uniqueness', a position which is untenable 'unless one wants to argue that theory cannot grasp the unique and hence the perception of the unique is theory-neutral – an idea which died at least twenty years ago with the demise of the concept of a theory-neutral observation language' (Sayer, 1989, page 303). Second, this argument continues the confusion between the dimensions concrete–abstract and local–general. Yet the fact that a phenomenon is 'more general' in the simple sense of being 'bigger' does not make it any more amenable to theoretical analysis. Third, this is true not only because both levels are the product of many determinations, but also because abstract analysis can be just as much about 'small' objects as it is about 'large' ones. As was pointed out above, the fact that in the debate about changes in the United Kingdom some of the conclusions being drawn at national level could not really be drawn at that level, because some of the key significant causal processes were also operative at smaller spatial scales, was one of the reasons behind the locality studies. 'The local' (meaning the small-scale) is no less subject to nor useful for theorisation than big, broad, general things. The counterposition of general and local is quite distinct from the distinction between abstract and concrete [see Sayer (1991) for expansion of these issues].

Indeed, when locality research came onto the agenda, new insights into understanding and *explanation* of concrete phenomena were central to the debate in human geography (Massey, 1984; Sayer, 1984), and these provided ways into the question of local variation. So the coappearance of an interest in methodology and studies of localities was mutually highly beneficial. But they are not equivalent to each other.

There is a difference, then, between the *reasons* for the importance of local studies at that time, and the *conditions provided* by theoretical developments which made such analysis more possible.

The same kind of argument must be spelled out (because it seems to be so widely misunderstood) in relation to that oft-quoted slogan 'geography matters'. Duncan (1989), for instance, rests a large part of his case against the adoption of a notion of locality precisely on the idea that locality research grew out of arguments about the effectivity of spatial form. It is certainly true, as I have already said, that this was a period in which it was increasingly argued that 'place' was important. Moreover, the methodologies adopted for the study of localities, for the explanation of uniqueness, emphasised the point in a different way. For it was stressed that, not only was the character of a particular place a product of its position in relation to wider forces (the more general social and economic restructuring, for instance), but also that that character in turn stamped its own imprint *on* those wider processes. There was *mutual* interaction (Massey, 1984). Moreover, the nature of the interaction, of the impact of local specificities on the operation of wider processes, may vary in kind. It may be that it occurs through self-conscious social activity. In the United Kingdom of the early 1980s, it was this which was the political focus of attention and enquiry. As the local political activists aimed to demonstrate in practice and as the localities projects showed in their research, there was a huge variety – of varying effectiveness – of local activity, resistance, and promotion (Cooke, 1989; 1990; Harloe et al, 1990). In these cases the focus of the 'local impact' was the local government, but it could of course be other agencies, social movements, or constellations of them. Moreover, the mutual conditioning of local and wider processes need not be a product of conscious social agency. Local impact may equally well, indeed more frequently, come about through the structural interaction of social processes without any deliberate local social agency. So studies of localities *may* certainly endorse the idea that geography matters, but it is an empirical question.

Finally, in this brief tour through things which locality-studies are *not*, or not necessarily, they are not *necessarily* part of the turn to the postmodern. That is to say, the debate about locality studies is in principle distinct from the debate about postmodernism. There are, of course, many apparent points of contact [Cooke (1990) has recently explored some of them], but many of these are more the result of the accidents of language than real connections, and none of them amount to real equivalences. Perhaps what a focus on localities can share with the shifts toward postmodernism is a recognition of, and a recognition of the potential significance of, both the local and variety. This, it seems to me, is unequivocally positive. Gregory, in another context, has argued tellingly that

one of the *raisons d'être* of the human sciences is surely to comprehend the 'otherness' of other cultures. There are few tasks more urgent in a multicultural society and an interdependent world, and yet one of modern geography's greatest betrayals was its devaluation of the specificities of place and of people (1989, page 358).

Even those who are critical of the philosophical arguments of the postmodernists also recognise at least this characteristic to be potentially progressive. Thus, Harvey writes

> How, then, should postmodernism in general be evaluated? My preliminary assessment would be this. That in its concern for difference, for the difficulties of communication, for the complexity and nuances of interests, cultures, places, and the like, it exercises a positive influence (1989, page 113).

The problem, of course, is that postmodernism in its current guise rarely lives up to the democratic potential opened by this move. On the other hand, as Harvey recognises, the recognition of difference is a characteristic which a reformed modernism could take on board.

However, there are other ways in which locality studies are sometimes thought to be closer to postmodernism than they are. One confusion arises over the term 'local' itself. The meaning of the term in the context of 'locality studies' is not the same as its meaning when used for instance by Lyotard in his arguments for 'local determinisms' and the abandonment of grander theories. [There seem, suitably enough, to be numerous confusions over words. The problems provoked by the multiple meanings of the term 'specific' were pointed to above, and Sayer (1991) follows up this issue further. Here it is the term 'local' which is at issue.] Neither a focus on the empirically local (in terms of geographical scale) nor an insistence that not all theorisable causal processes operate at the level of global accumulation, implies local determinism in the sense meant by Lyotard. 'Local' in locality is not opposed to 'meta' as in 'metatheory'. Once again, there is a potential confusion between the question of level in terms of geographical scale and level of abstraction in thought. Let us take one example where the confusion can arise. Harvey (1989, page 117) writes:

> Postmodernism has us accepting the reifications and partitionings . . . all the fetishisms of locality, place or social grouping, while denying that kind of metatheory which can grasp the political-economic processes (money flows, international divisions of labour, financial markets, and the like) that are becoming ever more universalizing in their depth, intensity, reach and power over daily life.

There are a number of points here. First, studying localities does not amount to fetishising them (I shall address this point again later); nor is Harvey necessarily saying it does. There is perhaps no disagreement here. Second, locality studies as I see them, most definitely do not deny the kind of theory which can grasp political-economic processes such as the international division of labour. The CURS programme was of course founded on precisely such concepts, and it was axiomatic that studying local areas necessarily required theories which were wider than their application to that area both in the sense that they had a broader spatial reach and in the sense of being more abstract. Such theories need not, though, only relate, as the quotation seems to imply, to economic phenomena.

Third, this seems to be, precisely, a misuse of the term 'metatheory', confusing the philosophical meaning with the question of scale. The same confusion arises later in the book when an acceptance of grand narratives is opposed to an emphasis on community and locality (page 351). Yet, if they are needed at all, grand narratives are needed just as much in the study of the local as of the international. Fourth, and more politically, it is difficult to reconcile this quotation's dismissive treatment of 'fetishisms of . . . social groupings' (women, for instance?) with the apparent commitment to a democratic recognition of the existence of difference cited above. Although post-modernism certainly has its difficulties in doing anything more democratic than recognising the existence of others, modernism seems to have problems in really, in the end, taking seriously the autonomy of others. Thus, just before this quotation, Harvey writes 'Post-modernist philosophers tell us not only to accept but even to revel in the fragmentations and the cacophony of voices through which the dilemmas of the modern world are understood' (page 116). He opposes such a position absolutely. I know what he means, and I have some sympathy. But I also have real reservations about this formulation. At one, very practical level, there seems to me to be not *enough* fragmentation at the moment. At least in the context of some political debates, there seems sometimes to be one megaphone – that of reaction. But that raises the second point in relation to Harvey's position: you cannot argue for the right to oppose when others are in power (and that includes being in power even within 'the left') if you will not allow it when the situation is reversed. Put together, all these quotations seem to say that it is OK to have a background orchestra of others, so long as you yourself are the conductor. Thus to return to the original issue of the meaning of the word 'local', the argument here is not that *some* local studies may not adopt a postmodern approach, indeed as Harvey points out (page 47), Fish (1980) *has* understood 'local determinism' to mean 'localities' such as interpretative communities and particular places; but it *is* to argue that the one does not necessarily entail the other.

Again, the debate about the postmodern has brought with it a sudden recognition of, indeed a revelling in, the importance of space and place. It is a realisation, a sudden discovery, which seems to have dawned on intellectuals across much of the social sciences. Jameson (1984) is only the most obvious theoretician to whom one could point. But yet again it is important to make distinctions. Although it emerged at the same period, the argument of the postmodernists about the importance of space and place is distinct, in its roots and in its nature, from the debate in geography which led to 'geography matters' and 'the difference that space makes'. For one thing, and most trivially, gratifying as it is to geographers, perhaps, to have the dimension which they have always treated as their own now accorded such centrality, it has to be said that some of the claims being made in the postmodern literature about the current importance of the spatial are grandly unsubstantiated. But, more significantly, the claim made in the debate about postmodernity is a historically specific one: it is that space and place are important *now*, and that this is something new. The arguments being made in the debate about 'geography matters' were rather different. They also involved a distinct, and I believe more constructive, engagement with and development of the form of Marxism which had been dominant in the previous decade. Here, the argument was not an empirical one in the sense that it was saying that the world had changed. More, it was an argument about our intellectual focus and about the complexity of the causal processes which we should recognise. This is not incompatible with the argument that space and place have great and real significance in these times, nor that this sig.:ificance may be increasing, but it is not the same argument.

Localities, reaction, and progressiveness

A wider argument has, however, been made by Harvey in the context of the debate over *The Condition of Postmodernity*. This argument is that a focus on place and the local is, by its very nature, antiprogressive. It is necessary to be clear here. Harvey is not saying that all foci on localities are necessarily reactionary; nor certainly am I saying that a focus on the local is necessarily progressive (far from it!), and even more certainly I am not saying that it is any more than one among many potential ways of studying for geographers (for a further development of this argument see Massey, 1990). So, in broad terms we probably agree. Nonetheless, Harvey's argument is interesting and important to consider.

There are two interweaving strands. The first begins from the philosophical arguments of such as Heidegger and Bachelard that whereas Time connotes Becoming (which is assumed, in modernist terms, to be progressive), Space connotes Being. And this in turn implies fixity, stasis. The second thread is that a concern with place leads inexorably to an aesthetic mode, and that in turn virtually inevitably to reaction. Both of these lines of argument are interesting. But they both also have weaknesses. First, both of them involve internal slippages and leaps of logic. Second, both singly and together they imply a concept of 'locality' which is certainly not the only one available and, I would argue, is at odds with the one which is implied by at least some locality studies.

The first argument, then, equates Space with Being; 'space contains compressed time. That is what space is for' (Harvey, 1989, page 217). The implication, immediately, is that spaces, such as localities, are essentially simultaneities. Moreover, and more importantly, they are static.

> Being suffused with immemorial spatial memory, transcends Becoming ... Is this the foundation for collective memory, for all those manifestations of place-bound nostalgias that infect our images of the country and the city, of region, milieu, and locality ... And if it is true that time is always memorialized not as flow, but as memories of experienced places and spaces, then history must indeed give way to poetry, time to space ... (page 218).

This notion is closely tied to what Harvey sees as important dilemmas, most particularly for capital ... 'the most serious dilemma of all: the fact that space can be conquered only through the production of space' (page 258). When placed in the context of capital accumulation this leads, of course, to the crucial contradiction of 'the spatial fix', and Harvey is here essentially generalising that concept to a wider field. But there are real difficulties in such an attempt at generalisation. The problem with the idea of spatial fix is that it really is about fixity, about immobility. The spatial fix is the physical forms of buildings and infrastructure; it is the prison-house of capital tied up. But this imagery is not transferable to wider fields nor, in particular, to localities. There is clearly a tension between trying on the one hand to capture a synchronicity and attempting to follow a process on the other. But localities, as I see them, are not just about physical buildings, nor even about capital momentarily imprisoned; they are about the intersection of social activities and social relations and, crucially, activities and relations which are necessarily, by definition, dynamic, changing. There is no stable moment, in the sense of stasis, if we *define* our world, or our localities, ab initio in terms of change. The CURS programme has 'Change' in the first word of its title. As was argued in the opening section of this paper, its empirical focus was precisely on the quite contrasting ways in which local sets of social relations were being transformed: *how* they were 'becoming'. It is an accepted argument that capital is not a thing, it is a process. Maybe it ought to be more clearly established that places can be conceptualised as processes, too.[3] If that were so, then it would be possible not only to agree that 'the present is valid only by virtue of the potentialities of the future' (Poggioli, 1968, page 73, cited in Harvey, 1989, page 359), but to apply it to localities as well.

The second thread of Harvey's argument is that place inexorably brings with it aesthetics and, in its turn, political reaction. One starting point for the staking out of this position is the close connection made between place and identity. The next step is to endow both place and identity with some kind of seamless coherence. A sense of identity is needed because of the unsettling flux of modern times (more on this later); a sense of identity means something stable, coherent, uncontradictory; places have already been identified as means of constructing identities, hence places are coherent, uncontradictory . . . a characterisation which is of course further reinforced by – indeed is integral to – the attribution of stasis already discussed.

Now, there are a number of comments to be made at this point in the argument. First, this *is* certainly one way in which the notion of 'place' is commonly used, and I would agree with Harvey that it has potential dangers (see below). The problem is that Harvey seems to elide this version of the concept 'place' with any and every notion of locality and the local. Second, this way of thinking of identity is curiously solid in an age of recognition of the decentred subject and of multiple identities. Individuals' identities are not aligned with *either* place *or* class; they are probably constructed out of both, as well as a whole complex of other things, most especially race and gender. The balance between these constituents, and the particular characteristics drawn upon in any one encounter or in any one period, may of course vary. And, third, this applies to places too. They do not have single, pregiven, identities in that sense. For places, certainly when conceptualised as localities, are of course *not* internally uncontradictory. Given that they are constructed out of the juxtaposition, the intersection, the articulation, of multiple social relations they could hardly be so. They are frequently riven with internal tensions and conflicts. Places are shared spaces: you could not think about London's Docklands at the moment without precisely that conflictual sharing and the conflict between interests and views of what the area is, and what it ought to *become*. This is not an idiosyncratic view, although there is horrendous terminological confusion. Thrift (1983), for instance, writes that 'the region, initially, at least, must not be seen as a *place*; that is a matter for investigation. Rather, it must be seen as made up of a number of different but connected *settings for interaction*'. In the argument in the present paper the term 'locality' could be substituted for 'region' in this quotation. Moreover, if the term 'place' is to have the extra endowment of meaningfulness implied in that quotation, then it must be understood (as Thrift makes clear) as different from other spatial terms, including locality. Chouinard (1989) argues that localities are not bounded areas but spaces of interaction. Sayer (undated) points out that spatial juxtaposition may mean that localities contain many quite unrelated elements:

> Yet, despite this lack of functional integrity, they may still be distinctive and even derive their identity from the lack of unity . . . . The awkward aspect of this property of localities is that people can actually be shaped by factors which, among themselves, are totally unrelated (page 3).

Moreover, this crucial aspect of internal differentiation, of articulation, and of potential contradiction and conflict applies even more strongly when analysis turns to how actors actively draw upon localities as a basis for interpretation. Wright (1985) [included as Chapter 8] has written of the variety of meanings and interpretations of Hackney, many of them implicitly if not always overtly, though indeed sometimes quite actively, in conflict with each other. For each the 'meaning' of Hackney is distinct – for the old white working class, for the variety of ethnic minorities, for the new monied gentrifiers. Each has its view of what the essential place is, each partly based on the past, each drawing out a different potential future. For the analyst of the locality this intersection is surely precisely one of the things which must be addressed. Hackney *is* Hackney only because of the coexistence of all those different interpretations of what it is and what it might be. There are, of course, many definitions of locality in the literature at the moment, but, given the argument in this paper, it would seem that any requirement that an intersection of social relations in a particular space can only graduate to locality status if there is a shared local consciousness is inordinately (and arbitrarily) restrictive (and also potentially more open to the arguments about reaction put by Harvey – see below). McArthur (1989), in contrast, argues strongly that any local consciousness, should it exist, will anyway be likely to vary widely in degree and nature between different groups in an area.

All of this relates strongly back to deeper issues of conceptualisation. Perhaps localities may be conceptualised as, in one aspect at least, the intersection of sets of (Giddens-type) locales. But, whatever else they are, localities are *constructions* out of the intersections and interactions of concrete social relations and social processes in a situation of copresence. Whether that copresence matters, and whether it leads to new emergent powers, is an open question which will not have an empirically generalisable answer. Moreover, the particular social relations and social processes used to define a locality will reflect the research issue (which in turn means that any locality so defined will *not* be the relevant spatial area for the investigation of all and every social process deemed in some way to have a local level of variation or operation). But all this does mean that localities are not simply spatial areas you can easily draw a line around. They will be defined in terms of the sets of social relations or processes in question. Crucially, too, they are about *interaction*. Such interaction, moreover, is likely to include conflict. Localities will 'contain' (indeed in part will be *constituted by*) difference and conflict. They may also include interaction between social phenomena which may not be 'related' in any immediate way in terms of social relations aspatially. It may be only the fact of copresence which makes them have quite direct impacts upon each other. Moreover, the constellations of interactions will vary over time in their geographical form (see Massey, 1984, pages 123, 196, 299). And the definition of any particular locality will therefore reflect the question at issue.

But all this returns us to the very originating view of Space-as-Being which Harvey adopts. This definition, and its counterposition with the equation Time – Becoming, is a curious mode of argument for him to follow. In most of the other major conceptualisations in the book there is a dynamic tension, sometimes a constructive contradiction. The initiating and powerful definition of modernism, which forms the framework for much of the argument in the book, is precisely of this nature. So why at this point relapse into this simple static dichotomy? Heidegger's is not the only approach to space which could have been adopted, and indeed in other parts of his argument (see below) Harvey is clearly critical of Heidegger precisely for his potentially romantic/reactionary views.

Indeed, the next steps in Harvey's argument are that 'The assertion of any place-bound identity has to rest at some point on the motivational power of tradition' (1989, page 303) and that such place-relating structures of feeling and action are (almost – it varies) always reactionary. 'Geographical and aesthetic interventions always seem to imply nationalist, and hence unavoidably reactionary, politics' (pages 282–283). Now, it has already been argued that the concept of locality is not, or need not be, the same as Harvey's concept of place in his argument here. So many points of potential disagreement between the lines of argument may simply evaporate if clear distinctions are made. Nonetheless, there are wider issues to consider. Harvey exemplifies his logic of place ⇒ aesthetic-isation ⇒ reaction at a number of points in his book, and the examples he gives, of reactionary nationalisms, most obviously of Nazism, or even the urban designs of Sitte, are very telling. But it is never quite clear just how *necessary* this chain of connections is supposed to be. Thus, of Sitte he writes

Under conditions of mass unemployment, the collapse of spatial barriers, and the subsequent vulnerability of place and community to space and capital, it was all too easy to play upon sentiments of the most fanatical localism and nationalism. I am not even indirectly blaming Sitte or his ideas for this history. But I do think it important to recognize the potential connection between projects to shape space and encourage spatial practices of the sort that Sitte advocated, and political projects that can be at best conserving and at worst downright reactionary in their implications. These were, after all, the sorts of sentiments of place, Being, and community that brought Heidegger into the embrace of national socialism (1989, page 277).

Yet if a reactionary outcome is not inevitable, but only a likely danger, still almost no examples of progressive possibilities are given by Harvey (Nicaragua gets a mention). But 'tradition' and an awareness of history can also be strengthening in an oppositional sense. Just within the United Kingdom, examples from the Little Moscows to Red Clydeside, Poplar, Clay Cross, and the 'Socialist Republic of South Yorkshire', show how local bases and traditions can be used. (The last of these was the name accredited to that part of the country for its attempts to combat national policies, most particularly over public transport!) It does not always only work for capital; we have our own traditions, too, and they are not simply to be sentimentalised, they are also to be built on. Moreover, building on traditions can also mean being critical of them. The labour-movement tradition of Sheffield, for instance, has been a strength in many ways, a resource to be drawn upon; but it has also delivered an understanding of gender relations and of the meanings of masculinity and femininity, which have had to be challenged head on for there to be any chance of maintaining a contemporary radical political culture in the local area. Localities, in that sense, are part of the conditions not of our own making. There are, of course, dangers even here. Even labour-movement history can be commodified, commercialised, romanticised, and sold off. Yes, it can, and it often is. But the consistency with which Harvey points to this kind of outcome (in the case of local history, in the case of local economic strategies, in relation to attempts to create spaces and places to celebrate the French revolution), indicates a wider problem. This is that, in Harvey's account, capital always wins and, it seems, only capital can ever win. Thus, in the discussions of locally based economic strategies most of the discussion is of capitalist strategies (trying to attract private capital, creating competitive images, etc, etc), and where 'municipal socialism' is referred to it is labelled 'defensive' (page 302) without any further explanation.

Moreover, if as I have argued there are indeed multiple meanings of places, held by different social groups for instance, then the question of which identity is dominant will be the result of social negotiation and conflict. In Wright's account of Hackney the different social groups had distinct interpretations, not just of Hackney's present, but also of its past, its 'traditions'. The past is no more authentic than the present; there will be no one reading of it. And 'traditions' are frequently invented or, if they are not, the question of which traditions will predominate can not be answered in advance. *It is people, not places in themselves, which are reactionary or progressive.* Unless, then, *any* notion of the past, any consciousness of any tradition, is ipso facto reactionary, the reactionary meaning of places focused on by Harvey is itself a result of conflict and not in principle necessary. Moreover, that means it must be opposed; it cannot simply be ignored.

Harvey writes ' "Regional resistances", the struggles for local autonomy, place-bound organization, may be excellent bases for political action, but they cannot bear the burden of radical historical change alone' (page 303). This is certainly correct in the sense that none of them are world revolution. But there are problems with the slippage of terms; place-*based* action gets conflated with place-*bound* action. And here the debate links back to that about postmodernism.

This is the progressive angle to postmodernism which emphasizes community and locality, place and regional resistances, social movements, respect for otherness, and the like .... . But it is hard to stop the slide into parochialism, myopia, and self-referentiality in the fact of the universalising force of capital circulation (Harvey, 1989, page 351).

This is true, but any strategy has its dangers. Harvey opposes to local-based action a very abstract universalism. The danger of *his* strategy is that one sits in one's university and urges the world proletariat to unite. Surely in these postmodern days we should, and could, be actively promoting a conceptualisation and a consciousness of place which is precisely about movement and linkage and contradiction. A sense of place which is extra-verted as well as having to deal with and build upon an inheritance from the past. That is surely the meaning of the joint existence of uniqueness and interdependence. Harvey argues (also page 351) that locally based action can lead to fragmentation. Again, it may. But is it not also a necessary condition for building real unity? We can only build unity if we have the confidence to face diversity without it frightening us and to analyse the *real* conditions for solidarity. This returns us again to the debate about difference, and how to conceptualise localities. At minimum we can say that localities are not internally introspective bounded unities. They have to be constructed through sets of social relations which bind them inextricably to wider arenas, and other places.

Conclusions

The point of this paper has in no way been to glorify the local level, either as object of analysis or as arena of political action. There are great dangers in an overemphasis on its importance, its significance to 'daily life', its relation to the constitution of identity (Massey, 1990). Nor is the issue whether we *only* do locality studies or *only* do something else. One of the problems with the current debate is that it has been understood by some as being about a new 'orthodoxy' on what geographers ought to be studying. By others, with equally little understanding, it has been dismissed as a fashion. Both of these positions are crippled by thinking of the development of foci of study as happening entirely as a product of events in the academic world or intellectual debate. But things are not (or should not be) so.

Other explanations of the current focus on localities do set the shifts in a wider, and historically specific, context. Thus Harvey, who is addressing a much wider issue than simply the current locality studies, interprets an increasing focus on place as deriving from the unsettling nature of the times in which we live, the current perturbations being a result of a heightened process of time–space compression. There are many paragraphs evoking the ephemerality, confusion, uncertainty, the shifting and the fragmentation, the disruption. 'In periods of confusion and uncertainty, the turn to aesthetics (of whatever form) becomes more pronounced' (page 328). Apart from the serious question of how one can begin to evaluate such a claim, there is a further point. If people *are* beginning to turn to localities in reactionary ways, then it may precisely be important to study them. Such phenomena are themselves – or should be – amenable to historical materialist analysis. To study something is not necessarily to glorify it; indeed it can be an important part of exposing myths, of locality and place as much as of anything else.

But I also find mystifying the idea, argued by many, that time–space compression is somehow psychologically disturbing. Such flux and disruption is, as Harvey says, part of modernity. Why should the construction of places out of things from everywhere be so unsettling? Who is it who is yearning after the seamless whole and the settled place? A global sense of place – dynamic and internally contradictory and extra-verted – is surely potentially progressive.

Nonetheless, it *is* true that the current programme of locality studies was proposed for reasons which were historically specific.[4] They arose from the situation *then* and *there*. And moreover that situation was not one only, nor even primarily, defined by academic or intellectual debate. It was a situation defined by what was happening in society more widely, and by important questions which were raised as a result of those changes. Such a history, in other words, does *not* imply that locality research, the study of particular places, should in some more general sense, always and everywhere, be the focus of human geographical enquiry. Sometimes we may want to study particular localities for particular, strategic, reasons. Most often, indeed, we may find that other foci of research will be more important.

Notes

1 The author was the initiator of the original proposal, and responsible for drawing up the original outline. The funders and the participants subsequently developed and implemented the programme in greater detail and as a product of their own ideas and research.

2 The argument here is directed to 'the left' because that is the debate which I am addressing. But parallel points could be made about relevance across the political spectrum. Government departments, for instance, displayed interest in the geographical variation in penetration of the 'enterprise culture'.

3 Pred (1984; 1989) has of course for a time argued something along those lines, although from a rather different perspective. Perhaps ironically given the context of the present paper, I would argue that he greatly overemphasises the significance of 'the local' (see Massey, 1990).

4 There are, of course, other more transhistorical claims made by some authors. These are discussed in Massey (1990).

References

Arrighi G., 1990, 'Marxist century, American century: the making and remaking of the world labour movement' *New Left Review* number 179, 29–63.

Chouinard V., 1989, 'Explaining local experiences in state formation: the case of cooperative housing in Toronto' *Environment and Planning D: Society and Space* 7 51–68.

Cochrane A. (Ed.), 1987 *Developing Local Economic Strategies* (Open University Press, Milton Keynes).

Cooke P. (Ed.), 1989 *Localities* (Unwin Hyman, London).

Cooke P., 1990 *Back to the Future: Modernity, Postmodernity and Locality* (Unwin Hyman, London).

Duncan S., 1989, 'What is locality?', in *New Models in Geography: The Political Economy Perspective* Volume II, Eds R. Peet, N. Thrift (Unwin Hyman, London) pp. 221–252.

Fish S., 1980 *Is There a Text in this Class? The Authority of Interpretive Communities* (Harvard University Press, Cambridge, MA).

Gregory D., 1989, 'The crisis of modernity? Human geography and critical social theory', in *New Models in Geography: The Political Economy Perspective* Volume II, Eds R. Peet, N. Thrift (Unwin Hyman, London) pp. 348–385.

Harloe M., Pickvance C., Urry J., 1990 *Place, Policy and Politics: Do Localities Matter?* (Unwin Hyman, London).

Harvey D., 1989 *The Condition of Postmodernity* (Basil Blackwell, Oxford).

Jameson F., 1984, 'Postmodernism, or the cultural logic of late capitalism' *New Left Review* number 146, 53–92.

Jonas A., 1988, 'A new regional geography of localities?' *Area* **20** 101–110.

Lane T., 1982, 'The unions: caught on an ebb-tide' *Marxism Today* September, pp. 6–13.

Massey D., 1983, 'The shape of things to come' *Marxism Today* April, pp. 18–27.

Massey D., 1984a *Spatial Divisions of Labour: Social Structures and the Geography of Production* (Macmillan, London).

Massey D., 1988, 'A new class of geography' *Marxism Today* May, pp. 12–17.

Massey D., 1990, 'L' "estudi de localitats" en geografia regional' *Treballs de la Societat Catalana de Geografia* **21** 73–87.

Massey D., Miles N., 1984, 'Mapping out the unions' *Marxism Today* May, pp. 19–22.

McArthur R., 1989, 'Locality and small firms: some reflections from the Franco-British project, "Industrial systems, Technical Change and Locality" ' *Environment and Planning D: Society and Space* 7 197–210.

Poggioli R., 1968 *The Theory of the Avant-garde* (Harvard University Press, Cambridge, MA).

Pred A., 1984, 'Place as historically contingent process: structuration and the time-geography of becoming places' *Annals of the Association of American Geographers* **74** 279–297.

Pred A., 1989, 'The locally spoken work and local struggles' *Environment and Planning D: Society and Space* 7 211–233.

Sayer A., 1984 *Method in Social Sciences: A Realist Approach* (Hutchinson, London).

Sayer A. (undated), 'Locales, localities, and why we want to study them', mimeo, School of Social Sciences, University of Sussex, Brighton, Sussex.

Sayer A., 1989, 'Dualistic thinking and rhetoric in geography' *Area* **21** 301–305.

Sayer A., 1991, 'Behind the locality debate: deconstructing geography's dualisms' *Environment and Planning A* **23** 283–308.

Smith N., 1987, 'Dangers of the empirical turn: some comments on the CURS initiative' *Antipode* **19** 59–68.

Soja E., 1989 *Postmodern Geographies: The Reassertion of Space in Critical Social Theory* (Verso, London).

Thrift N., 1983, 'On the determination of social action in space and time' *Environment and Planning D: Society and Space* **1** 23–57.

Walker R., 1989, 'What's left to do?' *Antipode* **21** 133–165.

Wright P., 1985 *On Living in an Old Country: The National Past in Contemporary Britain* (Verso, London).

23 Iris Marion Young
Together in Difference: Transforming the Logic of Group Political Conflict[1]

Excerpts from: J. Squires (ed.), *Principled positions*, pp. 121–50. London: Lawrence and Wishart (1993)

William J. Wilson, among others, has forcefully argued that race-focused political movements and policies to improve the lives of poor people of colour are misplaced. Race-focused explanations of black and Hispanic poverty divert attention from the structural changes in the US economy that account primarily for the unemployment and social isolation experienced by rapidly growing numbers of inner city Americans. Race-focused policies such as affirmative action, moreover, have benefited only already better off blacks, and fueled resentment among middle class and working class whites. The problems of poor people, whether white or black, male or female, are best addressed, he argues, through a strong class based analysis of their causes and the promotion of universal public programmes of economic restructuring and redistribution.[2]

Group focused movements and policy proposals, these arguments suggest, only continue resentment and have little chance of success. The more privileged white, male, able-bodied, suburban sectors of this society will not identify with economic and social programmes that they associate with blacks, or women, or Spanish speakers or blind people. Only a broad coalition of Americans uniting behind a programme of universal material benefits to which all citizens have potential access can receive the widespread political support necessary to reverse the 1980s retreat of the state from directing resources to meet needs – programmes such as national health service, family allowance, job training and public works, housing construction and infrastructure revitalisation.

Wilson is at least partly right, both about the causes of poverty and deprivation and the necessity of a broad based coalition of diverse sectors of society coming together to cure them. There are nevertheless basic problems with this approach to political organising and policy. Since the working class, broadly understood, is fractured by relations of privilege and oppression along lines of race, gender, ethnicity, age, ablement, and sexuality, experiential differences and group-based distrust will make it difficult to bring this coalition together unless the distrust is openly addressed and the experiential differences acknowledged. If a unified movement were to develop for a universal working class programme, moreover, it is liable to be led by and reflect the interests of the more privileged segments of each fracture. Wilson's own analysis, for example, is seriously male biased; he tends to perceive female-headed households as pathological, and recommends economic and social programmes that assume women's economic connections to men as the most desirable arrangement. Finally, without countervailing restrictions built in, any universal benefits policies are likely to benefit most those already more privileged. Any new job training programme must learn from the

CETA (Central Education and Training Agency) experience, for example, how to combat the tendency to train the already most trainable young white men through programmes that better target young single mothers, older people, and poor people of colour. If a political movement wishes to address the problems of the truly disadvantaged, it must differentiate the needs and experiences of relatively disadvantaged social groups and persuade the relatively privileged – heterosexual men, white people, younger people, the able bodied – to recognise the justice of the group based claims of these oppressed people to specific needs and compensatory benefits. Such recognition should not rule out that a programme for social change benefit these relatively more privileged groups, but justice will be served only if the programmes are designed to benefit the less privileged groups more, and in group specific ways. I believe that some theory and practice of socialist, feminist, black liberation and other group based social movements in the last decade has aimed to develop analysis, rhetoric and political practice along these lines.

I conclude from all this that both a unified working-class based politics and a group differentiated politics are necessary in mobilisations and programmes to undermine oppression and promote social justice in group differentiated societies. Given the above dialectic, however, it is not obvious how both kinds of politics can occur. This problem appears in many forms all over the world; societies, classes, social movements, are riddled with inequality, hatred, competition, and distrust among groups whom necessity brings together politically. This paper examines one aspect of this problem, specifically how political actors conceive group difference and how they might best conceive it. Historically, in group based oppression and conflict difference is conceived as otherness and exclusion, especially, but not only, by hegemonic groups. This conception of otherness relies on a logic of identity that essentialises and substantialises group natures.

Attempts to overcome the oppressions of exclusion which such a conception generates usually move in one of two directions: assimilation or separation. Each of these political strategies itself exhibits a logic of identity; but this makes each strategy contradict the social realities of group interfusion. A third ideal of a single polity with differentiated groups recognising one another's specificity and experience requires a conception of difference expressing a relational rather than substantial logic. Groups should be understood not as entirely other, but as overlapping, as constituted in relation to one another and thus as shifting their attributes and needs in accordance with what relations are salient. In my view, this relational conception of difference as contextual helps make more apparent both the necessity and possibility of political togetherness in difference.

Group difference as Otherness

Social groups who identify one another as different typically have conceived that difference as Otherness. Where the social relation of the groups is one of privilege and oppression, this attribution of Otherness is asymmetrical. While the privileged group is defined as active human subject, inferiorised social groups are objectified, substantialised, reduced to a nature or essence.[3] Whereas the privileged groups are neutral, exhibit free, spontaneous and weighty subjectivity, the dominated groups are marked with an essence, imprisoned in a given set of possibilities. By virtue of the characteristics the dominated group is alleged to have by nature, the dominant ideologies allege that those group members have specific dispositions that suit them for some activities and not others. Using its own values, experience, and culture as standards, the dominant group measures the Others and finds them essentially lacking, as excluded from and/or complementary to themselves. Group difference as otherness thus usually generates dichotomies of mind and body, reason-emotion, civilised and primitive, developed and underdeveloped.

Gender is a paradigm of this presumption that difference is otherness. Gender categorisation of biological and social group differences between men and women typically makes them mutually exclusive complementary opposites. Western culture, and other cultures as well, systematically classify many behaviours and attributes according to mutually exclusive gender categories that lie on a superior–inferior hierarchy modeled on a mind–body dichotomy. Men are rational, women emotional, men are rule-bound contractors, women are caretakers, men are right-brainers, women are left-brainers. Dichotomous essentialising gender ideologies have traditionally helped legitimate women's exclusion from privileged male places.

The oppressions of racism and colonialism operate according to similar oppositions and exclusions. The privileged and dominating group defines its own positive worth by negatively valuing the Others and projecting onto them as an essence or nature the attributes of evil, filth, bodily matter; these oppositions legitimate the dehumanised use of the despised group as sweated labour and domestic servants, while the dominant group reserves for itself the leisure, refined surroundings, and high culture that mark civilisation.[4]

Not all social situations of group difference have such extremely hierarchical relations of privilege and oppression. Sometimes groups are more equal than this model portrays, even though they may not be equal in every respect. Nevertheless, many such situations of group difference and conflict rely on a conception of difference as Otherness. Some of the conflicting groups in Eastern Europe seem roughly equal in this way and nevertheless see their relation as one of mutual exclusion. I will return to this later.

Whether unequal or relatively equal, difference as otherness conceives social groups as mutually exclusive, categorically opposed. This conception means that each group has its own nature and shares no attributes with those defined as other. The ideology of group difference in this logic attempts to make clear borders between groups, and to identify the characteristics that mark the purity of one group off from the characteristics of the Others.

This conception of group difference as Otherness exhibits a logic of identity. Postmodern critiques of the logic of identity argue that much Western thought denies or represses difference, which is to say, represses the particularity and heterogeneity of sensual experience and the everyday language immersed in it. Rational totalising thought reduces heterogeneity to unity by bringing the particulars under comprehensive categories. Beneath these linguistic categories, totalising thought posits more real substances, self-same entities underlying the apparent flux of experience. These substances firmly fix what does and does not belong within the category, what the thing is and is not. This logic of identity thereby generates dichotomy rather than unity, dichotomies of what is included and what is excluded from the categories. Through this dialectic initial everyday experience of particular differences and variations among things and events become polarised into mutually exclusive oppositions: light-dark, air-earth, mind-body, public-private, and so on. Usually the unifying discourse imposes a hierarchical valuation on these dichotomies, lining them up with a good-bad dichotomy.

The method of deconstruction shows how categories which the logic of identity projects as mutually exclusive in fact depend on one another. The essence of the more highly valued or 'pure' side of a dichotomy usually must be defined by reference to the very category to which it is opposed. Deconstructive criticism demonstrates how essentialised categories are constructed by their relations with one another, and bursts the claim that they correspond to a purely present reality. Deconstruction not only exposes the meaning of categories as contextual, but also reveals their differentiation from others as undecidable: the attempt to demarcate clear and permanent boundaries between things or concepts will always founder on the shifts in context, purpose and experience that change the relationships or the perspectives describing them. Allegedly fixed identities thus melt down into differentiated relations.

Defining groups as Other actually denies or represses the heterogeneity of social difference, understood as variation and contextually experienced relations. It denies the difference among those who understand themselves as belonging to the same group; it reduces the members of the group to a set of common attributes. In so far as the group categorisation takes one set of attributes as a standard in reference to which it measures the nature of other groups as complementary, lacking, excluded, moreover, it robs the definition of a group's attributes of its own specificity.

The method of deconstruction shows how a self-present identity – whether posited as a thing, a substantial totality, a theoretical system, or a self – drags shadows and traces that spill over that unity, which the discourse representing the identity represses at the same time that it relies on them for its meaning. This process of criticism that reveals the traces, exhibits the failure of discourse to maintain a pure identity, because it appears as internally related to what it claims to exclude. Attempts to posit solid and pure group identities in social life fail in just this way. The practical realities of social life, especially but not only in modern, mass, economically interdependent societies, defy the attempt to conceive and enforce group difference as exclusive opposition. Whatever the group opposition, there are always ambiguous persons who do not fit the categories. Modern processes of urbanisation and market economy produce economic interdependencies, the physical intermingling of members of differently identifying groups in public places and workplaces, and partial identities cutting across more encompassing group identities.

Think, for example, of the social disruptions of an oppositional gender dichotomy. Homosexuality is the most obvious problem here. Enforced heterosexuality is a cornerstone of the gender edifice that posits exclusive opposition between masculine and feminine. The essence of man is to 'have' woman and the essence of women is to depend on and reflect man. Men who love men and women who love women disrupt this system along many axes, proving by their deeds that even this most 'natural' of differences blurs and breaks down.[5] Thus the need to make homosexuality invisible is at least as much existential and ontological as it is moral.

The inability to maintain categorical opposition between social groups appears in examining any social group difference, however. Where there are racial, ethnic, or national group differences there is always the 'problem' of those who do not fit because they are of 'mixed' parentage. The effort to divide such racial or ethnic groups by territory is always thwarted by what to do with those frontier areas where opposing groups mingle residentially, or how to account for and reverse the fact that members of one group reside as a minority population in a neighbourhood, city or territory conceptually or legally dominated by another. The needs of capital for cheap labour increasingly exacerbate this problem of the 'out groups' dwelling in the territory from which they are conceptually excluded. Capital encourages despised or devalued groups defined as Other to accept low paying menial jobs which keep them excluded from the privileges of the dominant group, but cannot keep them physically excluded from land and buildings the privileged claim as theirs. In many parts of the world those defined as 'guest workers' have become a permanent presence, and survive as blatantly ambiguous groups, excluded by definition from the places where they live, but excluded as well from their supposed homelands.[6]

The method of deconstruction consists in showing how one term in a binary opposition internally relates to the other. Group difference conceived as Otherness exhibits a similar dialectic. Frequently the most vociferous xenophobia, homophobia, misogyny arises as a result of a logic that defines a self-identity primarily by its negative relation to the Other. Some gender theorists suggest, for example, that for many men masculinity is primarily defined as what women are not.[7] Racist discourses similarly articulate the purity and virtue of white civilization by detailed fascinated attention to the attributes of the coloured Others, and the white subjects thereby derive their sense of identity from this negative relation to the Other. A group identity formed as a negation of the Other in these ways is fragile and relatively empty, and perhaps for that reason often violently insists on maintaining the purity of its border by excluding that Other.[8] This negative dialectic of group identity denies the subjects so identified a positive specificity. If men were less worried about avoiding invasion by feminine attributes, they might better be able to consider whether there is anything positively specific about masculine experience that does not depend on excluding, devaluing and dominating women.

To challenge this conceptualisation of difference as otherness and exclusive opposition, I propose a conception of difference that better recognises heterogeneity and interspersion of groups. It makes explicit the relational logic I just articulated according to which even the most fixed group identities define themselves in relation to others. A more fluid, explicitly relational conception of difference need not repress the interdependence of groups in order to construct a substantial conception of group identity.[9]

This relational conception of difference does not posit a social group as having an essential nature composed of a set of attributes defining only that group. Rather, a social group exists and is defined as a specific group only in social and interactive relation to others. Social group identities emerge from the encounter and interaction among people who experience some differences in their way of life and forms of association, even if they regard themselves as belonging to the same society. So a group exists and is defined as a specific group only in social and interactive relation to others. Group identity is not a set of objective facts, but the product of experienced meanings.

In this conception difference does not mean otherness, or exclusive opposition, but rather specificity, variation, heterogeneity. Difference names relations of both similarity and dissimilarity that can be reduced neither to coextensive identity nor overlapping otherness. Different groups always potentially share some attributes, experiences, or goals. Their differences will be more or less salient depending on the groups compared and the purposes of the comparison. The characteristics that make one group specific and the borders that distinguish it from other groups are always *undecidable*.

A primary virtue of this altered conception of group difference is that from it we can derive a social and political ideal of togetherness in difference which I think best corresponds to the political needs of most contemporary situations of group based injustice and conflict. Continuing my previous work on this theme, I will call this the ideal of a heterogeneous public.[10] Perhaps the best way to explore the uniqueness of this ideal is to contrast it with the two ideals that tend to surface in contemporary political debate and strategies involving group oppression or group conflict: assimilation and separation.

The tradition of liberal individualism promotes an assimilationist ideal. It condemns group based exclusions and discriminations, along with the essentialist ideologies of group superiority and objectification that legitimate these oppressions. Liberal individualism not only rightly calls these conceptions of group identity and difference into question, it also claims that social group categorisations are invidious fictions whose sole function is to justify privilege. In fact there are no significant categorical group based differences among persons, this position suggests. People should be considered as individuals only, and not as members of groups. They should be evaluated on their individual merits and treated in accordance with their actions and achievements, not according to ascribed characteristics or group affinities which they have not chosen.

Liberal individualism thus proposes an assimilationist ideal as a political goal. The assimilationist ideal envisions a society where a person's social group membership, physical attributes, genealogy, and so on, make no difference for their social position, the advantages or disadvantages that accrue to them, or how other people relate to them. Law and other rules of formal institutions will make no distinction among persons and will assume their moral and political equality. In a society which has realised this assimilationist ideal, people might retain certain elements of group identity, such as religious affiliation or ethnic association. But such group affiliation would be completely voluntary, and a purely private matter. It would have no visible expression in the institutional structure of the society. Workplaces, political institutions, and other public arenas would presume every person as the same, which is to say a free self-making individual.[11]

The assimilationist ideal properly rejects any conception of group difference as otherness, exclusive opposition, and rightly seeks individual freedom, equality and self-development. It wrongly believes, however, that the essentialist substantialising conception of group identity and difference is the only conception. The liberal individualist position associates group based oppression with assertions of group differences as such; eliminating group oppression such as racism, then, implies eliminating group differences.

There are several problems with this assimilationist ideal. First, it does not correspond to experience. Many people who are oppressed or disadvantaged because of their group identity nevertheless find significant sources of personal friendship, social solidarity, and aesthetic satisfaction in their group based affinities and cultural life. While the objectifying, fixed conception of group identity is false, it does not follow that group identity is false altogether. Some group affinities that mean a great deal to people are not tied to privilege and oppression; even among presently privileged groups one can find positive group affinity networks and culturally specific styles that help define people's sense of themselves without being tied to the oppression or exclusion of others. The assimilationist ideal exhibits a logic of identity by denying group difference and positing all persons as interchangeable from a moral and political point of view.

Second, the assimilationist ideal also presumes a conception of the individual self as transcending or prior to social context. As Sandel and others have argued, however, a conception of the self as socially constituted and embedded in particular communities is much more reasonable.[12] The assimilationist ideal carries an implicit normative requirement that the authentic self is one that has voluntarily assumed all aspects of her or his life and identity. Such a voluntarist conception of self is unrealistic, undesirable, and unnecessary. We cannot say that someone experiences injustice or coercion simply by finding themselves in social relationships they have not chosen. If unchosen relationships do not produce systematic group inequality and oppression, and also allow individuals considerable personal liberty of action, then they are not unjust.

The strategy for undermining group based oppression implied by the assimilationist ideal, finally, is not likely to succeed under circumstances where there are cultural differences among groups and some groups are privileged. If particular gender, racial or ethnic groups have greater economic, political or social power, their group related experiences, points of view, or cultural assumptions will tend to become the norm, biasing the standards or procedures of achievement and inclusion that govern social, political and economic institutions. To the degree that the dominant culture harbours prejudices or stereotypes about the disadvantaged groups, moreover, these are likely to surface in awarding positions or benefits, even when the procedures claim to be colour-blind, gender-blind, or ethnically neutral. Behaviourally or linguistically based tests and evaluations cannot be culturally neutral, moreover, because behaviours and language cannot be. When oppressed or disadvantaged social groups are different from dominant groups, then, an allegedly group-neutral assimilationist strategy of inclusion only tends to perpetuate inequality.[13]

Contemporary oppressed or disadvantaged social groups with these criticisms of an assimilationist strategy have often envisioned only one alternative to it, the separatist strategy. Understood in its purist form, separatism says that freedom and self-development for an oppressed or disadvantaged group will best be enacted if that group separates from the dominant society, and establishes political, economic and social autonomy. For many separatist movements this vision implies the establishment of a separate sovereign state with a distinct and contiguous territory. Some movements of cultural minorities and oppressed groups, especially those with residentially dispersed populations, do not find such a state feasible. Radical separatist policies nevertheless call for maintaining and establishing group based political, cultural, and especially economic institutions through which members of the group can pursue a good life as much as possible independently of other groups.

I think that the separatist impulse is an important aspect of any movement of oppressed or disadvantaged groups in a society. It helps establish cultural autonomy and political solidarity among members of the group. By forcing dominant groups who have assumed themselves as neutral and beneficent to experience rejection, moreover, separatism also often threatens and disturbs dominant powers more than other political stances. Sometimes the separatist impulse results in the construction of institutions and practices that do make life better for the oppressed or disadvantaged group, and/or give it more political leverage with which to confront dominant groups. A separatist inspired philosophy of 'women helping women' has established rape crisis centres and battered women's shelters in North America, for example, or women-based economic co-operatives in places such as Chile or India (though in the latter cases the organisers may not call themselves feminist separatists).

Separatism asserted by oppressed groups is very different from the processes of enforced separation, segregation and exclusion perpetrated by dominating groups that assert their superiority. Dominant groups depend for their sense of identity on defining the excluded group as Other and keeping the border between themselves and the Other clear. Separatism is inward looking where chauvinism looks outward; separatism is a positive self-assertion where racism and anti-Semitism are negative; separation of the oppressed is voluntary where their exclusion is coerced. Nevertheless, separatism also submits to a logic of identity structurally akin to that underlying a conception of difference as Otherness. It aims to purify and enclose a group identity and thereby avoid political conflict with other groups.

The aim of self-determination and autonomy propelling separatist movements might be sensible if it were not the case that almost everywhere the groups in conflict are already together, their histories intertwined with mutual influences as well as antagonisms. As I discussed earlier, most social groups today currently reside in patterns interspersed with other groups; where the groups are relatively segregated geographically, there are usually mixed border areas, or some members of a relatively separated group are dispersed elsewhere. Groups that perceive themselves as very different in one context, moreover, often find themselves similar when they together encounter another group. But most important, processes of economic centralisation and diversification, along with urbanisation, have created a necessary economic interdependence among many groups and regions that would much prefer to be separate.

The logic of identity expressed by separatist assertions and movements, finally, often tends to simplify and freeze the identity of its group in a way that fails to acknowledge the group differences within a social group. A strong nationalist separatist movement, for example, may reinforce or increase its domination of women or a religious minority, because it wrongly essentialises and homogenises the attributes of members of the group.

I conclude from these arguments that social movements of oppressed or disadvantaged groups need a political vision different from both the assimilationist and separatist ideals. I derive such a vision from the relational conception of group difference. A politics that treats difference as variation and specificity, rather than as exclusive opposition, aims for a society and polity where there is social equality among explicitly differentiated groups who conceive themselves as dwelling together without exclusions.

What are the elements of such an ideal? First, the groups in question understand themselves as participating in the same society. Whether they like it or not, they move within social processes that involve considerable exchange, interaction, and interdependency among the groups. Their being together may produce conflicts, division, relations of privilege and oppression that motivate their *political* interaction.

Thus, second, to resolve these conflicts the group must be part of a single polity. The polity should foster institutions and procedures for discussing and deciding policies that all can accept as legitimately binding, thereby creating a public in which the groups communicate.

But, third, this public is *heterogeneous*, which means that the social groups of the society have a differentiated place in that public, with mutual recognition of the specificity of the groups in the public. Political processes of discussion and decision-making provide for the specific representation of those groups in the society who are oppressed or disadvantaged, because a more universal system of representation is unlikely to include them in manner or numbers sufficient to grant their perspective political influence. The primary moral ground for this heterogeneous public is to promote social justice in its policies. Besides guaranteeing individual civil and political rights, and guaranteeing that the basic needs of individuals will be met so that they can freely pursue their own goals, a vision of social justice provides for some group related rights and policies. These group institutions will adhere to a principle that social policy should attend to rather than be blind to group difference in awarding benefits or burdens, in order to remedy group based inequality or meet group specific needs.

Conclusion

I began this paper by referring to a dilemma faced by those seeking change in social policy in the United States toward economic restructing that will be oriented to meeting needs. The dilemma appears to be that a unified class-based social movement is necessary to achieve this change, on the one hand, but justice within and as a result of such a movement requires differentiating group needs and perspectives and fostering respect for those differences, on the other. All over the world group based claims to special rights, to cultural justice and the importance of recognising publicly different group experiences and perspectives have exploded, often with violence.

In many places the claims, debates and conflicts of this ethnopolitics seem to dwarf other political issues. When they do it is generally no less true there than in the United States that economic structures are a primary case of group disadvantage, where it exists. Some socialists, and some liberals as well, might rightly claim that focus on political group difference and conflict diverts attention from these issues of economic structure.

In this paper I have suggested that the context of economic interdependence provides an important basis for the necessity of groups who define each other as different to maintain a single polity, both in oppositional social movements of civil society or in legislative and other governmental institutions. I have also suggested, however, that the subject of discussion in such a polity should be not restricted to distribution and redistribution of economic resources, or even to issues of control over the means of production and distribution. The way differently identifying groups understand themselves and each other, as well as how their group specific needs and interests intersect with policies and institutions of political decision-making must also be an explicit part of political discourse.

Notes

1 This paper was first presented to the Center for Social Theory and Comparative History at the University of California at Los Angeles in June 1991. Thanks to the participants for a lively discussion. Thanks also to David Alexander and Robert Beauregard for comments on an earlier version. Thanks to Carrie Smarto for research assistance.

2 W. J. Wilson, *The Truly Disadvantaged*, University of Chicago Press, Chicago 1987.

3 See Homi K. Bhabha, 'Interrogating Identity: The Postcolonial Prerogative', in David Goldberg (ed), *Anatomy of Racism*, University of Minnesota Press, Minneapolis 1990.

4 For more development of how the conception of difference as otherness connects with disgust, see Chapter 5 of my book, *Justice and the Politics of Difference*, Princeton University Press, Princeton 1990.

5 For an interesting analysis of the structures of enforced heterosexuality and its relation to gender structuring, see Judith Butler, *Gender Troubles*, Routledge, London and New York 1989.

6 Aysegul Bayakan develops a fascinating account of the identity crisis this process has produced among Turks and those of Turkish descent in Germany. 'The Narrative Construction of the Turkish Immigrant in Germany', unpublished manuscript, Women's Studies, University of Pittsburgh.

7 Nancy Chodorow developed this idea in her famous paper, 'Family Structure and Feminine Personality', in Rosaldo, R. and Lamphere, L. (eds), *Women, Culture and Society*, Stanford University Press, 1974; see also Nancy Hartsock, *Money, Sex and Power*, Longman, New York 1983.

8 I have elaborated a description of this process using Julia Kristeva's theory of the abject in Chapter 5 of my book already cited, *Justice and the Politics of Difference*.

9 For a development of this relational conception of difference, see Martha Minow, *Making All the Difference*, Cornell University Press, Ithaca 1990.

10 See my 'Impartiality and the Civic Public: Some Implications of Feminist Critiques of Moral and Political Theory', in Cornell and Benhabib (eds), *Feminism as Critique*, U. Minnesota, 1987; and *Justice and the Politics of Difference*, Chapters 4 and 6.

11 For a well articulated expression of this assimilationist ideal, see Richard Wasserstrom, 'On Racism and Sexism', in *Philosophy and Social Issues*, Notre Dame University Press, Notre Dame 1980.

12 Michael Sandel, *Liberalism and the Limits of Justice*, Cambridge University Press, Cambridge 1982.

13 For an extended argument that principles and procedures of 'merit' evaluation cannot be group neutral, see *Justice and the Politics of Difference*, Chapter 7.

INDEX